DATA STRUCT

FOR
CODING INTERVIEWS

by

KAMAL RAWAT

MEENAKSHI

BPB PUBLICATIONS

FIRST EDITION 2018

ISBN: 9789387284500

LIMITS OF LIABILITY AND DISCLAIMER OF WARRANTY

Distributors:

BPB PUBLICATIONS
20, Ansari Road, Darya Ganj
New Delhi-110002
Ph: 23254990/23254991

BPB BOOK CENTRE
376 Old Lajpat Rai Market,
Delhi-110006
Ph: 23861747

DECCAN AGENCIES
4-3-329, Bank Street,
Hyderabad-500195
Ph: 24756967/24756400

MICRO MEDIA
Shop No. 5, Mahendra Chambers, 150
DN Rd. Next to Capital Cinema, V.T.
(C.S.T.) Station, MUMBA -400001
Ph: 22078296/22078297

Published by Manish Jain for BPB Publications, 20, Ansari Road, Darya Ganj, New Delhi-110002 and Printed by Repro India Ltd., Mumbai

Dedication

This book is dedicated to a yogi

MASTER MAHA SINGH

who mended relations and excelled in each role he lived

SON | BROTHER | FRIEND | HUSBAND | FATHER | FATHER-IN-LAW | TEACHER | FARMER | CITIZEN | DEVOTEE

the most upright, conscientious, and able man I know

We miss you in every possible way Papa.

Acknowledgements

To write a book, one needs to be in a certain state of mind where one is secure from inside so that he can work with single-minded focus.

A guru helps you be in that state. We want to thank our ideological and spiritual gurus.

Time spent on this book was stolen from the time of family and friends. Wish to thank them, for not logging an FIR for stolen time.

We also wish to thank each other, but that would be tantamount to one half of the body thanking the other half for supporting it. The body does not function that way.

Preface

Data structure and algorithms go hand-in-hand. It is almost meaningless to discuss Binary Search Tree without discussing the algorithm to search in the tree.

In this book, we discuss all major data structures, related algorithms and interview questions based on each data structure. As in the other books of "Coding Interviews" series, our focus is on learning data structures from interviews point of view. After reading this book, you will feel confident in solving unknown problems asked in programming interviews that are directly or indirectly based on any of the given data structure.

Each chapter in this book focuses on one data structure. It first introduces the concepts and theory related to a data structure, followed by interview questions based on that data structure.

Sometimes, the usage of a data structure is not intuitive in solving a particular problem, but as we will see, there are templates that can be used to identify if a data structure fits in the solution of a particular problem.

Our focus is on problem-solving and not on technology (programming language). That's why we have used freedom to code in multiple programming languages. In this book, the primary language of code is C and C++, but at times, we have written the code in Java as well. When the code is in Java, it is explicitly mentioned. If nothing is mentioned, assume the code to be in C++.

Strategy to master data structure for problem-solving is much like the strategy used for mathematics, which is:

- Understand the theorems and concepts (Understand each data structure separately).
- Solve simple problems on those concepts (Write simple programs using that data structure).
- Even if you know the concept and have practiced simple problems, you may not be fit to crack the IIT entrance exam. For that, you have to solve (and keep solving) difficult problems. (Now code difficult problems).
- Remember, while solving difficult problems, we are usually not learning new theorems (same for data structure), but when we see a pattern, we try to create our own rule and note down the pattern. In Data Structures you will find many problems following the same pattern of the solution. Make notes of such patterns.

One small note, Data structure does not exist in isolation. You code an Algorithm, that works on a data structure. The scope of this book is limited to data structures and we are not discussing algorithm paradigms like Sorting algorithms, Recursion, Dynamic Programming, Backtracking, etc. which are not based on any specific data structure. We have written separate books algorithms under the "Coding Interview" series.

Table of Contents

Chapter 1

Introduction

In this chapter, we discuss some of the topics that we should know before start learning individual data structure in detail. I suggest you read this chapter, even if you are an experienced programmer. It will only take few minutes.

Data

In computer science, Data is just a sequence of 0's and 1's. Unless it is converted into some meaningful information, it does not make much sense. If a memory byte stores `00000110` we may not be able to conclude anything from it, but if we know that this byte stores student's roll number, then we know that student's roll number is `6`. That's the difference between **data** and **information**.

Computer programs store, retrieve, and process data. If the data is stored in a well-organized way, then it can be accessed quickly for processing. To process the data, our program should understand the meaning of stored data. Now, we are not just talking about data and information, we are also talking about the operations that can be performed on the data. All this is together called **Data Structure**.

Data structure is more of a logical concept than physical. Essentially, it addresses two fundamental questions:

1. **How is the data stored?**

   It talks more about the logical storage than physical storage. For example, when we define a Linked list, we talk about the concept of Node, we care less about whether the list is stored in RAM or Hard disk or a cache.

2. **What operations can be performed on it?**

   Different data structures support different operations. For example, a stack usually does not support accessing the middle element.

The functional definition of a data structure is known as Abstract Data Type (ADT) and is independent of its implementation. For example, when we talk about the ADT of Stack data structure, we talk about the push and pop operations. But whether the actual implementation store stack elements in an Array or Linked List is not in the scope of ADT discussions.

Actual problem solving involves Data structure and Algorithms, where an algorithm is a step-by-step process to solve a problem. In this book, we focus on data structures and algorithms related to each data structure, for example, Binary Search Tree; data structure cannot be discussed without discussing the algorithms to insert, search and remove data from the tree. So, those algorithms form part of the data structure itself.

This book does not cover pure algorithm concepts like Dynamic Programming, Sorting algorithms, etc.

Data Type and Data Structures

Data type usually refers to the type of data. For example, the data type of literal `3.5` is `double` in C language. The Data type of below array literal:

```
{1, 2, 3, 4, 5}
```

is an "Integer-Array-Of-Size-Five".

Data structure is a way of organizing, storing and processing a particular type of data. A user-defined class is one data structure and File can be another data structure.

Every language comes with a predefined set of data types that are available as a part of the language. The data types available in C language are the following:

```
char, short, int, long, long long, float, double.
```

Note that these data types also come along with the operations that can be performed on variables and literals of these types. For example, we cannot perform bit-wise operations on single or double precession real numbers.

Recursion

Recursion is one of the most important concept that will be used in solving lot of problems.

In computer programming, "*when a function calls itself either directly or indirectly it is called a Recursive Function and the process is called Recursion*".

Typically, a function performs some part of overall task and rest is delegated to recursive call(s) of same function. There are multiple instances of function's stack frame (also called Activation record) in the stack and a function may use return value of recursive calls. Function stops calling itself when it hits a terminating condition.

Writing recursive code is not difficult, in fact, in most cases it is relatively simpler because we are not solving the complete problem. Writing recursive function code is a two-step process, it:

1. Visualize recursion by defining larger solution in terms of smaller solutions of exact same type, and,
2. Add a terminating condition.

The solution to find sum of first n natural numbers is defined recursively as:

$$\sum_{i=0}^{n} i = n + \sum_{i=0}^{n-1} i$$

writing it in terms of function call:

```
Sum(n) = n + Sum(n-1)
```

Terminate this recursion when n=1. Following is the recursive function:

```
int sum(unsigned int n)
{
  // FIRST TERMINATING CONDITION
  if(n == 0)
    return 0;

  // SECOND TERMINATING CONDITION
  if(n == 1)
    return 1;

  return n + sum(n-1);
}
```

Following are important points in a recursion:

1. Recursion always have a **terminating condition** (else it will be infinite recursion). In `sum` function, condition, if `n=1`, then, stop recursion and return 1, this is a terminating condition.
2. Recursive function **performs some part of the overall task and delegate rest of it to recursive call**(s). `sum` function adds n with return value of `sum(n-1)` but delegate computation of `sum(n-1)` to recursive call.

Important point while writing recursive code is, *never miss the terminating condition(s), else your function may fall into infinite recursion.*

Sum of first n natural numbers can be computed non-recursively by using a simple loop as shown below:

```
int sum(int n)
{
  int s= 0;
  for(int i=1; i<=n; i++)
    s += i;
  return s;
}
```

Question 1.1: Recursion to compute n^{th} power of a number is defined below:

$$x^n = \begin{cases} x * x^{n-1} & \\ 1 & \text{if } n=0 \end{cases}$$

Write a function that accets x and n, and return x^n.

Solution: Following is the code:

```
int power(int x, int n)
{
    if(0 == n || 1 == x){ return 1; }
    return x * power(x, n-1);
}
```

Function `power` receives two parameters. One of them remains fixed, other changes and terminates the recursion. Terminating condition for this recursion is

IF (n **EQUALS** 0) **THEN** return 1

But we have used two terminating conditions:

IF (n **EQUALS** 0) **THEN** return 1

IF (x **EQUALS** 1) **THEN** return 1

This is to avoid unnecessary function calls when `x` is `1`. Every function call is an overhead, both in terms of time and memory. The four things that we should focus on while writing a function (in this order) are:

1. It should serve the purpose. For every possible parameter, our function must always return expected result. It should not be ambiguous for any input.
2. Time taken by function to execute should be minimized.
3. Extra memory used by function should be minimized.
4. Function should be easy to understand. Ideally code should be self-explanatory to an extent that it does not even require any documentation (comments).

During coding or while in the interview we do not usually care about how many lines of code does particular function runs into as long as the length of code is justified (you are not writing a duplicate piece of code).

Please note that an optimal implementation of `power` function takes O(lg(n)) time. The purpose here is not to write optimal code, but to demonstrate recursion.

Recursive code usually takes more time to execute and also uses high memory. Still, recursion is one of the most powerful and rampant problem-solving tools, because sometime, a solution that otherwise is very complex to comprehend, can be very easily visualized recursively. We just need to solve the problem for the base case and can leave rest of the problem to be solved by recursion.

Some algorithms like merge sort have a simple recursive implementation and are difficult to code non-recursively. Algorithms like binary search can be coded both recursively and non-recursively with equal ease. Algorithms like Linear search or bubble sort are better implemented and understood iteratively than recursively. There is no blanket statement that can relate easy code with recursion.

With this minimal groundwork, let us start discussing one data structure at a time.

Chapter 2

Arrays and strings

"*An array is an* ***indexed collection*** *of* ***homogeneous*** *elements.*"

There are three keywords in the above definition.

1. **Indexed:** Array elements are indexed and can be accessed by using their position in the array. In most programming languages, index starts from zero. If array name is `arr` then, `arr[3]` represents the fourth element in array `arr`.
2. **Collection:** An array is a collection of elements. There can be more than one element in an array.
3. **Homogeneous:** All elements in an array are of the same data type. It is only because of this homogeneity that we can access the array elements by using indexes.

In C language, data type, total number of elements and name of the array are provided while defining the array.

```
int arr[5];   // arr IS AN ARRAY OF 5 INTEGERS
```

We may also (optionally) provide the initialization list that specify initial values of array elements

```
int arr[5] = {1, 2, 3, 4, 5};
```

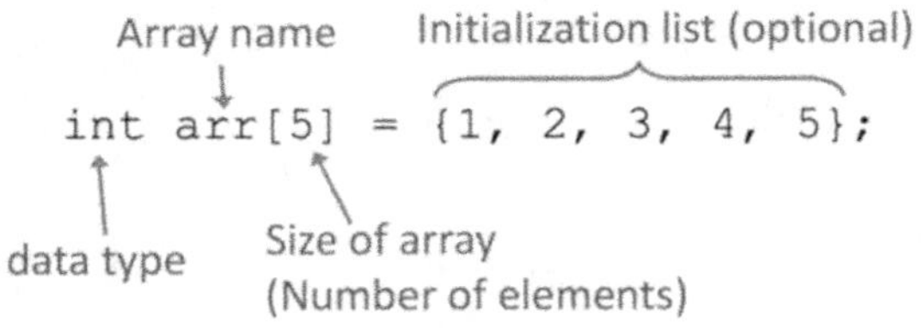

Array name
Initialization list (optional)

```
int arr[5] = {1, 2, 3, 4, 5};
```

data type
Size of array
(Number of elements)

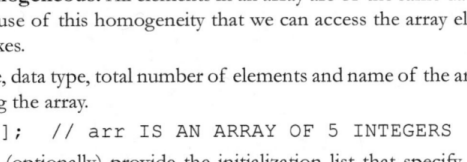

Size of the array can be skipped if initial values of all the array elements are given. Size of `arr` below is 5 (Number of values in the initialization list).

```
int arr[] = {1, 2, 3, 4, 5};
```

If size of the array is more than the number of values in the initialization list, rest of the array gets initialized with zero's of the corresponding data type. In below declaration, entire array is zero.

```
int arr[1000] = {0};
```

Code 2.1 print all elements of given array from left to right. Function `printArray` may be used to debug code while executing.

```
void printArray(int *arr, int n){
  for(int i=0; i<n; i++)
    printf("%d", arr[i]);
}
```

Code 2.1

We cannot have an array of type `void`. Below code is an error.

```
void arr[5]; // ERROR
```

If such a definition is allowed, it is impossible to determine the amount of memory to be allocated to `arr`. However, an array of void-pointers is valid.

```
Void* arr[5]; // OK
```

An array of void-pointers is also used as a generic array

```
void* arr[3];

// MAKING ELEMENTS POINT TO DIFFERENT TYPE'S MEMORY
arr[0] = malloc(sizeof(char));
arr[1] = malloc(sizeof(int));
arr[2] = malloc(sizeof(double));

// STORING VALUES OF DIFFERENT TYPES IN ARRAY
*((char*)&(arr[0])) = 'A';
*((int*)&(arr[1])) = 10;
*((double*)&(arr[2])) = 5.3;
```

Code 2.2

Each element of array in Code 2.2 points to memory of different data type. In a way, it is a generic array storing `char`, `int` and `double` data.

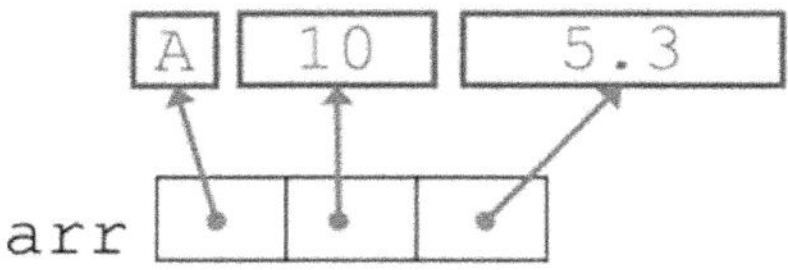

To dereference such an array, we need to know the data type that individual element points to.

```
printf("Element at arr[1] : %d", *(int*)arr[1]);
```

C language does not have a true generic programming framework like templates in C++. Macros provide generic programming in the true sense but they have limitations and are also prone to errors. In the previous declaration, all elements of `arr` are of the exact same type, `void*`. In Java, an array of `Object` data type is generic, because `Object` is the parent class of all.

```
Object arr[] = new Object[5];
```

Allocating array on the heap

In C language, `malloc` function does not know the type or nature of memory it is allocating (that's why it cannot initialize the array), it just allocates a contiguous chunk of memory. In the below allocation, `malloc` does not (and cannot) know if the memory it is allocating is used to store one `int` or an array of `4` characters.

```
int *ptr = (int*)malloc(sizeof(int));
```

If we assign this memory to a character pointer.

```
int *chPtr = (char*)malloc(sizeof(int));
```

Then memory looks like below:

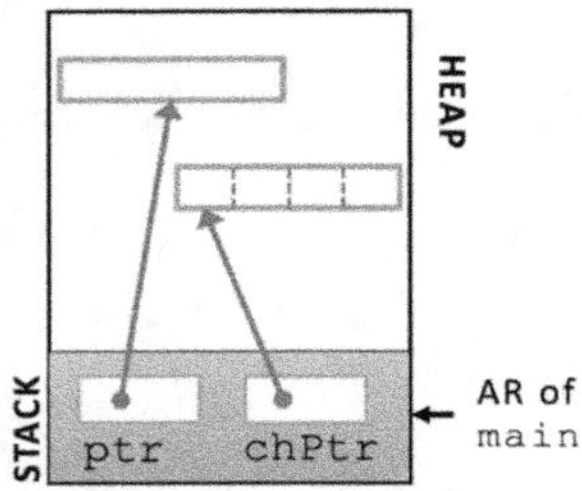

The memory pointed to by `ptr` is interpreted as one integer memory, but same amount of memory pointed to by `chPtr` is interpreted as an array of four characters. Memory allocated by using `malloc` (or `calloc`) is a contiguous stream of bytes, we need to interpret it as an array.

```
void fun(){
  int arr[5] = {1, 2, 3, 4, 5}; // ARRAY ON STACK
  int *ptr = NULL;
  ptr = (int*) malloc(5 * sizeof(int));
}
```

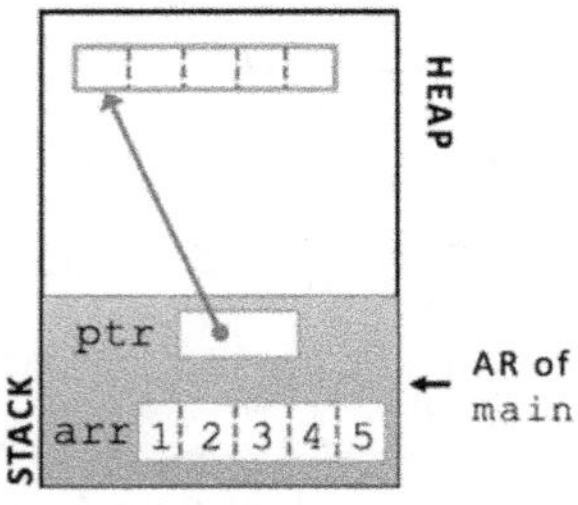

Array `arr` reside on the stack in the activation record (Stack frame) of function `fun`. When `fun` function returns to the calling function, memory allocated to `arr` will be deallocated. The array allocated by using `malloc` function resides on the heap and remain there unless explicitly revoked by using `free` function. When function returns the pointer, `ptr` will go out of scope, creating a memory leak in the heap.

In Java, all arrays are allocated on heap. Fortunately, when a memory on heap is not pointed by any variable, it is automatically garbage collected thus, avoiding any memory leaks.

A pointer variable, pointing to an array can access its elements like an array itself (and vice-versa). If `ptr` points to the first element of `arr`, the below expressions are same and access the third element of `arr`:

Array	Pointer	Comment
`arr[2]`	`ptr[2]`	
`*(arr+2)`	`*(ptr+2)`	
`2[arr]`	`2[ptr]`	Subscript operator is commutative.
`*(2+arr)`	`*(2+ptr)`	Add operator is also commutative.

While passing an array to a function, we pass a pointer to the first element of array (along with the number of elements) and receive it in a pointer variable.

```
void printArray(int *a, int n){
  for(int i=0; i<n; i++)
    printf("%d", a[i]);
}
int main(){
  int arr[5] = {1, 2, 3, 4, 5};
  printfArray(arr, 5);
}
```

Output: `1 2 3 4 5`

When function `printArray` is called, the memory looks like below:

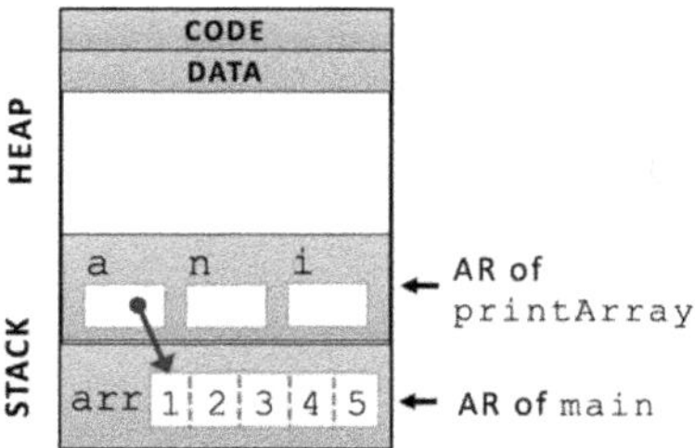

The array resides on the activation record of `main`. If any change in the array is made inside function `printArray`, it will change the original array `arr` of `main`.

Array name (`arr`) has a special meaning. In most cases, it acts as a constant pointer to the first element of array. Both pieces of following code compute the sum of elements of array `arr`. Both are same in terms of performance also.

`int arr[]={1,2,3,4,5};`
`int iSum=0;`
`for(i=0; i<10; i++)`
  `iSum += *(arr + i);`
	`int arr[]={1,2,3,4,5};`
	`int iSum=0, *ptr=arr;`
	`for(i=0; i<10; i++)`
	  `iSum += ptr[i];`

Deallocating 1-dimensional array on heap

A one-dimensional array on the heap is a chunk of memory allocated by a single `malloc` function call. The memory can be deallocated by calling function `free` with the initial address of memory.

```
free(ptr);
ptr = NULL;
```

The assignment of `NULL` to `ptr` is important, because otherwise, `ptr` will become a dangling pointer.

Arrays in Java

All arrays in java are dynamically allocated on the heap as shown below.

```
int[] arr = new int[5];
```

Length of arrays in java can be found by using property `length`. An array of user-defined objects is defined in the same way as that of a primitive type.

```
Student[] arr = new Student[5];
```

The good thing about Java or C++ is that type information, property or functionality can be maintained inside the class itself.

Strings

A string in C language is a character sequence stored in a character array and is terminated with a `NULL` character.

```
char str[10];
```

`str` is a string of size `10` that can hold up to 9 characters, one place is left for `NULL` character. Initialization of string can be done as below:

```
char str[10] = "Hello";
```

The way `str` is stored in the memory is:

H	e	l	l	o	\0				

Characters after '\0' are garbage (this is different from how numeric arrays are initialized).

All character arrays may not be strings. Look at the amount of memory allocated to below two strings:

```
char s1[]="YOGO";                    //sizeof(s1)is 5
char s2[]={'Y', 'O', 'G', 'A'}; // sizeof(s2)is 4
```

`s1` is a string and is null-terminated. `s2` is just a character array and no extra `NULL` character is appended at the end.

While passing strings to a function we do not need to pass length of the string because `NULL` character marks end of the string. Consider below implementation of `strlen` function

```
size_t strlen(const char *s){
  int cnt = 0;
  while(s != NULL && *s != '\0')
    {cnt++; s++;}
  return cnt;
}
```

String Length v/s Size of String

Strings are null-terminated character arrays. A special format specifier, `%s` is dedicated for reading/writing strings. A `char` array is used to hold a string.

```
char str[10];
```

There are two lengths associated with a character array, size of memory allocated to the array and length of string stored in that memory.

```
char str[10] = "Hello";
```

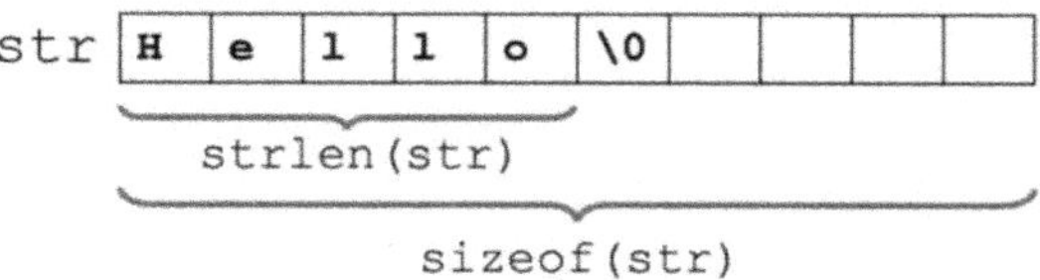

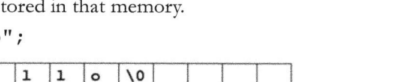

`10`-character memory is allocated to `str`, but it stores a string of length `5` only. `sizeof(str)` returns length of memory in bytes allocated to `str`, in this case `10`. `strlen(str)` returns length of the string stored in `str`.

```
printf("%d : %d", sizeof(str), strlen(str));
```

Output: `10 : 5`

Another difference is that, `size of` is an operator and `strlen` is a function. Also, value of `size of` remain same for a variable, but `strlen` may change.

```
char str[10] = "Hello";
printf("\n%d : %d", sizeof(str), strlen(str));

strcpy(str, "Bye");
printf("\n%d : %d", sizeof(str), strlen(str));
```

Output: `10 : 5`

`10 : 3`

Strings in Java

Java strings are objects of `String` class that represents a sequence of characters. The most important thing about strings in Java is that they are immutable, which literally means that an object of `String` class cannot change after it is created. If `str` is a string defined as:

```
String str = new String("Shiva");
```

Then, `str.toLowerCase()` does not change the value of String `str`. A new `String` object (with value `Shiva`) is created. The previous object ("`Shiva`") will be free by the garbage collector. Similarly, all the below functions of `String` class create a new `String` object.

```
String to UpperCase()

// REMOVES WHITE SPACES FROM BOTH ENDS
String trim()

// CREATE NEW String WITH oldChar REPLACED BY newChar
String replace(char oldChar, char newChar)

// SAME AS this String + str2
String concat(String str2)

// CREATES A char ARRAY FROM STRING
char[] to CharArray()
```

`String` class in `java.lang` package is very rich in methods. You think of an operation on an immutable string, and there is a high chance of that operation already present in the `String` class.

There are two ways of creating a string object:

```
String str1 = "Krishna";
String str2 = new String("Shiva");
```

The first `String` object pointed to by `str1` is created in the string pool, but object pointed to by `str2`, created using `new String`, is created on the Heap.

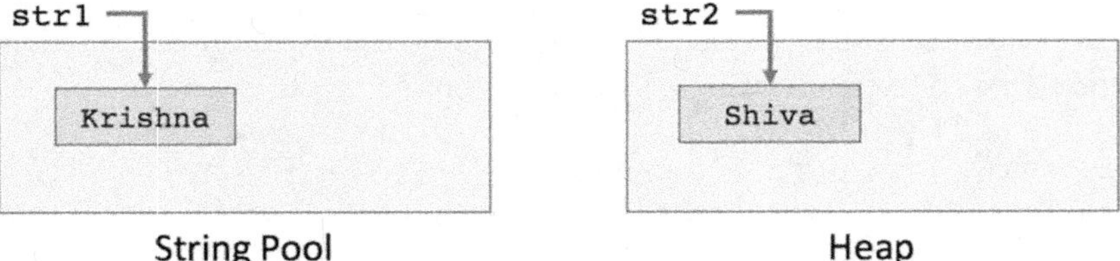

String pool is like a pool of strings where all string literals are unique. When user tries to create a new String in the string pool, the compiler will look to see (using `equals` method) if the same string already exists in the pool or not. If the same string already exists, then no new string is created. Both string objects will point to the same string in the string pool.

But, in case of heap, a fresh object is always created on the heap whenever `new String` is used. Consider below definitions:

```
String str3 = "Krishna";
String str4 = new String("Shiva");
```

After the above four declarations, the memory looks like below:

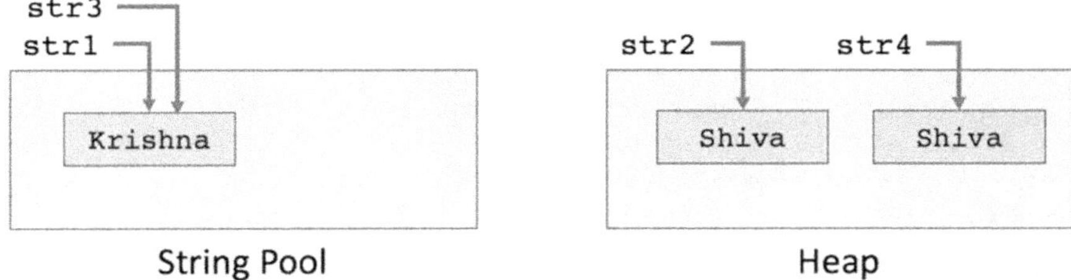

`str1` and `str3`, both points to the exact same object. Output of both the statements below is `true`:

```
System.out.println(str1 == str3);    // true
System.out.println(str1.equals(str3)); // true
```

`str2` and `str4` has the same string values, but they are different objects and their hashcodes are different, therefore `(str2==str4)` is `false`. But the library function `equals` return `true` (because it compare the values).

```
System.out.println(str2 == str4);    // false
System.out.println(str2.equals(str4)); // true
```

When we store only one copy for distinct string objects, it is called interning. Java's `String` class also privately maintains a pool of strings and also has a `public` method `intern()`.

When the `intern()` method is invoked on a `String` object, it looks for the string contained by this String object in the pool, if the string is found there then the string from the pool is returned. Otherwise, this `String` object is added to the pool and a reference to this newly created `String` object is returned.

```
String str1 = "Krishna";
String str2 = new String("Krishna");
String str3 = str2.intern();

System.out.println(str1 == str3);   // true
```

In this case, the heap and string pool looks like below:

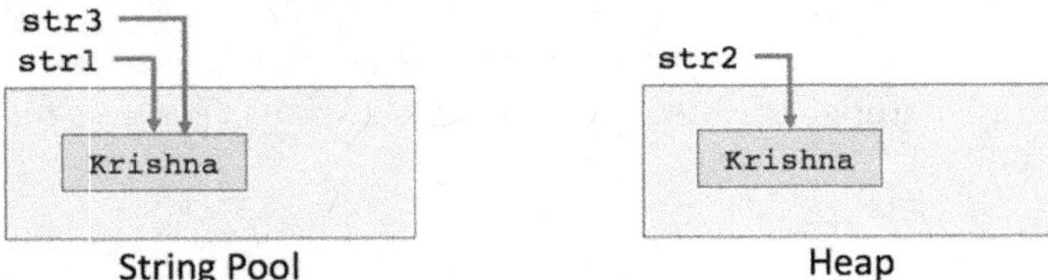

The below java function receives a string and reverse it.

```
public String reverse(String str){
  int len = str.length();
  char[] charArray = new char[len];

  for (int i = 0; i < len; i++)
    charArray[i] = str.charAt(len-i-1);

  return  new String(charArray);
}
```

The `reverse` operation above takes O(n) memory, where as reversing character arrays in C language is in-place.

Practice Questions

Question 2.1: How will you reverse an array?

Note: *This is a too simple question to be asked in an interview. Interviewers usually do not ask such simple problems, but a variation of this can be asked, like, how will you reverse words in a string.*

Solution: One way to reverse an array is to have two variables, `low` and `high` initialized to `0` and `n-1` respectively. Swap the elements at `low` and `high` and after swapping, increment `low` and decrement `high`. Keep doing this as long as `low` is less than `high`.

```
reverse(int *arr, int n){
  int low=0, high=n-1;
  while(low<high){
    swap(&arr[low], &arr[high]);
    low++;high--;
  }
}
```

We can do the same by using one variable also as shown below

```
void reverse(int *arr, int n){
  for(int i=0; i<n/2; i++)
```

```
    swap(&arr[i], &arr[n-i-1]);
}
```

It is not an improvement over the previous code, but interviews are conversations, at the end of which interviewer has to take a call on whether to hire you or not. It is always good to know multiple options for performing the same task.

The `swap` function used above is shown in Code 2.3

```
void swap (int *a, int * b){
  int temp  =*a;
  *a = *b;
  *b = temp;
}
```

Code 2.3

The recursive code takes more time and more memory, but recursion is such a powerful problem-solving tool that every developer should be very comfortable using it. If an interviewer asks you to write a recursive code for above problem here it is.

```
void reverse(int *arr, int low, int high){
  if(low>=high){ return; } // TERMINATING COND.
  swap(&arr[low], &arr[high]);
  reverse(arr, low+1, high-1);
}
```

Recursion usually replaces a loop. Function performs work of one iteration of the loop and rest is left for recursion. If there are `10` elements in the array, the function swap the first and last elements and call itself recursively for the reduced size.

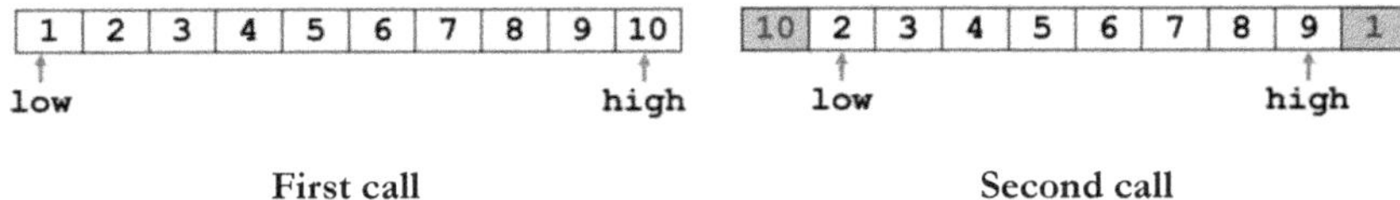

First call **Second call**

Recursive implementation takes O(n) extra memory.

Question 2.2: Given an array and a number `d`, how will you rotate the array by `d` positions. For example, if the array is

```
int arr[] = {1, 2, 3, 4, 5, 6, 7, 8};
```

and `d=2`, rotate `arr` around index `2`. The array should change to:

`{3, 4, 5, 6, 7, 8,` **`1, 2`**`};`

Solution: There can be multiple methods. Let us look at some of them:

Method-1: Rotate one element at a time

```
// d IS THE POINT OF ROTATION
for(i=1; i<=d; i++){
  int temp = arr[0];
  for(int cnt=1; cnt<n; cnt++)
    arr[cnt-1] = arr[cnt];

  arr[n-1] = temp;
}
```

This logic takes `O(n*d)` time.

Method-2: Use a temporary array

1. Use a temporary array of size `d`.
2. Store first `d` elements of original array in the temporary array.
3. Shift all elements of original array left by `d` positions.
4. Copy elements of temporary array at the end of original array

```
void rotate(int *arr, int n, int d){
  int* tempArr = (int*)malloc(sizeof(int)*d);
  int i=0;

  // MOVE d ELEMENTS TO tempARR
  for(; i<d; i++){ tempArr[i] = arr[i]; }

  // SHIFT n-d ELEMENTS
  for(i=0; i<n-d; i++){ arr[i] = arr[i+d]; }

  // COPY FRONT d ELEMENTS AT END
  for(; i<n; i++){ arr[i] = tempArr[i-n+d]; }

  free(tempArr);
}
```

This function takes linear time, but also takes `O(d)` extra memory. Please note that the memory allocated to `tempArr` is deallocated at the end.

Method-3: Reversal algorithm

There is an interesting algorithm that use the reverse function discussed in the previous question. Let us modify the `reverse` function to reverse sub array of an array from `start` index to `end` index (both inclusive).

```
void reverse(int *arr, int start, int end);
```

Function call,

```
reverse(arr, 0, n-1);
```

reverse the entire array. Array can be rotated in three steps:

```
reverse(arr, 0, d-1);   // REVERSE FIRST d ELEMENTS
reverse(arr, d, n-1);   // REVERSE LAST n-d ELEMENTS
reverse(arr, 0, n-1);   // REVERSE ENTIRE ARRAY
```

If input array is `{1,2,3,4,5,6,7,8}`and d=3, then

```
reverse(arr, 0, 2);   // {3,2,1,4,5,6,7,8}
reverse(arr, 2, 7);   // {3,2,1,8,7,6,5,4}
reverse(arr, 0, 7);   // {4,5,6,7,8,3,1,2}
```

This method takes linear time and constant extra memory.

Question 2.3: Print all possible permutations of an Array elements. For array
`int arr[] ={1, 2, 3}.`

the possible permutations are:
`123, 132, 213, 231, 312, 321`

Solution: Think of it recursively. If the given array is {1, 2, 3, 4, 5}. Remove one of these five values and find all permutations of the remaining four values. Then put the removed value in front of each of these permutations.

```
permutation( {1,2,3,4,5} )
{
  put 1 in front of eachpermutation({2,3,4,5} )
  put 2 in front of each permutation({1,3,4,5} )
  put 3 in front of each permutation({1,2,4,5} )
  put 4 in front of each permutation({1,2,3,5} )
  put 5 in front of each permutation({1,2,3,4} )
}
```

While coding, we keep the removed element at first position as shown below:

```
void permutation(int * arr, int n, int l){
  if(l==n-1){
    printArray(arr, n);
    return;
  }

  for(int i=l; i<n; i++){
    swap(&arr[i], &arr[l]);
    permutation(arr, n, l+1);
    swap(&arr[i], &arr[l]);
  }
}
```

Parameter `l` holds the starting index of the array for a particular instance of the function call. Initially `l` is zero. When `l` is equal to `n-1`, there is only one element in the array to be processed. Single element has only one permutation, so it is our terminating condition.

Function `printArray` is discussed in Code 2.1. We are calling `swap` function twice to avoid disturbing the order of elements in original array in the calling function.

What if there are repeating elements? If array is `{1,2,2}` output should be `122, 212, 221`. Duplicate permutations should not be printed.

This can be done by putting a check in the `for` loop to not swap the elements if they are same.

Question 2.4: For a given string, print the pattern as shown below. If the input string is `hat`.

- Replace one character at a time with numeric one(`1`):

  `1at, h1t, ha1`

- Replace two characters with `'1'` each:

  `11t, h11, 1a1`

- Replace three characters with `'1'` each:

  `111`

In all above patterns, if two numbers are consecutive, add their numeric value. So, `11t` becomes `2t`. Similarly, `111` becomes `3`. The final output for input string `hat` should be
`hat ha1 h1t h2 1at 1a1 2t 3`

Solution: This looks like a permutation problem, but it is not. The arrangement of characters is not changing. If `h` is present in the pattern it will always come at the first position.

Let us break the solution in modules. First, let us write a helper function to check if the given character is a numeric digit.

```
bool isDigit(char ch){
  return (ch >='0' && ch<='9');
}
```

Let us divide the entire solution into two parts. The first part will generate the patterns that has `1` placed at specific positions (without consecutive ones being added). The second part will receive such a pattern and print it, while printing the pattern it makes sure to add the consecutive ones and print the sum in place of individual digits.

Let's first write the printing logic. This function receives a string whose each character is either an alphabet or a `'1'`. The function prints alphabets as it is and add all consecutive `'1'` and print the sum instead.

```
void printPattern(char* str){
  for(int i=0; str[i]!='\0'; i++){
    if(!isDigit(str[i])){
      printf("%c", str[i]);
```

```
    }else{
      int sum = 0;
      int j=i;
      while(str[j] != '\0' && isDigit(str[j])){
        sum++;
        j++;
      }
      i=j-1;
      printf("%d", sum);
    }
  }
}
```

Generating pattern is simple. For each position, there are exactly two possibilities, when the corresponding character comes in that position and when `1` comes in that position.

```
void generatePatterns(char *str, int i)
{
  if(str[i] == '\0')
    printPattern(str);
  else{
    char temp = str[i];
    generatePatterns(str, i+1);
    str[i] = '1';
    generatePatterns(str, i+1);
    str[i] = temp;
  }
}
```

Question 2.5: Given a string of bits and a number `k`. In one flip, you can toggle exactly `k` consecutive bits. With only this flip operation available, change the string to all ones. For example:

`Input String: 0000110000` `k: 3`

Following are four flips using which we can change all bits to 1:

FLIP-1: **111**0110000 FLIP-2: 1110110**111**

FLIP-3: 111**100**0111 FLIP-4: 1111**111**111

If it is not possible to set all bits in the string, print `IMPOSSIBLE`. For example, it is not possible to change all bits of input string `01010101` to `1`for `k=3`. (Flipping less than `k` bits is not allowed).

Solution: If there are `n` bits, the total number of places that can be flipped are `n-k+1`. Let us say **Flipping at Bit-0** means flipping `k` bits starting from bit at index-`0`. Each of these `(n-k+1)` positions can either be flipped or not. Total number of possible operations are 2^{n-k+1}. After all these operations, check if the string is in its final state (all 1's) or not.

This solution takes exponential time. It is also giving the two indicators to apply Dynamic Programming or Greedy Approach, i.e. Optimal Substructure and Overlapping Sub problems. Consider the following greedy approach:

In each iteration flip at position of left-most 0.

If Input string is 0001000000001110 and k=4, first zero is at left-most position, **0**001000000001110. After flipping 4 bits starting from this position, the string becomes, 1110000000001110. Now, look for next zero-bit (111**0**000000001110) and flip the k bits starting from that left-most zero-bit: 1111111000001110

Again, find the left-most zero and flip next k bits. The string now becomes: 1111111111101110

After repeating the same thing again, the string becomes: 1111111111110000

And finally, after flipping the last 4 bits, we get a string with all 1's 1111111111111111

If we are flipping last k bits and the string is still not converting to the final string, it means the string cannot be converted to all 1's (See Code 2.4).

Question 2.6: In the previous question, find the minimum number of flips required to produce the final String.

Solution: The greedy approach used in solution of the previous question actually takes minimum number of flips only. Function countFlips in Code 2.4 takes linear time and returns the minimum number of flips.

```
int countFlips(char* str, int k){
  int flips = 0, i=0;
    unsigned long n = strlen(str) - k; // MAX FLIPS

  for(; i<=n; i++){
    if(str[i] == '0'){
      // FLIP next k bits
      for(int j=0; j<k; j++)
        str[j+i] = (str[j+i] == '0') ? '1':'0';
      flips++;
    }
  }

  while(str[++n] != '\0')
    if(str[i] == '0')
```

```
        return -1;
    return flips;
}
```

Code 2.4

Question 2.7: Given an array containing both positive and negative integers, write code to rearrange the elements so that positive and negative numbers come alternatively. If they are not equal, extra positive (or negative) numbers should come at the end of the array. For example:

```
Input: {1, 2, -2, -5, 6, 7, -8}
Output: {1, -2, 2, -5, 6, -8, 7}

Input: {-1, 2, 3, -5, -6, -7, -8}
Output: {-1, 2, -5, 3, -6, -7, -8}
```

Solution: Check for each position.

If sign of the number at that position is same as the one required for that position, move forward.

If a positive number is required, and a negative number is present, find the first positive number after that position and put it at this position by shifting all numbers in between by one position each. See the following code:

```
void shiftAndInsert(int *arr, int fromIdx, int toIdx)
{
  int temp = arr[toIdx];
  for(int j = toIdx; j > fromIdx; j--)
    arr[j] = arr[j-1];
  arr[fromIdx] = temp;
}
```

Function `shiftAndInsert` moves `arr[toIdx]` at `fromIdx` by shifting elements from `fromIdx` till `toIdx-1` one position forward. Below is the main function that does the work

```
void arrange(int* arr, int n)
{
  if(n<1){ return; }

  // KEEP CHECK OF SIGN EXPECTED AT arr[1].
  int positive = false;
  if(arr[0] < 0){ positive = true; }

  int fromIdx = 1;
  for( int i = 1; i < n; i++){
    if(positive){
```

```
      if(arr[i] >= 0){

        // IF EXPECTING +VE AND HAVE -VE
        if(fromIdx != i){
          shiftAndInsert(arr, fromIdx, i);
          i = fromIdx;
        }
        positive = false; // FOR NEXT ELEMENT
        fromIdx = i + 1;
      }
    }
    else{
      if(arr[i] < 0){

        // IF EXPECTING -VE AND HAVE +VE
        if( fromIdx != i){
          shiftAndInsert(arr, fromIdx, i);
          i = fromIdx;
        }
        positive = true;
        fromIdx = i + 1;
      }
    }
  }
}
```

The good thing about this code is that it maintains the order in which element are present in the original array. But it is an `O(n²)` time algorithm.

Method-2: This solution uses the partition logic of quick sort algorithm. First separate the positive and negative numbers by bringing all negative numbers on one side and all positive numbers on another side. This can be done by using the partition logic of QuickSort algorithm with `pivot=0`. Element less than pivot (negative numbers) will come on left side and element greater than pivot (positive numbers) will come on the right side.

After partitioning, start from first negative number, and swap every alternate negative number with next positive number.

```
// HELPER FUN. SWAP TWO ARRAY ELEMENTS
void swap(int*arr, int x, int y){
  int temp = arr[x];
  arr[x] = arr[y];
  arr[y] = temp;
}

// BRING ALL -VE NUMBERS AT START
```

```
int partition(int *arr, int n){
  int low = 0, high = n-1;
  while(low<high){
    while(arr[low] < 0) low++;
    while(arr[high] > 0) high--;

    if(low<high)
      swap(arr, low, high);
  }
  return low;
}

void arrange(int *arr, int n){
  // INDEX OF FIRST -VE AND POSITIVE NUMBERS
  int neg = 0;
  int pos = partition(arr, n);

  if(pos >= n){ return; }

  while( (pos<n)&&(neg<pos)&&(arr[neg]<0) ){
    swap(arr, pos, neg);
    pos++;neg += 2;
  }
}
```

This logic takes O(n) time and constant extra space, but it does not maintain the order of positive and negative numbers in original list.

Question 2.8 : Given an array of integers, create another array whose `i`th element is the product of all elements in the original array except the `i`th element.

```
Input Array:a[]= {1, 2, 3, 4};
Output Array: b[] = {24, 12, 8, 6};
```

Notice that, `b[1] = a[0]*a[2]*a[3];` (Product of all, but `a[1]`).

Solution: The first attempt to answer this question is usually to multiply all numbers in the input array and then divide the product by `a[i]` to get `b[i]`.

```
void multiplyMat(int* a, int* b, int n){
  int prod = 1;
  for(int i=0;i<n;i++){ prod *= a[i]; }
  for(int i=0;i<n;i++){ b[i] = prod/a[i]; }
}
```

Try running this code for array, `{0, 1, 2, 3}` and it will crash because of division by zero,

```
b[0] = prod/a[0];
```

It can be fixed by putting a condition

```
if(a[i] != 0)
b[i] = prod / a[i];
```

But then the computation of `b[0]` is skipped all together. Moreover, the `prod` is zero (and rightfully so).

It need to be fixed. While you are trying to fix it, the interviewer may ask you to solve it without using division operator. This happens if the interviewer has a solution in his mind and he wants you to come up with the same solution (not my type of interviewers).

Method-2 (Without using division operator): Consider below algorithm:

```
STEP-1: SET ALL ELEMENTS OF ARRAY b SUCH THAT
    b[i] = b[i-1] * a[i].
```

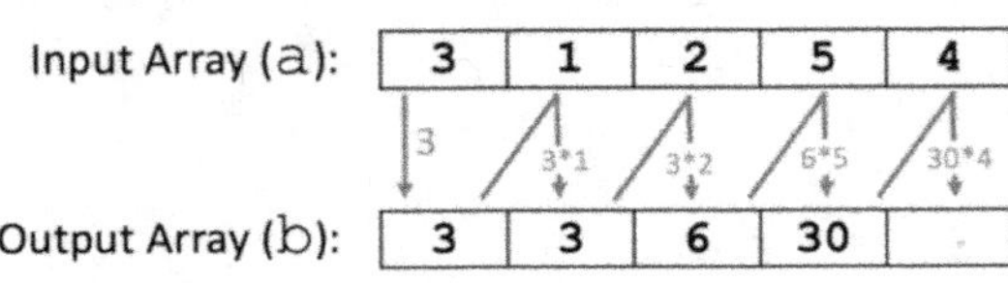


```
        (The last element can be left)

STEP-2. prod = 1;

STEP-3. FOR(i=n-1; i>0; i++)
            b[i] = b[i-1] * prod;
            prod = prod * a[i];
```

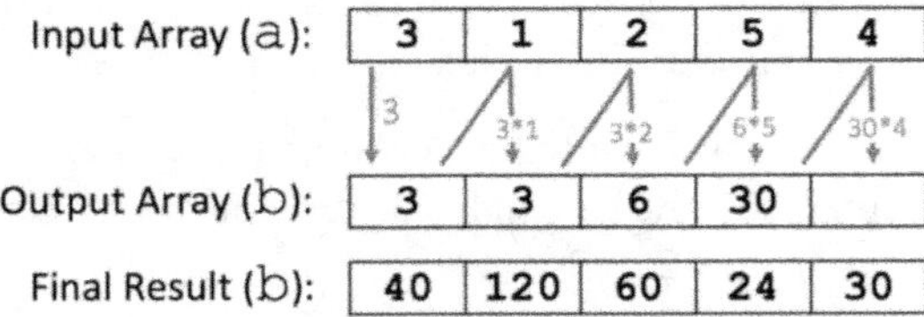

Code for this algorithm is as following:

```
void multiplyMat(int* a, int* b, int n){
  b[0] = a[0];
  for(int i=1;i<n-1;i++){
    b[i] = b[i-1] * a[i];
  }

  int prod = 1;
  for(int i=n-1; i>0; i--){
    b[i] = b[i-1] * prod;
    prod *= a[i];
  }
  b[0] = prod;
}
```

This solution takes `O(n)` time and constant extra memory (memory of `b` array is part of input).

Question 2.9: Given a string with white spaces. Write code to remove all white spaces from the string.

```
Input String: "C INTERNALS FOR CODING INTERVIEWS"
Output String: "CINTERNALSFORCODINGINTERVIEWS"
```

Solution: In ASCII, white space is either a space, tab, vertical tab, carriage return, newline or formfeed. We are only considering space and tab, but the function `isWhiteSpace` can be easily extended to include all of them.

```
int isWhiteSpace(char ch){
  return (ch == ' ' || ch == '\t');
}
void removeWhiteSpace(char *str){
  int i=0, j=0;
  while(str[j] != '\0') {
    while(str[j] != '\0' && isWhiteSpace(str[j]))
      j++;
    str[i] = str[j];
    i++; j++;
  }
  str[i] = '\0'; // ADDING NULL CHAR AT END
}
```

There are two variables, `i` and `j`. `i` is always before `j` and there (may be) is a window between them. Size of this window may vary, but `j` is always greater than or equal to `i`.

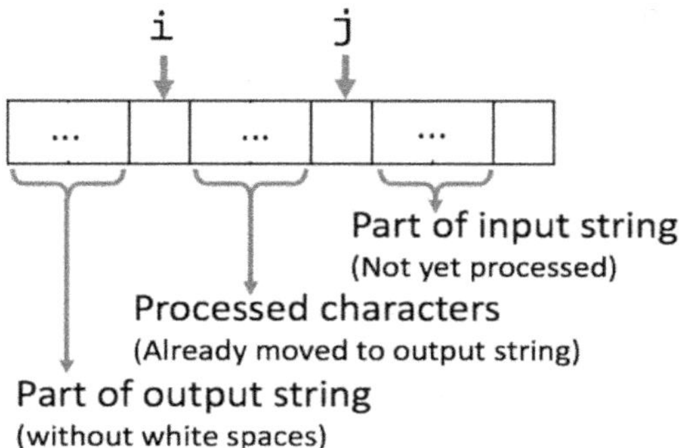

`j` holds the index of current character in the input array. `i` holds the position where the next character should go. If current character is whitespace, then it is ignored and `j` is moved forward. Non-white characters are copied at index `i`, after which both i and j are moved forwad. This algorithm traverses the array only once and takes O(n) time.

If we consider the part of string starting from i to j as a window, then this window is moving from start to end of the string. The length of window also increases with time. Initially length of the window is zero, with both i and j being same, later, when a white space is encountered, this length increases. If there is no white space in the string, then length of window remains zero till the end. This concept of using a window that moves is called **Sliding Window Technique** and it is used in many questions.

Question 2.10: Given an array of n integers and a number k. Find the maximum sum of k consecutive elements in the array.

Solution: The brute-force way of doing it is to use two nested loops

```
int maxConsecutiveSum(int *arr, int n, int k){
int maxSum = INT_MIN;

  for (int i=0; i<n-k+1; i++){
    // CHECK SUM OF k VALUES STARTING FROM i
    int temp = 0;
    for (int j=0; j<k; j++)
      temp += arr[i+j];

    if(maxSum < temp)
      maxSum = temp;
  }

  return maxSum;
}
```

This code takes O(nk) time because of nested loops. In the solution, we are performing some of the additions multiple times. When i=0, the sum of elements from arr[0] to arr[k-1] is computed, and when i=1, sum of elements from arr[1] to arr[k] is computed. These two sum values overlap and both of them are computing sum of values from arr[1] to arr[k-1].

An alternate solution is to use a sliding window of width k, initially starting from 0 to k-1. If k=3 and given array is

```
int arr[] = {2, 6, 3, 9, 1, 5, 2, 3, 14, 0}
```

The initial positioning of the sliding window has i=0 and j=k-1.

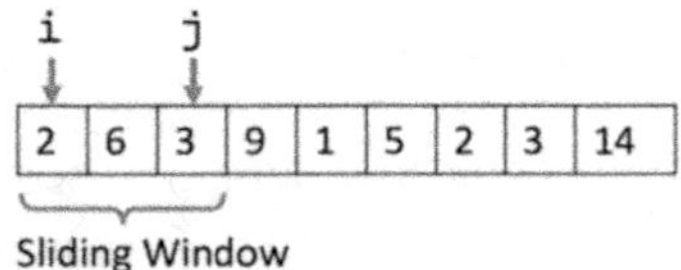

Compute the sum of numbers from `i` to `j` (both inclusive). To compute the sum of next `k` consecutive numbers we shift the window right by one position and compute new sum value in constant time:

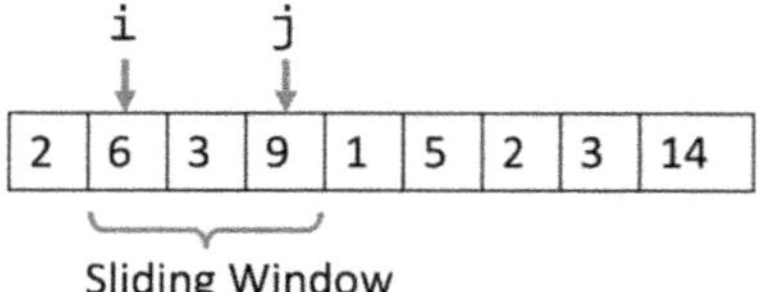


```
New_Sum = Old_Sum - arr[0] + arr[k];
```

All sum values can be computed in a single loop in linear time:

```
int maxSumSlidingWindow(int *arr, int n, int k){
  int sum = 0;

  // COMPUTING THE INITIAL SUM
  for(int cnt=0; cnt<k; cnt++)
    sum += arr[cnt];

  int i=0, j=k; // BOUNDS OF SLIDING WINDOW
  int maxSum = sum;
  while(j<n){
    sum = sum - arr[i] + arr[j];
    if(sum > maxSum)
      maxSum = sum;

    i++; j++; // MOVING THE WINDOW
  }

  return maxSum;
}
```

We have not checked for error cases like if `k>n`, to keep it simple.

Sliding window concept can be used to solve multiple such problems. Some of the examples are:

1. Find the character that is repeating maximum number of times consecutively in a given string.
   ```
   Input String: "Lettere"       Output: t
   Input String: "Kaaaamdsfff"   Output: a
   ```
   The question is not to consider the overall count, but the count of consecutive characters.

2. Given a sorted string of characters. Write code to remove duplicate values from the string.
   `Input String: "aaabbccgg"   Output: "abcg"`
   `Input String: "aaaadfffss"  Output: "adfs"`

Question 2.11: Count words in a given string.

`Input String: "C INTERNALS FOR CODING INTERVIEWS"`

`Output String: 5`

Solution: This is a very simple problem to comprehend, and reasonably easy to solve. The only thing to note is that there can be multiple white spaces between any two words. So, do not put a logic that accept only one space in between two words. In fact, two words may be separated by a tab (and not by a space) or a new line, with first being the last word of a line and second beign the first word of the next line. The logic has to ignore all consecutive white spaces and look for a non-white character. Once you find the first non-white character increment the word count and ignore other consecutive white characters spaces.

```
int wordCount(char* str){
  int numWords = 0;
  while(*str != '\0')
  {
    if(isWhiteSpace(*str)){
      // CURRENT GROUP OF CHAR ARE WHITE SPACES
      while(*str!= '\0' && isWhiteSpace(*str))
      str++;
    }else{
       // CURRENT GROUP OF CHAR IS A WORD
       numWords++;
       while(*str!= '\0' && !isWhiteSpace(*str))
         str++;
    }
  }
  return numWords;
}
```

This function uses the `isWhiteSpace` function of Question 9.

Question 2.12: Given a string, write code that reverses the string, but does not reverse individual words.

`Input String: "C INTERNALS FOR CODING INTERVIEWS"`

`Output String: "INTERVIEWS CODING FOR INTERNALS C"`

Solution: We have seen the logic to reverse an array or a string. Let us write a variation of the reverse function that reverse part of the string starting from one position to another.

```
void reverse(char *str, int startpos, int endpos) {
  int i = startpos;
  int len = startpos + (endpos-startpos)/2;

  for (; i<len+1; i++){
    char temp = str[i];
    str[i] = str[endpos-i + startpos];
    str[endpos-i + startpos] = temp;
  }
}
```

This function takes a string and two indices and reverse the string from `startpos` to `endpos` (both inlusive). Using this function, we first reverse the entire string and then reverse each word in the string. If input string is:

```
char str[] = "C INTERNALS FOR CODING INTERVIEWS"
```

Then after reversing the entire string, we get
`"SWEIVRETNI GNIDOC ROF SLANRETNI C"`

Now when we reverse each word in the string, we get the desired output
`"INTERVIEWS CODING FOR INTERNALS C"`

Below is the code that use `reverse` and `isWhiteSpace` functions:

```
void reverseWords(char* str)
{
  int len = strlen(str)-1;
  reverse(str, 0, len);   // REVERSE ENTIRE STRING

  int startpos = 0, i = 0;

  // MOVE TO FIRST NON-WHITE CHARACTER
  while( isWhiteSpace(str[i])&& (str[i] != '\0'))
    i++;

  while(i<=len){
    i++;
    if(isWhiteSpace(str[i]) || str[i] == '\0')
    {
      reverse(str, startpos, i-1);
      // MOVE i TO NEXT NON-WHITE CHARATER
      while(isWhiteSpace(str[i]) && str[i] != '\0')
        i++;
      startpos = i ;
    }
  }
}
```

Reversing a string is a linear time operation. The function reverses the array twice and takes O(n) time.

Multi-Dimensional arrays

A two-dimensional array is defined as:

```
int arr[3][2] = {{1, 2}, {3, 4}, {5, 6}};
```

It represents the array with three rows and two columns with given values:

1	2
3	4
5	6

`arr` is an array of three elements, each of which is an array of two `int`s. `arr[1]` represent the second element of the array (which itself is an array). `arr[1][0]` represent first integer in the second array that is 3.

Passing 2-dimensional array to functions

While passing a one-dimensional array to a function, we pass (and receive in the called function), the address and number of elements separately.

```
int func(int *arr, int n){ … … }
```

Even if we mention empty brackets, it means the same

```
int func(int arr[], int n){ … … }
```

But only first dimension can be left like this, all the other dimensions has to be specified explicitly. The below code is an error

```
int func(int arr[][], int n, int m){ … … } // ERROR
```

If we mention the second dimension explicitly, then it is fine:

```
int func(int arr[][5], int n){ … … }
```

The second dimension has to be specified. There are two popular ways of passing a multi-dimensional array to a function

If dimensions are available globally, either as macro, constant or variable.

```
#define N 4
#define M 5
void fun(char data[N][M]);
```

As stated above, the first dimension can be skipped.

Second way is to pass size as function parameter.

```
int func(int n, int m, int arr[n][m]){ … … }
int func(int m, int arr[][m]){ … …}
```

This is variable length array, needless to say, variable length arrays work only if your compiler is compatible with C99 standard.

In some specific cases, we can also pass two-dim array type casted as one dimensional and receive it in a single pointer array as shown below:

```
void print(int *arr, int m, int n){
  int i, j;
  for (i = 0; i < m; i++)
    for (j = 0; j < n; j++)
      printf("%d ", *((arr+i*n) + j));
}

int main(){
  int arr[][3]={{1, 2, 3}, {4, 5, 6}, {7, 8, 9}};

  print((int *)arr, 3,3);
  return 0;
}
```

We are just reinterpreting the 2-dimensional array as one-dimensional chunk of memory and take advantage of its contiguous memory allocation. Let us see if we can do the same for two-dimensional arrays on heap?

2-dimensional array on heap

As seen above, each element of a two-dimensional array of integers is an `int` array (pointer-to-pointer-to-integer and not just pointer-to-integer).

```
int **p;
```

An integer array with 3 rows having 5 columns each is allocated memory on heap in a two-step process:

1. Allocate 1-dim array of 3 `int*` and store its address in p.

```
p = (int**)malloc(3 * sizeof(int*));
```

2. Allocate 1-dim memory of 5 integers thrice and store their addresses in `p[0]`, `p[1]` and `p[2]` respectively.

```
for(int i=0; i<3; i++)
  p[i] = (int*) malloc(5 * sizeof(int));
```

Memory allocated to p looks like: (blocks has index of individual element)

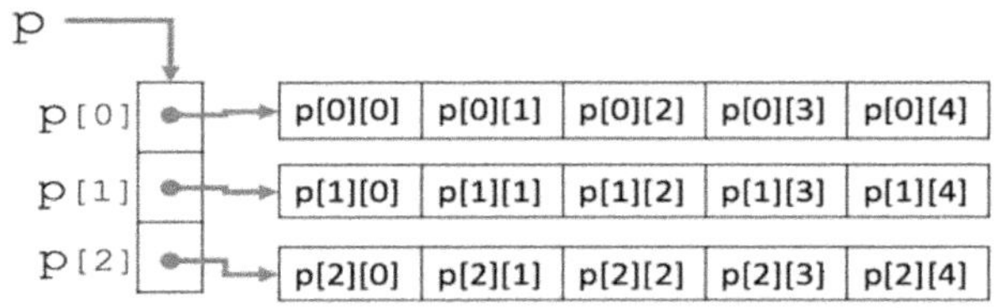

p
p[0] → p[0][0] p[0][1] p[0][2] p[0][3] p[0][4]
p[1] → p[1][0] p[1][1] p[1][2] p[1][3] p[1][4]
p[2] → p[2][0] p[2][1] p[2][2] p[2][3] p[2][4]

Array on the left side is used to store meta data. It is required because the array is on heap. If this 2-dim array is on stack only 15 contiguous integers blocks of memory are allocated. But on heap total memory required is:

```
- 15 blocks of int memory location
- 3 blocks of int* memory location
- 1 int** memory location
```

While accessing `(i,j)`th element of a 2-Dimensional array we have to dereference two levels to reach the actual element. All the four expressions below refer to element `p[i][j]` in the array:

```
p[i][j], *(p+i)[j], *(p[i] + j), *(*(p+i)+j)
```

Probably the biggest advantage of having a two-dim array on heap is that each row can have different number of columns, as in figure below:

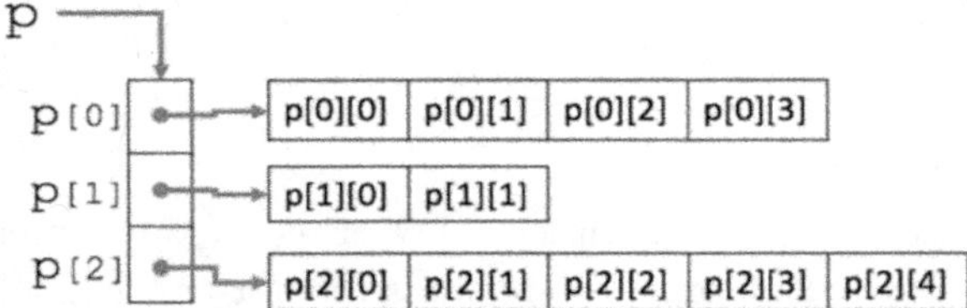

A statically allocated two-dimensional array provides the number of columns in the declaration itself

```
int arr[3][5];
```

If we are having a sparse array, then storing it on heap may be a good design choice.

Deallocating a two-dimensional Array on heap

To deallocate the two-dimensional array allocated on heap, we have to first deallocate each element of the one-dimension array on left side (array pointing to by pabove) before deallocating pointer `p` that points to the two-dim array

```
for(int i=0; i<m; i++)
  free(p[i]);

free(p);
p = NULL;
```

Note that if we first free `p` without freeing individual `p[i]`, then all `p[i]`'s become memory leaks.

Note that we did not assign `NULL` to individual `p[i]` elements. This is because they are deallocated in the very next statement.

N-dim Array

Arrays with more than two dimensions are difficult to visualize, but process of allocating them on heap remains same with one more added level of abstraction. For example: The address of a three-dimension array is `int***` and we need multiple loops. Following is the C language code to allocate a 3-dim array of order `A*B*C`

```
int ***p = (int***)malloc(A * sizeof(int**);

for(int i=0; i<A; i++)
  p[i] = (int**) malloc(B* sizeof(int*));

for(int i =0; i<A; i++)
  for(int j=0; j<B; j++)
    p[i][j] = (int*) malloc(C * sizeof(int));
```

and element at index (`i,j,k`) is accessed like `p[i][j][k]` or `*(*(*(p + i) + j) + k)`.

It can be further generalized for N-dimensional arrays.

Deallocating N-dimensional array

Deallocating a `3`-dimensional array is an intuitive generalization of the way we de-allocate two-dimensional array.

```
for(int i=0; i<A; i++)
  for(int j=0; j<B; j++)
    free(p[i][j]);

for(int i=0; i<A; i++)
  free(p[i]);

free(p);
p = NULL;
```

The process of deallocating memory is opposite to that of allocating memory. In C++, you may notice, the sequence of calling destructors is opposite to the sequence of calling constructors. While making a wall, you will first put the bricks and then apply plaster over it. But, while breaking the same wall, you have to first break the plaster.

Practice Questions

Question 2.13: Given a matrix, write code to print the diagonal elements.

1	2	3	4
5	6	7	8
9	10	11	12
13	14	15	16

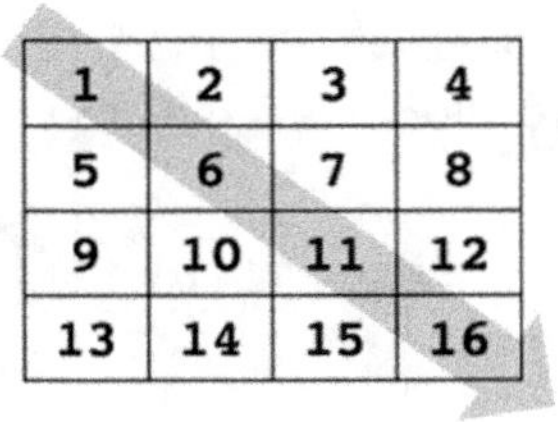

Output: 1 6 11 16

Solution: An element arr[i][j] is diagonal elementiffi and j are equal. Consider the below code:

```
for(int i=0; i<N; i++)
  for(int j=0; j<N; j++)
  if(i==j)
      printf(" %d", arr[i][j]);
```

This is a working code, that produce the desired output. But, while answering questions in an interview, make sure you are giving your best.

This code takes O(n^2) time. We are unnecessarily traversing the entire array. Following code jumps from one diagonal element to another and print the diagonal elements in O(n) time.

```
for(int i=0; i<N; i++)
  printf(" %d", arr[i][i]);
```

Question 2.14: Given a matrix, write code to print the elements in the anti-diagonal as shown below:

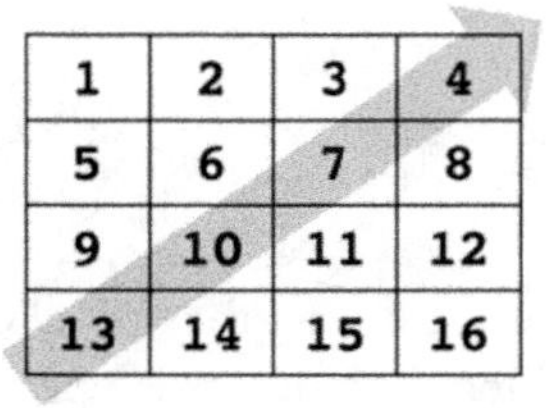

1	2	3	4
5	6	7	8
9	10	11	12
13	14	15	16

Output: 13 10 7 4

Solution: Like the previous question, we want to give an O(N) time solution. Below code, initialize the value of i and j to the bottom-left element and then move one element up in the anti-diagonal (decrease current-row and increase current-column).

```
int i = N-1, j = 0;
while(i>=0 && j<N){
  printf(" %d", arr[i][i]);
  i--; j++
}
```

Question 2.15: Given a square matrix of order `N*N`, write code to print all the elements in the order of their diagonal. For example, in the below matrix, the elements should be printed in the marked order, and the final output should be:

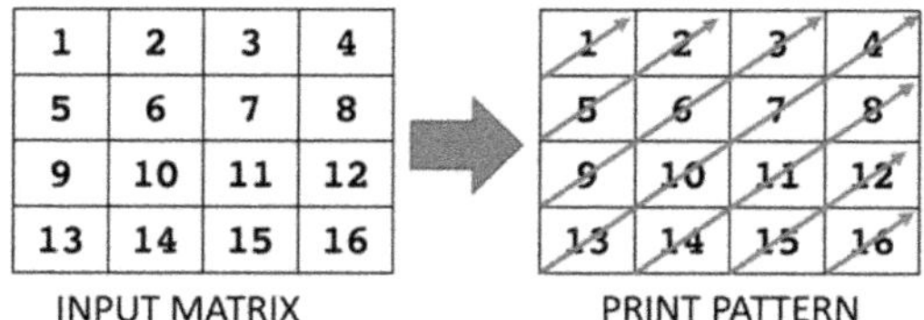

Output: 1 5 2 9 6 3 13 10 7 4 14 11 8 15 12 16

Solution: Code to print all the elements is similar to the previous question. Printing one small diagonal at a time. Total number of diagonals are `2*N-1` (each arrow represents one diagonal in the above diagram).

```
void printAllDgl(int N, int a[][N]){
  int i=0, j=0;

  // LOOP TO PRINT EACH DIAGONAL
  for(int cnt=0; cnt<2*N-1; cnt++){
    if(cnt<N) {
      i = cnt;
      j = 0;
    }
    else {
      i = N-1;
      j = (cnt+1)%N;
    }

    while(i>=0 && j<N){
      printf(" %d", a[i][j]);
        i--;j++;
    }
  }
}
```

The code takes $O(N^2)$ time, which is optimal because total number of elements printed are also $O(N^2)$.

Question 2.16: Given a two-dimensional array, write code to traverse it in spiral order. For example, for the below matrix, the spiral order traversal is:

1	2	3	4
5	6	7	8
9	10	11	12
13	14	15	16

Given Matrix

Spiral Order

```
1 2 3 4 8 12 16 15 14 13 9 5 6 7 11 10
```

Solution: In one loop iteration, print one complete circle of elements. Hence, we will have 4 loops inside a bigger loop to print the

1. **Top Row** from Left to Right
2. **Last Column** from Top to bottom
3. **Last Row** from Right to Left
4. **First Column** Bottom to Top

In the first pass of the outer loop the greyed elements get printed and in the next pass, inner circle is printed.

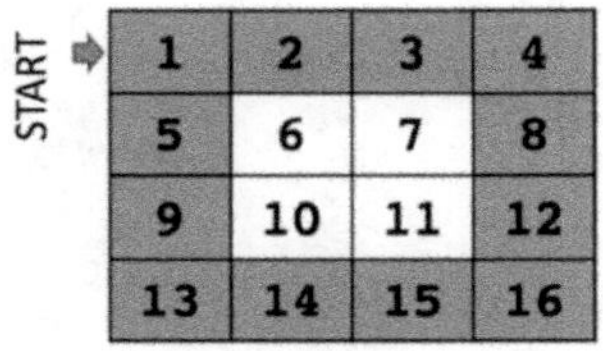

1	2	3	4
5	6	7	8
9	10	11	12
13	14	15	16

First Pass

START

6	7
10	11

Second Pass

And so on… Below is the code for above logic:

```
#define M 4
#define N 4
void spiralTraversal(int a[M][N]){
  int rs=0, cs=0;      // Row-Start AND Column-Start
  int re=M-1, ce=N-1; // Row-End AND column-End

  while(rs<=re && cs<=ce){     // OUTER LOOP
    int i=rs, j=cs;

    for(j=cs; j<=ce; j++)   // TOP ROW
      printf(" %d", a[i][j]);
    for(i=rs+1, j--; i<=re; i++) // LAST COLUMN
      printf(" %d", a[i][j]);
     for(j=ce-1, i--; j>=cs; j--) // LAST ROW
      printf(" %d", a[i][j]);
```

```
    for(i=re-1, j++; i>=rs+1; i--) // FIRST COLUMN
      printf(" %d", a[i][j]);

    rs++; cs++; re--; ce--;
  }
}
```

Question 2.17: Consider a situation where data reside in a symmetric matrix (two-dimensional square array). Element stored in index (`i,j`) and (`j,i`) is always same. If a value is assigned to index `arr[i][j]`, then `arr[j][i]` also gets updated with the same value maintaining the symmetry.

1	2	3	4
2	5	6	7
3	6	8	9
4	7	9	10

Values in the `i`th row are same as values in the `i`th column. How will you implement this data structure in a memory-efficient way?

Solution: If an interviewer asks such a question, he is expecting us to:

1. Give a storage structure. Whether we will continue to store values in a matrix or in a one-dimensional array or in some other data structure.
2. The method to update value at index (`i,j`). It can be as simple as `arr[i][j] = arr[j][i] = new_value;`
3. The method to get the value stored at index (`i,j`).

Since the matrix is symmetric, duplicate values are stored and we can save some space by storing one value only once. Let us store the matrix in a one-dimensional array as shown below:

Original Matrix

1	2	1	3	2
2	3	5	1	4
1	5	6	7	2
3	1	7	2	8
2	4	2	8	9

Stored in 1-dimensional array

1	2	1	3	2	3	5	1	4	6	7	2	2	8	9

The duplicate values of matrix are not stored. Now, from user point of view, it is a two-dimensional matrix, but actually the values are stored in a one-dimensional array.

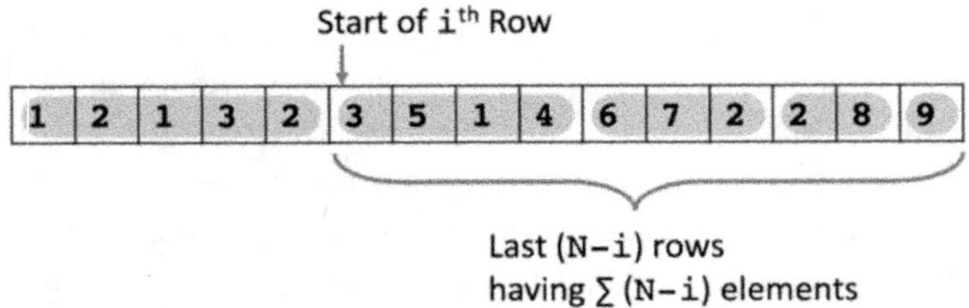

We have saved some memory without loosing data. Next, we should create a mapping between the index (i,j) of two dimensional matrix (that user use to access elements) and index of the actual one-dim array where values are stored. Total number of elements stored in the array are:

$$\sum_{i=0}^{N} i = \frac{N * (N+1)}{2}$$

Index (i,j) means jth element in the ith row. Elements at index (i,j) and (j,i) are same. Below function map the index of matrix with that of array.

```
int mapMatToArray(int i, int j){
  if(i<=j)
    // POSITION OF (i,j)
    return N*(N+1)/2 - (N-i)*(N-i+1)/2 + j-i;
  else
    // POSITION OF (j,i)
    return N*(N+1)/2 - (N-j)*(N-j+1)/2 + i-j;
}
```

If i<j, then we look for element at (j,i) even when asked for element at (i,j), because both are same and we are storing it only once.

Below are the getter and setter functions used to read and write values from the matrix.

```
// RETURN ELEMENT AT INDEX (i,j) IN MATRIX
int getValue(int i, int j){
  return arr[mapMatToArray(i,j)];
}
```

```
// PUT value AT INDEX (i,j) IN THE MATRIX
void setValue(int i, int j, int value){
  arr[mapMatToArray(i,j)] = value;
}
```

The below function prints Array as a two-dim matrix (with duplicate values):

```
void printMatrix(){
  for(int i=0; i<N; i++){
    for(int j=0; j<N; j++)
      printf("%d ", getValue(i, j));
    printf("\n");
  }
}
```

Output:
```
1 2 1 3 2
2 3 5 1 4
1 5 6 7 2
3 1 7 2 8
2 4 2 8 9
```

Question 2.18: Glasses are arranged in the form of a triangle (on top of each other) as shown below:

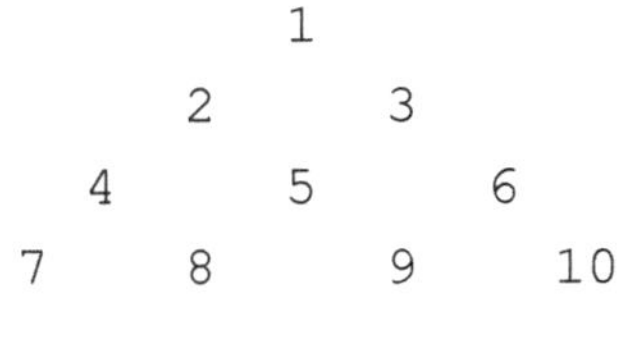

Glasses shown as numbers

Actual arrangement

The liquid is poured into Glass-1 (top glass). When it is full, extra liquid flows down into Glass-2 and Glass-3 in equal quantities. When Glass-2 and Glass-3 are full, liquid flows into Glasses 4, 5 and 6. Amount of liquid pouring into Glass-5 is twice the amount of liquid that goes into Glass-4(or Glass-6) because from Glass-2 liquid comes down into Glass 4 and 5 and from Glass-3 liquid flows into Glass 5 and 6. Glass-5 receives twice the amount of liquid received by Glass-4 because it gets from both the glasses.

If capacity of each glass is `C` and we pour `N` liters of liquid in Glass-1 from the top, how much liquid will be there in each glass?

Solution: The solution to this problem can be though in terms of Pascal's triangle. On the top water is coming directly into Glass-1. At level-2 water will be equally divided into two glasses.

Pascal Triangle

At level-3, there are three glasses, ratio of water coming in these glasses is 1:2:1. i.e. middle glass gets twice as much water as the left and right (this does not change the capacity, just the water flowing in it). and so on…

If 7 liters of water is poured from the top, then:

- At level-1, 7 liters of water flow in and 1 liter is stored.
- At level-2, 6 liters of water flow in and 2 liters is stored.
- At level-3, 4 liters of water flow in and 3 liters is stored.
- At level-4, 1 liters of water flow in (thru Glass-2 of level-3).

  This 1 liter flows equally into 2nd and 3rd glass at level-4.

The solution becomes simple if we visualize it and take a sample input. For simplicity, use a two-dimensional array data structure (and not tree, which it may look like).

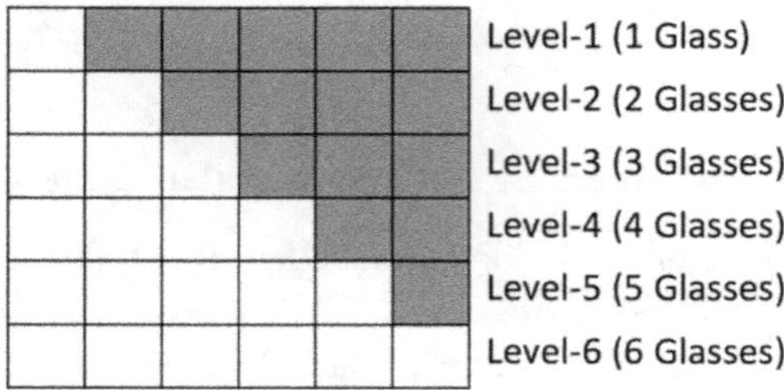

Glasses at level-`i` are represented by cells in Row-`i` as shown in the diagram above (`i` th level has `i` glasses). In the first row only one cell is used (because there is only one glass at that level). In the second row two cells are used and so on….

Almost half of the arrays pace is wasted, so its not efficient in terms of memory consumption and we can think of optimizing it by using method shown in previous question, but for the time being let us focus on the algorithm.

If N liters of liquid is poured from top, we store N in place of Glass-1 (arr[0][0]). Iterate on each row and for each cell check if the value stored in it is more than C (capacity of the glass). If it is more than C, add the extra value to its children (two glasses below it). The two glasses below arr[0][0] are arr[1][0] and arr[1][1].

```
            arr[0][0]
        arr[1][0]  arr[1][1]
  arr[2][0]  arr[2][1]  arr[2][2]
     ..................................
```

From cell arr[i][j] water will flow into cells arr[i+1][j] and arr[i+1][j+1].

In below code, function fillWaterInGlasses, accept the amount of water poured in from the top and returns total number of levels till which we will have water in the glasses.

```
#define MAX_LEVEL 100
// MATRIX OF GLASSES
double glasses[MAX_LEVEL][MAX_LEVEL];

int fillWaterInGlasses(double C, double N){
  glasses[0][0] = N;  // POUR FULL WATER IN TOP GLASS

  int level=0;
  bool waterInLevel = true;
  while(waterInLevel){
    waterInLevel = false;

    // FOR EACH GLASS IN THIS LEVEL.
    for(int j=0; j<=level; j++){

      // IF GLASS HAS MORE LIQUID THAN IT CAN STORE.
      if(glasses[level][j] > C){
        double extraWater = glasses[level][j] - C;
        glasses[level][j] = C;
        glasses[level+1][j] += extraWater / 2;
        glasses[level+1][j+1] += extraWater / 2;
        waterInLevel = true;
      }
    }
    level++;
  }
```

```
  return level-1;
}
```

Call function `fillWaterInGlasses` for `N=10`, and `C=1` (capacity of each glass is `1` liter, and `10` liters of water is poured from top) is as below:

```
int main(){
  int maxLevel = fillWaterInGlasses(1, 10);
  for(int i=0; i<=maxLevel; i++){
    for(int j=0; j<=maxLevel; j++)
      printf("%lf ", glasses[i][j]);
    printf("\n");
  }
}
```

Output is:

```
1.0   0.0    0.0    0.0    0.0
1.0   1.0    0.0    0.0    0.0
1.0   1.0    1.0    0.0    0.0
0.375 1.0    1.0    0.375  0.0
0.0   0.3125 0.625  0.3125 0.0
```

Which actually means (writing output till 2 decimal places):

```
                1
           1         1
       1        1         1
   0.37    1         1        0.37
 0      0.31     0.62     0.31     0
```

Question 2.19: Given a matrix of integers of order `M*N` and coordinates of a rectangular region in the matrix, write code to find the sum of all the elements in that rectangular region.

1	2	0	3	4	1	5	8	1	0
1	5	5	2	4	9	1	2	0	1
3	8	1	3	3	7	2	1	4	9
5	2	8	6	1	0	8	4	2	3
1	4	2	5	6	3	0	1	8	1
8	2	3	5	4	1	7	2	9	3
1	7	1	0	0	1	2	7	4	3
8	5	5	9	1	2	0	3	4	2

For example, if the input matrix is

And co-ordinated given are (3,4) and (6,8), then they represent the below rectangular region whose Sum is 70.

Hence the output should be 70.

How will your logic change if the matrix remains same but, we need to compute sum of rectangular regions many times?

1	2	0	3	4	1	5	8	1	0
1	5	5	2	4	9	1	2	0	1
3	8	1	3	3	7	2	1	4	9
5	2	8	6	1	0	8	4	2	3
1	4	2	5	6	3	0	1	8	1
8	2	3	5	4	1	7	2	9	3
1	7	1	0	0	1	2	7	4	3
8	5	5	9	1	2	0	3	4	2

Solution-1: The brute-force solution is very simple, we just need to traverse in the rectangular region and keep adding each value to compute the total sum.

```
int regionSum(int startI, int startJ, int endI, int endJ)
{
  int sum = 0;
  for(int i=startI; i<=endI; i++)
    for(int j=startJ; j<=endJ; j++)
      sum += arr[i][j];
  return sum;
}
```

This code takes $O(n^2)$ time in the worst case. The problem is that every time we want to compute the sum of a region, we have to do the entire work again and will take $O(n^2)$ time in each case.

We may precompute the sum of rectangles and store them in a cache to reduce this time complexity of calculating region sum again and again. We will be incurring a one-time cost of $O(n^2)$ time.

Solution-2: Precompute the cache

Take one more matrix of the same order in which cell `(i,j)` will store the sum of all the cells in the rectangular region starting from `(0,0)` till `(i,j)`, both included of the original matrix. For the given matrix, this cache will look like:

1	3	3	6	10	11	16	24	25	25
2	9	14	19	27	37	43	53	54	55
5	20	26	34	45	62	70	81	86	96
10	27	41	55	67	84	100	115	122	135
11	32	48	67	85	105	121	137	152	166
19	42	61	85	107	128	151	169	193	210
20	50	70	94	116	138	163	188	216	236
28	63	88	121	144	168	193	221	253	275

Now, for each cell we know the sum of the rectangular region starting from `(0,0)` to that cell.

Suppose we want to compute the sum of elements in the rectangular region in the below matrix:

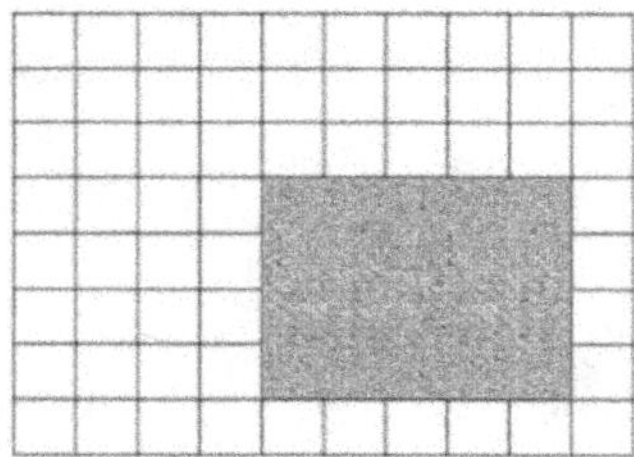

We do not know the sum of elements in this rectangle, but we know the sum of elements in all the below four rectangles, because their top-left cell is starting from `(0,0)`

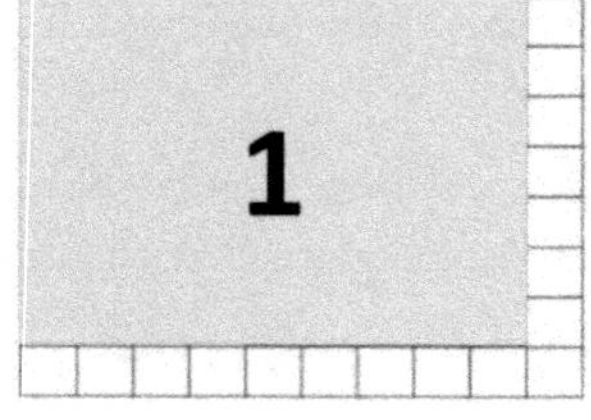

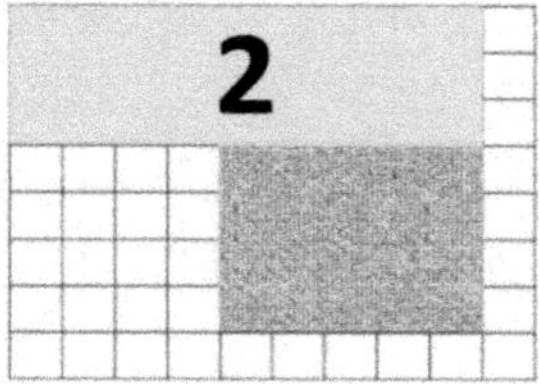

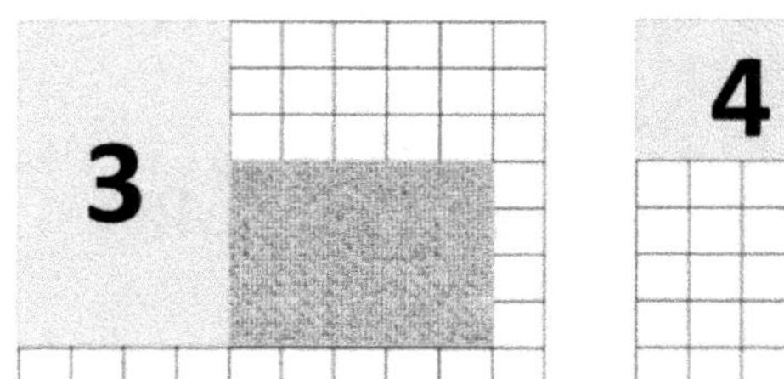

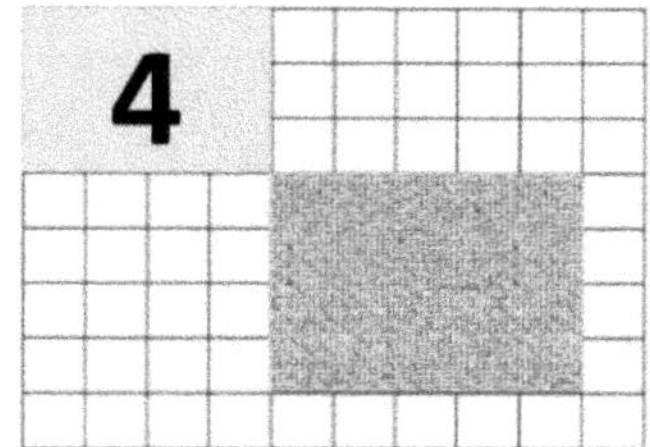

Sum of elements in the asked rectangle is equal to **1** – **2** – **3** + **4**. And it will be a constant time operation.

Chapter 3

Structures and Linked List

Linked list, like an array, is a homogeneous collection of elements, but, these elements are not contiguous and hence not indexed. Because they are not contiguous, next element cannot be reached by just incrementing the pointer. Each element of list holds a link to next element (in addition to data) and we move forward in the list by using that link.

This combination of data and link is called **Node**. A Node is defined as a structure in C and class in Java. Code 3.1 shows the structure of Node whose data is an integer.

```
struct Node
{
  int data;
  Node *next;
};
```

<u>**Code: 3.1**</u>

A variable of `struct Node` type is defined as:

```
struct Node obj;
```

Note that individual fields of a structure are not homogeneous, but they are contiguous, and each element of the list is of the same type (please do not confuse struct type with the linked list). The memory allocated to variable `obj` looks like below.

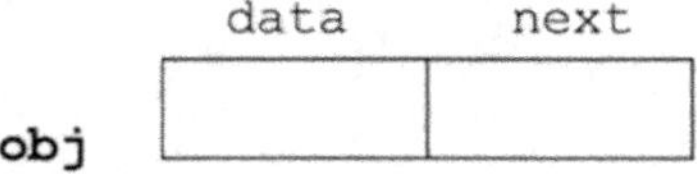

Individual field of a structure is accessed by using structure name with dot operator

obj.data = 5; // Assign 5 to data field
obj.next = NULL; // Assign NULL to next field

If `ptr` is a pointer pointing to the structure,

```
struct Node * ptr = &obj;
```

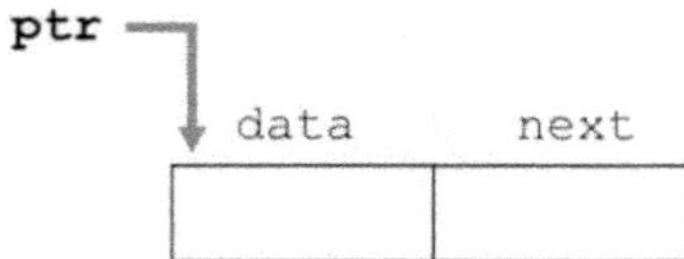

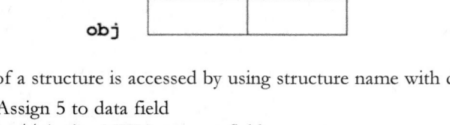

Then `(*ptr).data` can be written more conveniently by using arrow operator `ptr->data`. They both represent the l-value of field `data` of structure `Node`.

In C++, `struct` keyword in the declaration can be omitted.

```
Node obj;
```

In C language, developers usually `typedef` the structure type to the name.

```
typedef struct node
{
  int data;
  struct node *next;
} Node;
```

If you do not find the `struct` keyword in a declaration in this book, assume that it to be a `typedef.` Let us define three variables of structure `Node` and initialize them as below:

```
struct Node a, b, c;

a.data = 5; a->next = &b;
b.data = 10; b->next = &c;
c.data = 15; c->next = NULL;
```

If the address of `a`, `b` and `c` are `100`, `200` and `300` respectively, after assigning these values, the memory looks like:

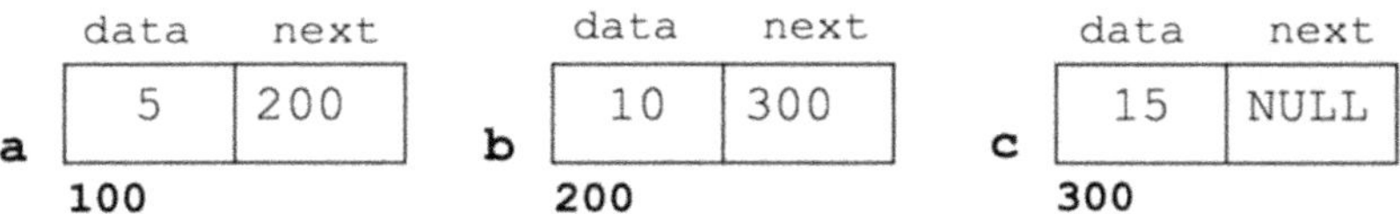

Physical addresses of these memories are not important, the fact is that `a.next` stores address of `b` and `b.next` stores address of variable `c`, and this fact is important. In next execution, physical addresses of these variables may be different, but this fact will always be true. A better picture of the memory, therefore is:

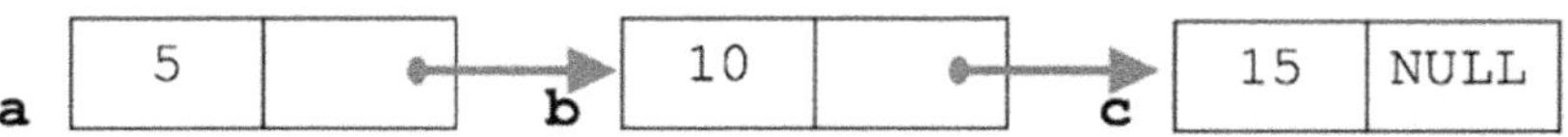

This is a linear homogeneous arrangement of data, where all the elements are of same type (there are three elements, all are of type `Node`). However, the elements are not contiguous and second element may physically be present before the first. The value in `next` field of the first node gives us the address of the second element. So to get to the kth element, we have to come to $(k-1)^{th}$ element (from where we will get the address of the kth element). The actual physical address of nodes is not important.

In this type of arrangement, even variable names are not required to access elements in the list. If we just have a pointer to the first element of the list.

```
struct Node *ptr = &a;
```

We can access all the elements

```
printf("%d ", ptr->data); // PRINT 5
printf("%d ", ptr->next->data); // PRINT 10
printf("%d ", ptr->next->next->data); // PRINT 15
```

Below function receives a pointer to the first element (called `head`) of the list and prints the value in each node in the list.

```
void printList(Node *head){
  while(head != NULL){
    printf("%d ", head->data);
    head = head->next;
  }
}
```

Code 3.2

Almost always, memory to a linked list is allocated on the heap and a pointer pointing to the head of list is used to maintain the list. Let us call this pointer `head`, because it points to the head (first element) of the list. I have never seen a linked list or Binary tree or any other data structure that use a linked list (Stack implemented using linked list) allocated memory on stack instead of heap memory.

Linked list is a collection of nodes where each node stores the data and pointer to next node. `next` field of the last node is NULL. Below figure shows same data residing in an array and a linked list.

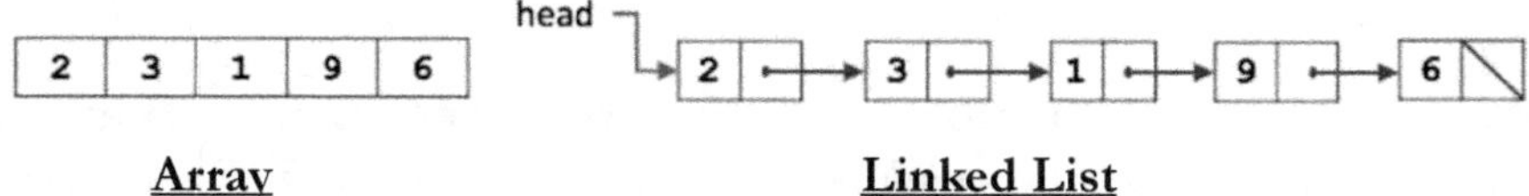

Array **Linked List**

All elements in an array are physically placed one after another. If we have address of the first element, we can traverse the entire array by using pointer arithmetic. Elements (Nodes) of linked list are scattered. We keep the address of first element (Node), first element stores address of the second element and so on.

Obviously, linked list takes more memory than an array. Problem with an array is that its size is fixed when memory is allocated to it. A Linked list, on the other hand, can grow and shrink dynamically.

Inserting an element in an array requires shifting of elements to make room for the new element. Inserting in a linked list does not require any such shifting and can be achieved by adjusting pointers. If we are maintaining the elements in sorted order, array and list are like below:

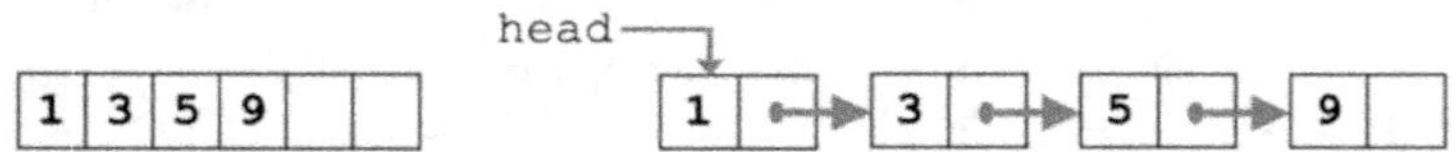

To allow insertion of elements in the array, we have to keep extra room while allocating the array (two empty spaces at the end). Now, if we want to insert value 4 in these two collections, then after insertion, they look like below (Note that inserting elements in an array require us to shift elements):

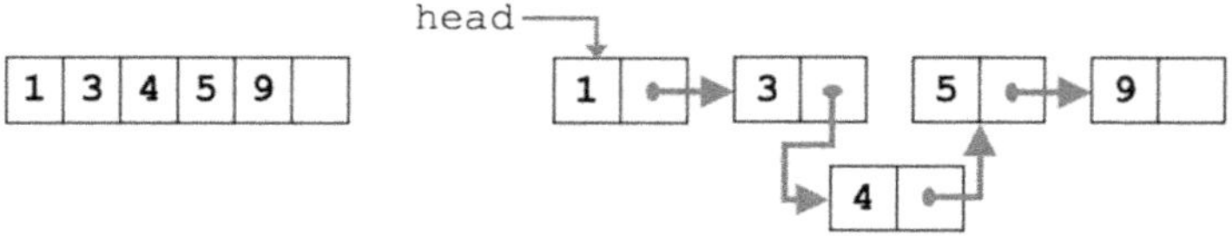

Let us look at the code to insert in a sorted array (assuming enough space at the end).

```
void insertInSortedArray(int* arr, int n, int value){
  int i=n-1;
  while(i>=0 &&(arr[i] >value)){
    arr[i+1] = arr[i];// SHIFT ELEMENT FORWARD

    i--;
  }
  arr[i+1] = value; // INSERTING data
}
```

While inserting an element in a linked list, there are three possibilities:

1. To insert the element as first element.
2. To insert the element as last element.
3. To insert the element in between.

If we are inserting at the head of list (inserting smallest element), then it's a special case. Otherwise, we first need to move to the node after which the new element need to be inserted. Let `ptr` points to the previous node,

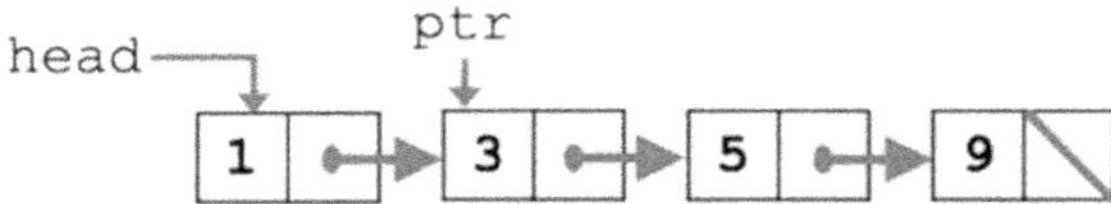

The process of inserting `value` after node pointed to by `ptr` is:

```
// ALLOCATE A NEW NODE.
Node * temp = (Node*) malloc(sizeof(Node));
```

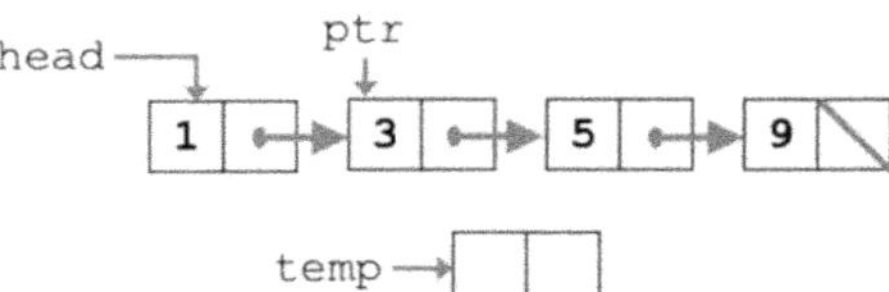

```
// STORE value IN THE NEW NODE
temp->data = value;
```

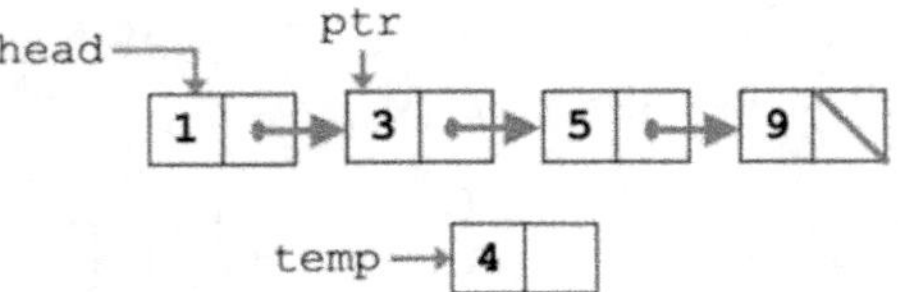


```
// ADJUST THE POINTERS TO BRING NEW NODE IN BETWEEN
temp->next = ptr->next;
ptr->next = temp;
```

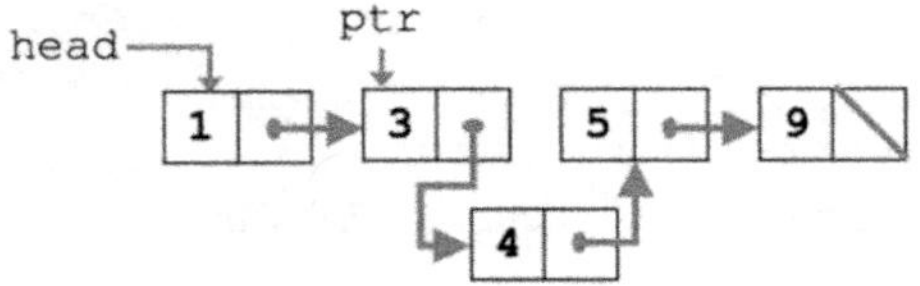

You may check that this logic also works for inserting node at the end of list. To insert a value as the first node, `head` pointer need to be changed:

```
temp->next = head;
head = temp;
```

Below is the complete program to insert an element in a sorted linked list. The function receives `head**` because `head` pointer itself may need to be changed (if inserting at the head of list).

```
// HELPER FUNCTION. ALLOCATE & INITIALIZE NODE
Node * newNode(int value){
  Node* temp = (Node*)malloc(sizeof(Node));
  temp->data = value;   temp->next = NULL;
  return temp;
}

void insertInSortedList(Node** hp, int val){
  Node *temp = newNode(val);
  Node* head = *hp;

  if(head == NULL || head->data>val){
    // INSERTING AT HEAD
    temp->next = head;
    *hp = temp; // CHANGING head OF main FUNC
  }else{
    // MAKE head POINT NODE AFTER WHICH TO INSERT
    while(head->next!=NULL && head->next->data<val)
      head = head->next;
```

```
      temp->next = head->next;
      head->next = temp;
    }
}

int main(){
  Node* head = newNode(1);
  head->next = newNode(3);
  head->next->next = newNode(5);
  head->next->next->next = newNode(9);

  printf("BEFORE INSERTION :"); printList(head);
  insertInSortedList(&head, 4);
  printf("\nAFTER INSERTION :"); printList(head);
}
```

Code 3.3

Output of this program is:

```
BEFORE INSERTION :1 3 5 9
AFTER INSERTION :1 3 4 5 9
```

The insertion logic takes `O(n)` time. This time taken is to identify the position at which the value need to be inserted. Once the point of insertion is identified, the actual insertion takes constant time.

If the new value in a list is always inserted at the head (e.g. List implementation of Stack data structure), then insertion is a constant time operation.

We need to make similar adjustments of pointers to delete anode from the list. Deleting first element is special because it changes the `head` pointer. Function `deleteFromList` search and delete the node that holds the value `val`. If `val` is not found in list, the function returns `false`.

```
bool deleteFromList(Node** hp, int val)
{
  if(hp == NULL || *hp == NULL) // EMPTY LIST
    return false;

  // DELETING FIRST ELEMENT
  Node* head = *hp;
  if(head->data == val){
    Node *temp = head;
    *hp = head->next; // ADJUST head OF main
    free(temp);
  }else{
    // SETTING PREVIOUS POINTER
    Node * prev = head; head = head->next;
```

```
        while(head != NULL && head->data != val){
            prev = head;
            head = head->next;
        }

        if(head == NULL) // val NOT FOUND
            return false;

        // NOW head POINTS TO NODE TO BE DELETED
        // AND prev POINT TO PREVIOUS NODE.
        prev->next = head->next; // DETACH FROM LIST
        free(head);
    }
    return true;
}
```

We first move to the node that need to be deleted. `head` pointer points to the node to be deleted, and `prev` pointer points to the previous node. First the node is logically deleted from the list by bypassing it by using:
`prev->next = head->next;`

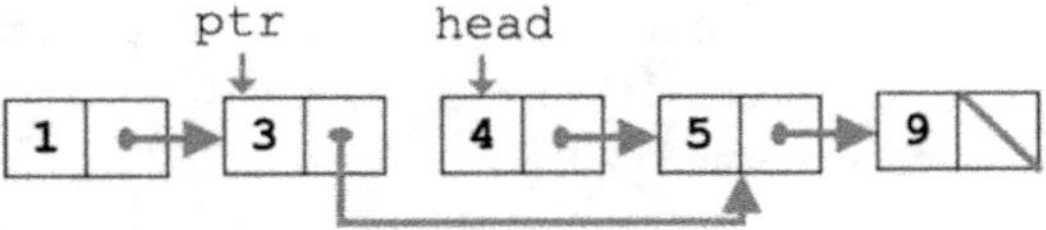

If list is traversed at this point, node with value 4 will not come in the traversal. The node is logically removed from the list, but it is still present in the memory. The `free` function returns this memory back to the system.

To remove a node from the list, we need a pointer to the previous node. It is not possible to delete a node if only a pointer to that node is given and there is no way to reach to the previous node[1]. Let us take an example

Question3.1: Remove all the duplicate nodes from a sorted linked list.
`Input: 1->3->4->4->6->6->6->7`
`Output: 1->3->4->6->7`

Solution: The logic for this is simple

```
For each node in the list
  IF next node value is same as the current node
      delete next node
```

---

1 There is an indirect way to remove the node when only pointer to that node is given.

- Copy data of next node to current node.
- Remove the next node.

But even this trick will not work when we are at last node.

In this algorithm, we are always deleting the next node. This function does not change the head pointer

```
void removeDuplicates(Node* head){
  while(head != NULL){
    if(head->next != NULL){
      // WHILE NEXT NODE'S DATA IS SAME AS THIS NODE
      while(head->next != NULL &&
            head->next->data == head->data)
      {
        Node* temp = head->next;
        head->next = temp->next;
        free(temp);
      }
    }
    head = head->next;
  }
}
```

Deleting entire linked list, means deleting all the node one-by-one and setting the head pointer to `NULL`. The task of setting `head` pointer to `NULL` can be left to calling function. Below function deallocate the entire list.

```
void deleteList(Node* head){
  while(head != NULL){
    Node *temp = head;
    head = head->next;
    free(temp);
  }
}
```

Note that the `head` pointer is set to the next node before deal locating the current node (`temp` pointer). The order of these statements is important, because if we first free the node, then its next pointer becomes dangling. Below is the recursive implementation of above function. It first calls the recursive function to remove rest of the list and then free the current node.

```
void deleteList(Node* head){
  if(head == NULL){ return; } // TERMINATING COND.
    deleteList(head->next);   // FREE REST OF THE LIST.
  free(head);                 // FREE CURRENT NODE
}
```

> Most interview questions on linked list are not tough on logic, but they require a very clever manipulation of pointers and meticulous attention to details and boundary conditions.

While coding for a linked list (and other dynamic data structures) take care of boundary conditions.

Practice Questions

Question 3.2: Given a linked list, write code to split it into two lists each containing alternate nodes from the original list. If the original list is:

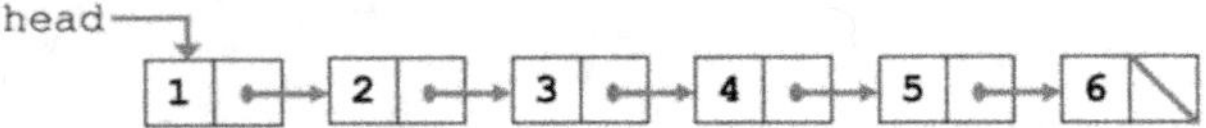

Then, your code should split into two lists like below:

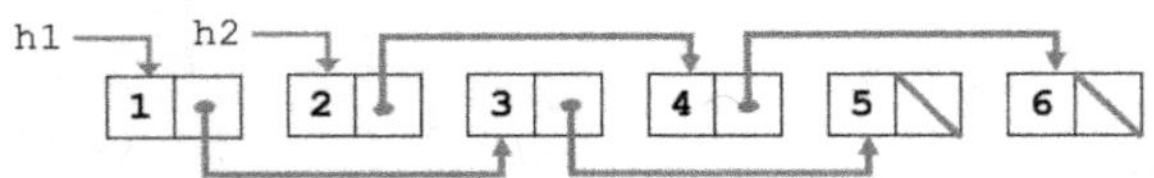

The two output lists are:

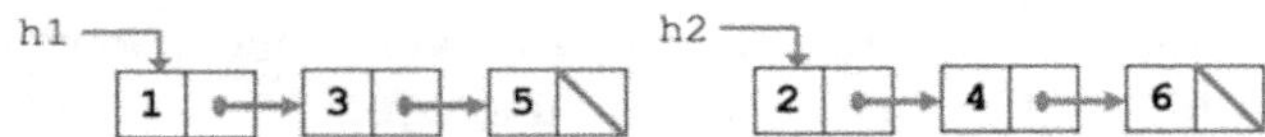

Solution: Traverse the list, remove each node and add it to the alternate list.

First let us decide the signature of the function, because a function cannot return two values directly, we need to populate reference variables:

```
// head: POINT TO HEAD OF ORIGINAL LIST
// pH1, pH2:ADDRESS OF h1 AND h2 IN CALLING FUNCTION.
void splitList(Node* head, Node**pH1, Node** pH2);
```

But before submitting your code to the interviewer, make sure you have checked it for the following:

- If the input list is empty, both output lists should be empty (they should not be garbage).
- If the list has only one element, second output list should be empty.
- The logic should work for even number of nodes.
- The logic should work for odd number of nodes.

They may look like normal test cases, but believe me, even the most expert programmers self-review their code for errors.

```
void splitList(Node* head, Node** pH1, Node** pH2){
  if(head==NULL){
    *pH1 = *pH2 = NULL; return;
  }
```

```
  *pH1 = head; *pH2 = head->next;// SETHEAD PTRs

  if(head->next == NULL)
    return; // LIST HAS ONLY 1 ELEMENT

  // START FROM 3rd ELEMENT

  head = head->next->next;
  Node* end1 = *pH1, *end2 = *pH2;
  int flag=1; // To mark alternate
  for(; head != NULL; head=head->next, flag*=-1){
    if(flag == 1){
      end1->next = head;
      end1 = end1->next;
    }else{
      end2->next = head;
      end2 = end2->next;
    }
  }
  end1->next = end2->next = NULL;
}
```

Two pointers, `end1` and `end2` pointing to last node of each list are kept because we are always inserting at the end of lists. Let us also write the `main` function that calls the above code.

```
int main(){
  Node* head = newNode(1);
  head->next = newNode(2);
  head->next->next = newNode(3);
  head->next->next->next = newNode(4);
  head->next->next->next->next = newNode(5);
  head->next->next->next->next->next = newNode(6);

  Node* h1 = NULL, *h2 = NULL; // FOR OUTPUT LIST
  printf("ORIGINAL LIST :"); printList(head);
  splitList(head, &h1, &h2);
  printf("\nFIRST LIST :");  printList(h1);
  printf("\nSECOND LIST :");  printList(h2);
}
```

Function `printList` is defined in Code 3.2. Note that `head` pointer inside `main` does not change and continue to point to node it was pointing earlier, `h1` and `h2` are set appropriately by function `splitList`.

Question 3.3: Given a singly linked list, write a function to reverse the list.

Solution: This is one of the most used operations on a linked list. As we will see in other questions also, many linked list problems require to reverse the list as a part of their solution. Even a simple code to print the list in reverse order may require us to first reverse the list.

Reversing a list will change the `head` pointer. There are two ways to do it. First is to accept a `Node**` and change the head pointer of calling function within `reverse` function itself, similar to how we update the `head` pointer while inserting and deleting nodes in the list.

```
void reverse(Node** headPtr);
```

Another way is to accept a `Node*`, reverse the list and return address of the head of reversed list from the function

```
Node*reverse(Node*head);
```

We are using the second signature below:

```
Node* reverse( Node * head){
  Node* a; Node* b = NULL;

  while(head != NULL){
    a = head->next;
    head->next = b;
    b = head;
    head = a;
  }

  return b;
}
```

`reverse` function has three pointers `a`, `b` and `head`. At any point

- **`head`**: points to the original list.
- **`b`**: points to the reverse list.
- **`a`**: used as temporary pointer.

Initially, all nodes are in the original list and reverse list is empty. In each iteration of `while` loop, we break the first node from original list and insert it at the head of reverse list. When all the nodes are removed from the original list, `b` points to the head of the reversed list.

This function takes `O(n)` time and constant extra memory. You should be very comfortable with the logic and code of reverse function because it is used in many problems.

Question 3.4: Write a function to check if a linked list is a palindrome or not. For example, following linked lists are palindromes.

```
2->3->4->4->3->2
M->A->L->A->Y->A->L->A->M
```

But following lists are not palindromes

```
2 -> 3 -> 4 -> 5 -> 4 -> 6 -> 2
M -> O -> K -> S -> H -> A
```

The function should take O(n) time in worst case.

Solution: A simpler solution to this problem uses `reverse` function from the previous question. Consider the below algorithm:

STEP-1: Move to the middle of the list (see Question 8).

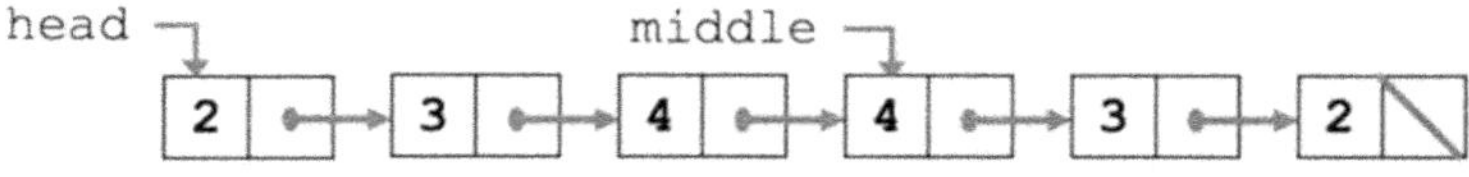

STEP-2: Reverse the second half of list (either count the nodes or use fast and slow pointers)

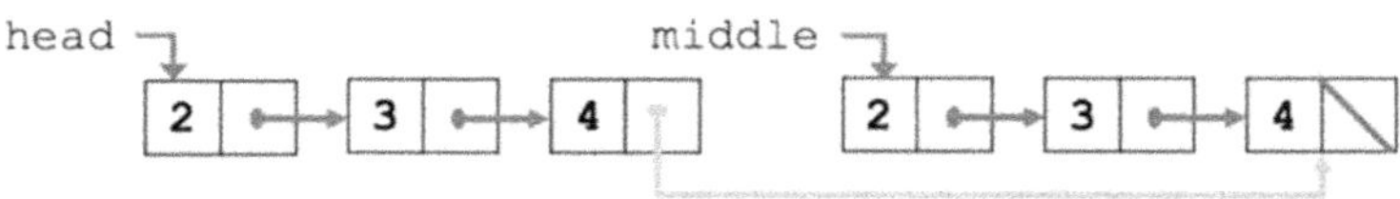

Note that the `next` field of the node before middle node is not changed. This is not required because we will ultimately `restore` the original list.

If there are odd number of elements, then let the first half, have that extra element.

```
head ┐                               middle ┐
     M -> A ->L -> A -> Y -> A -> L-> A-> M
```

STEP-3: Check if corresponding elements of lists pointed to by `head` and `middle` are equal

```
bool isPalindrome = true;
while(middle != NULL){
  if( middle->data != head ->data){
    isPalindrome = false;
    break;
  }
  head = head->next;
  middle = middle->next;
}
```

In the while loop we cannot replace middle with head, because there may be more elements in the list pointed to by `head` (in fact last element of second list is part of first list also).

STEP-4: Reverse the second half of list again to restore the original list.

Another (not so optimized) way can be to use recursion. We keep two pointers, one pointing to the first list and another pointing to the end. The end pointer is moved backward (through recursion stack) and front pointer is moved forward. Below is the code:

```
// WRAPPER FUNCTION - CALLED FROM OUTSIDE
bool isPalindrome(Node* head){
  return isPalindromeRec(&head, head);
}

bool isPalindromeRec(Node **l, Node *r){
  if (!r) // TERMINATING CONDITION
    return true;

  //IF SUBLIST IS NOT PALINDROME. DON'T CHECK FURTHER
  if (! isPalindromeRec(l, r->link) )
    return false;

  bool flag = (r->data == (*l)->data);
  *l = (*l)->next; // Move l forward

  return flag;
}
```

Note that the first `Node*` in the function is passed by reference.

Question 3.5: Write a function to print a linked list in reverse order (without actually reversing it).

Solution: One simple way to print the list in reverse order is to first reverse the list, and then print it in forward order (see Code 3.2) and then reverse it again to get the original list. This logic takes `O(n)` time.

```
void printReverse(Node *head){
  head = reverse(head);
  printList(head);
  head = reverse(head);
}
```

Another way to print list in reverse order is by using recursion as shown below:

```
void printReverseRec(Node *head){
  if(head != NULL){ return; } // TERMINATING COND.
  printReverseRec(head->next);
  printf(" %d", head->data);
}
```

Logic of `printReverse` function is simple. First print the list starting from second node in reverse order (using same function) and then print value of current node. This function also take `O(n)` time asymptotically, but absolute time taken is more than non-recursive (consider time off unction call/return). It also takes `O(n)` time memory also.

Question 3.6: Given a linked list and a number `n`. Reverse each group of `n` nodes in the list. The last group may have less than `n` nodes

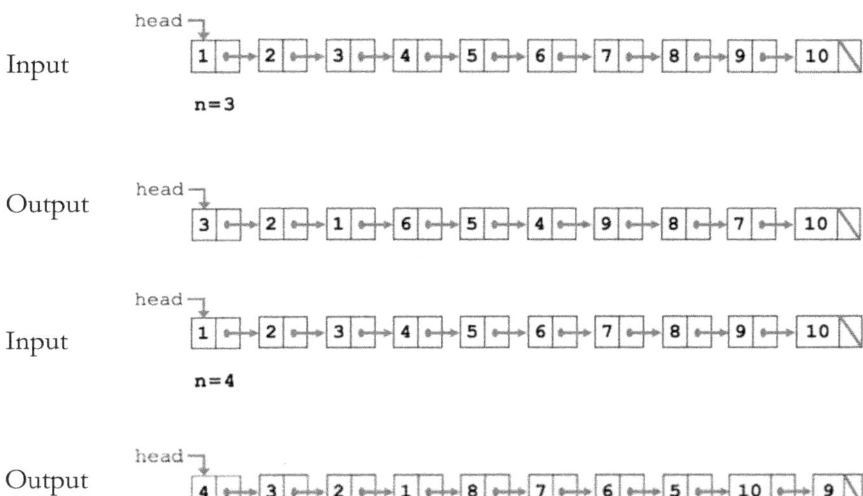

Solution: This problem is logically not difficult to solve, but because we are dealing with a lot of pointers, there is a good chance of error, especially if we miss connecting the end of first group with the start of second group after reversing both the groups.

We have to build on the top of reversal logic. We may have to modify the reversal logic for this problem. For the time being, let us assume, we have a reverse function that reverse first `n` nodes of the list pointed to by its first argument and populate the reference variables to point to the last node of current group and first node of next group. It also returns the new head of current group after reversing it.

```
Node* reverseFirstN(Node * head, int n,
Node** end, Node** nextHead);
```

With this function in place, let us write the actual code

```
// head - HEAD OF CURRENT LIST.
Node* reverseInGroup(Node* head, int n)
{
  if(head == NULL){ return NULL; }

  Node *end, *nextHead;

  // HEAD OF NEW LIST
  head = reverseFirstN(head, n, &end, &nextHead);
  while(nextHead != NULL){
```

```
        Node* temp = end;
        // JOIN TWO CONSECUTIVE GROUPS
        temp->next=reverseFirstN(nextHead, n, &end,
        &nextHead);
    }
    return head;
}
```

Now let us try to modify the reverse function we have already seen to implement our `reverseFirstN` function

```
// n - NUMBER OF NODES TO REVERSE.
// end - END OF REVERSED GROUP.
// nextHead - HEAD OF NEXT GROUP.
// RETURN head OF CURRENT REVERSED GROUP.
Node* reverseFirstN(Node * head, int n, Node** end, Node**
nextHead){
        Node * a;
        Node * b = NULL;
        int cnt = 0;

        *end = head;
        while(head != NULL && cnt<n){
            a = head->next;
            head->next = b;
            b = head;
            head = a;
            cnt++;
        }
        *nextHead = a;
        return b;
}
```

The entire logic takes O(n) time and constant extra memory.

Question 3.7: Given two linked lists that merge at a point as shown below:

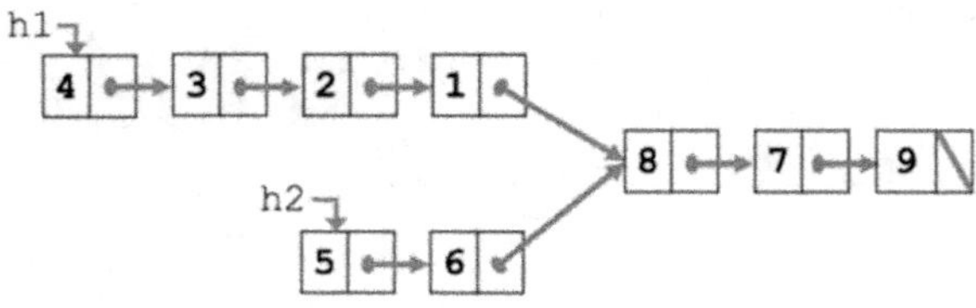

Find the number of unique elements in each list and number of nodes common in both the lists. For the above lists, the number of unique elements in the two lists are 4 and 2 respectively and 3 nodes are common.

Solution: Let the number of nodes in the first list are `(m+p)` and number of nodes in the second list are `(n+p)`, last `p` nodes being common as shown below:

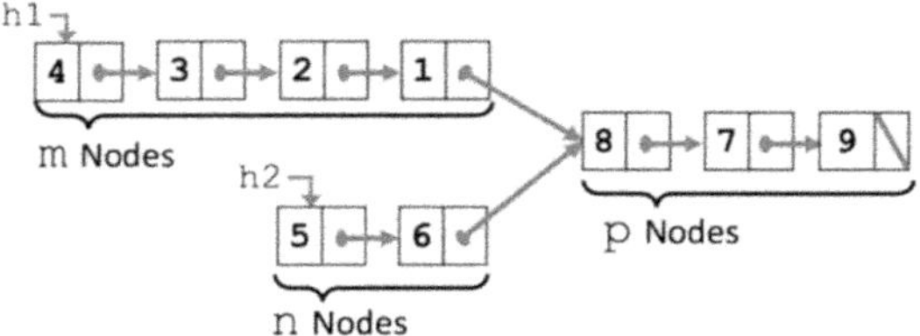

If we are able to find the number of common nodes, we can find all three values by traversing the two lists.

The most intuitive solution is to traverse the first list and for each node in the list, see if that node is also present in the second list. The first node that is common between the two nodes marks the start of common list. This logic takes $O(n^2)$ time and constant extra memory.

Another way is to store node addresses of first node in a hash table. Traverse the second list and for each node in the second list, see if it is present in the hash. This method uses `O(n)` extra space.

A better way is to calculate the number of nodes in two lists by traversing them linearly. They have `(m+p)` and `(n+p)` nodes respectively.

`ABS((m+p)-(n+p))` gives the number of excess nodes in larger list. Move head of the larger list forward by this difference, such that number of nodes in both the lists after that point are same:

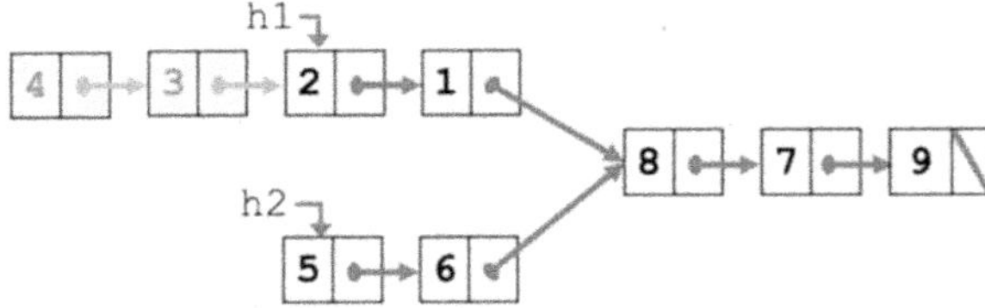

Now move both the heads forward in a loop and compare address of nodes they are pointing to as shown below:

```
Node* findFirstCommonNode(Node *h1, Node *h2){
  int mp = 0; Node *origH1 = h1;
  int np = 0; Node *origH2 = h2;

  while(h1 != NULL){
    h1 = h1->next; mp++;
  }
  while(h2 != NULL){
    h2 = h2->next; np++;
  }
```

```
    h1 = origH1; h2 = origH2;
    if(mp>np)
      for(int i=0; i<mp-np; i++)
        h1 = h1->next;
    else
      for(int i=0; i<np-mp; i++)
        h2 = h2->next;

    while(h1 != h2 && h1 != NULL && h2 != NULL){
      h1 = h1->next;
      h2 = h2->next;
    }
    return h1;
}
```

Once we get the pointer to first common node, we can calculate p. After calculating p, values of m and n can be calculated by traversing the lists.

Let us discuss one more linear-time solution.

There are three variables, m, n and p. To find the value of these three variables, we need three unique equations of these variables. We can get two equations by traversing the two lists:

```
m + p = 7                                   ... ... ... (i)
n + p = 5                                   ... ... ... (ii)
```

After having these two equations, reverse one of the two lists. If we reverse the second list, then it becomes:

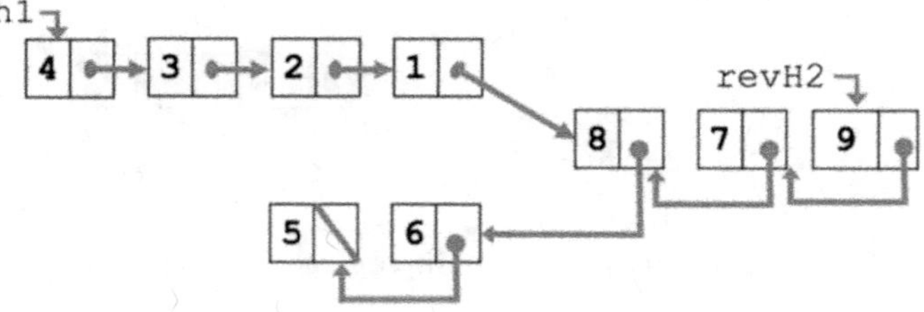

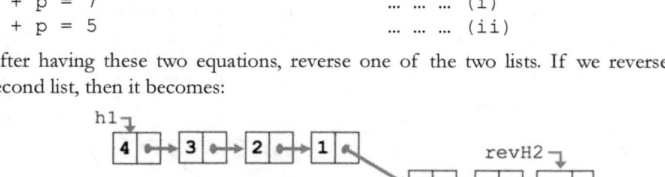

Now, if we traverse the first list, the nodes visited are shown below:

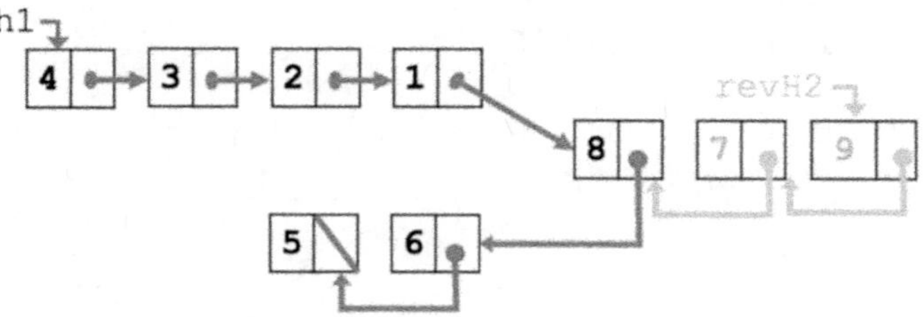

It gives us our third equation:

```
m + n + 1 = 7                                  ... ... ... (iii)
```

Solving these three equation, we can compute the values of m, n and p. Later, reverse the second list again to restore original.

Question 3.8: In a single loop, find k^{th} node from end of the list. If list is

```
2 -> 4 -> 6 -> 3 -> 7
```

and k=2, output should be 3 (2nd node from end).

Solution: The challenge in a linked list is that we cannot move backward from the end of list. Neither can we access an element directly. There are many question like this, that are solved by taking two pointers and move them at the same time (may be at a different pace).

Take two pointers ptr1 and ptr2, set ptr2 to point to k^{th} element from start of the list.

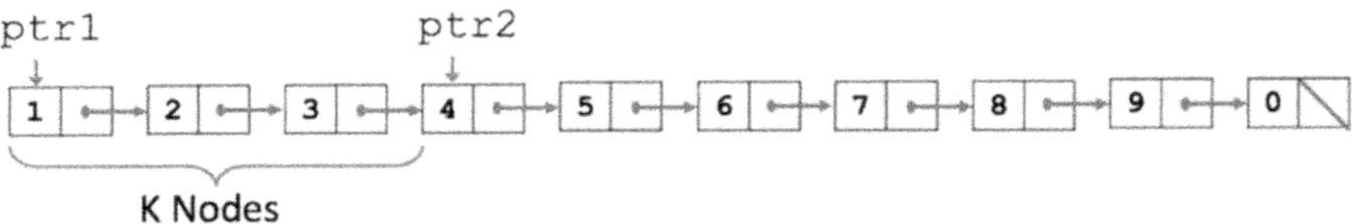

Move both of them forward at same pace, when ptr2 reach the end of list, ptr1 will be k positions before it pointing at the kth element from end.

```
Node * findFromEnd(Node* head, int k)
{
    // head POINTER ACTS AS ptr2
    Node* ptr1 = head;
    for(int i=0; i<k; head = head->next, i++)
        if(head == NULL)
            return NULL;    // LIST HAS <k NODES

    while(head != NULL){
        ptr1 = ptr1->next;
        head = head->next;
    }
    return ptr1;
}
```

Question 3.9: Find middle element of the linked list in a single traversal. If list is 1->2->3->4->5 then output should be 3. If there are even nodes, return any of the two middle elements.

Solution: This is very similar to the previous question. Take two pointers, initially both pointing to the first node, let us call them `slowPtr` and `fastPtr`. In each iteration of the loop move `slowPtr` forward by one position and `fastPtr` by two positions. When `fastPtr` reach the end, `slowPtr` will be at the middle node.

```
Node* middleElement(Node* head)
{
  Node* slowPtr = head;
  Node* fastPtr = head;

  while(fastPtr != NULL && fastPtr->next != NULL ){
    slowPtr = slowPtr->next;
    fastPtr = fastPtr->next;
    if(fastPtr != NULL)
     fastPtr = fastPtr->next;
  }
    return slowPtr;
}
```

This method is also called **Tortoise & Hare Method**.

Question 3.10: Given a linked list which may contain a loop as shown below:

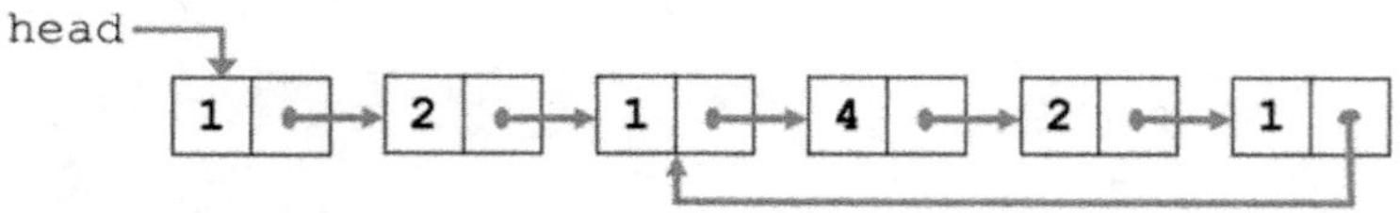

Write code to check if the list has a loop or not.

Solution: If there is a loop in the list, then there is no end. The usual traversal of the list may not get us the desired solution because we do not know when to stop.

Sometimes, we keep a pointer at the head and traverse the list by using a second pointer untill it becomes null or becomes equal to the head pointer. This will not work because the loop may not involve the first node (as shown in the given figure).

Method-1: Marking visited Nodes

When a node is traversed, it is marked as visited. While moving forward, if we encounter a visited node again, it means there is a loop in the linked list. Marking can be done in two ways

1. Use hash table to store addresses of visited nodes.

```
For each Node starting from the first
   If Node is present in the Hash table
     Loop detected
```

```
  Else
    Store Node in the Hash Table and move forward
```

2. Keeping an extra field per node. Structure of the `Node` will be:

```
struct Node
{
  int data;
  Node* next;
  bool visited;
};
```

Initialize `visited=false` for all nodes. Traverse the list and for each Node,

```
If visited field of the Node is true
  Loop detected
Else
  set node.visited=true and move forward
```

This method can detect the loop in linear time but, requires O(n) extra space.

Method-2: Reversing the list

Keep a pointer at the first node of the list and try to reverse the list. If there is a loop in the list, you will reach the first node while reversing the list. If you don't reach the first node, there is no loop in the list. Later reverse the list again to restore the original state.

This is an `O(n)` time algorithm and requires constant extra space. But you need to reverse the list twice, first to detect the loop and then to restore the original list.

Method-4: Using Two Pointers (Best Algorithm)

Use the "Tortoise & Hare" approach discussed in Question 3.8. Traverse the list simultaneously by using a fast pointer and a slow pointer. If there is a loop, the two pointers will become equal. Below is the code for this logic:

```
bool detectLoop(Node *head){
  Node* slowPtr = head;
  Node* fastPtr = head;

  while(fastPtr != NULL && fastPtr->link != NULL ){
    slowPtr = slowPtr->link;
    fastPtr = fastPtr->link->link;

    if (slowPtr == fastPtr)
      return true;
  }
  return false;
}
```

Question 3.11: Two numbers are given as linked lists (with one node storing one digit). Write an algorithm to compute sum of numbers represented by these two linked lists. The result should also be a linked list. For example, if given numbers are `N1 = 13625` and `N2=485`, given as:

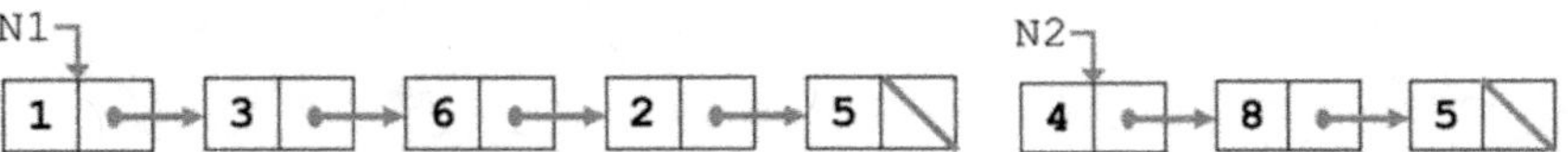

Then output should be a linked list of number `13625+485 = 14110.`

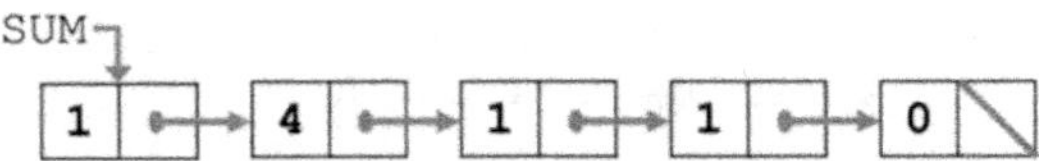

Solution: In mathematics, we add two numbers starting from the least significant digits.

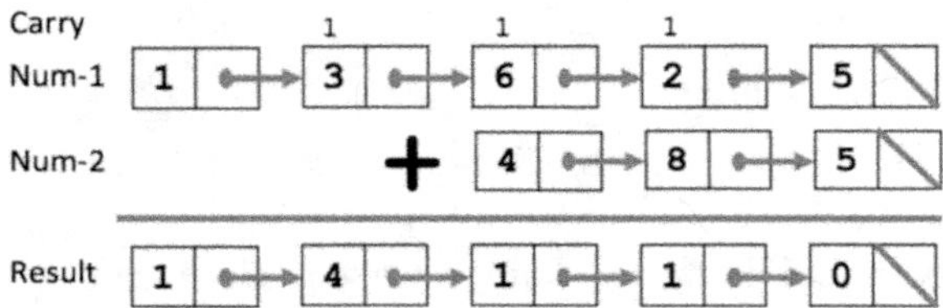

The problem with linked list is that we cannot traverse it backward from least significant digit to most significant digit. Hence, the computation of sum may not be linear time. One way of solving it is to reverse the two input lists, and then add nodes in forward order:

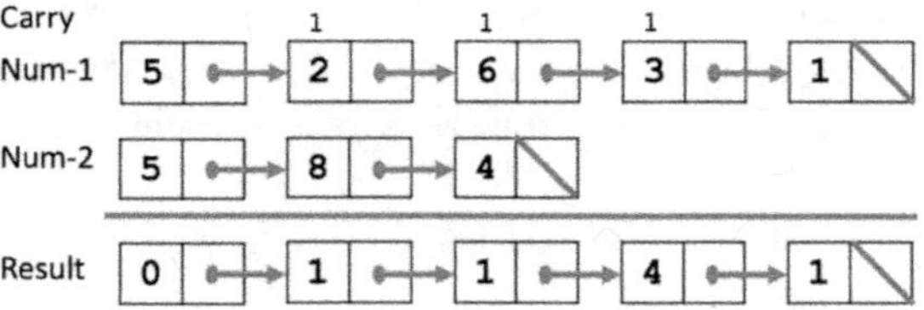

The result list is the reverse of the actual result. Reverse this list to get the final output. We can avoid reversing at the end if we insert new node at the head of result list. Function to insert node at head of a list is given below

```
void insertNodeAtHead(Node** hp, int val){
  Node* temp = newNode(val);
  temp->next = *hp;
  *hp = temp;
}
```

Below function has the main logic to add two numbers given as linked list.

```
Node* addNumbers(Node* n1, Node* n2){
  // BOUNDARY COND. IF ANY NUMBER IS 0
  if(n1 == NULL){ return n2; }
  if(n2 == NULL){ return n1; }

  // REVERSE THE TWO LISTS
  n1 = reverse(n1);  n2 = reverse(n2);

  Node* resList = NULL; int carry = 0;
  while(n1 != NULL || n2 != NULL)
  {
    int sum = carry;
    if(n1 != NULL){
      sum += n1->data;
      n1 = n1->next;
    }
    if(n2 != NULL){
      sum += n2->data;
      n2 = n2->next;
    }
    carry = sum / 10;
    sum = sum % 10;

    insertNodeAtHead(&resList, sum);
  }
  if(carry != 0) // ADDING LAST CARRY
    insertNodeAtHead(&resList, carry);

  // RESTORING LIST BEFORE RETURNING
  n1 = reverse(n1);  n2 = reverse(n2);

  return resList;
}
```

Another way of doing this is to use two Stacks and follow below algorithm:

```
Step-1: Push all nodes of Num1 in Stack1
Step-2: Push all nodes of Num2 in Stack2
Step-3: Keep popping elements from two stacks, adding them
and inserting them at front of the result list.
```

Also, keep track of carry in Step-3. This method will take O(n) extra space.

Question 3.12: Given a number represented in the form of a linked list (as in the previous question). Write code to increment the number in-place. Consider the given inputs and corresponding outputs below:

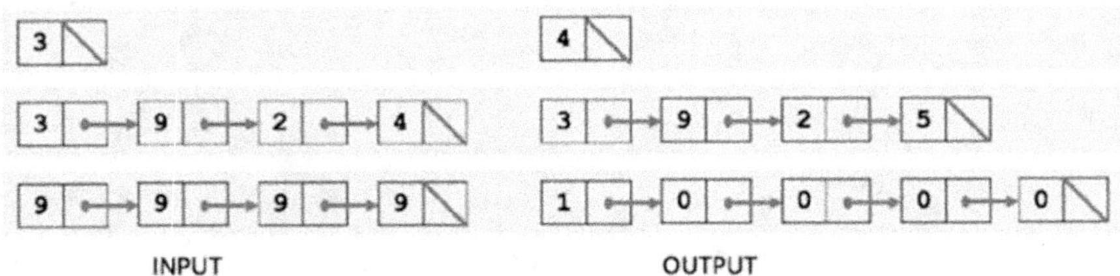

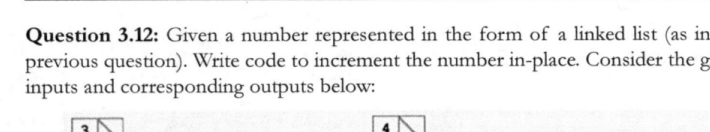

Solution: One of the solutions to this problem is a minor modification of the previous question's solution.

```
Step-1: Reverse the list.
Step-2: Increment first element of the list. If there is
        a carry, keep adding it to the next node till
        either, the carry becomes zero or end of the list
        is reached. If we reach the end of the list and
        carry is still not zero, allocate a new node with
        value 1 and append it to the list (carry cannot be
        more than 1).
Step-3: Reverse the list to get the original order.
```

Note that we are just incrementing a number and not adding two numbers. A carry can be generated only if the least significant digit is 9 and it can propagate to next place only when the current digit is 9. When a digit less than 9 is encountered, the carry will be consumed and nothing will change beyond that point.

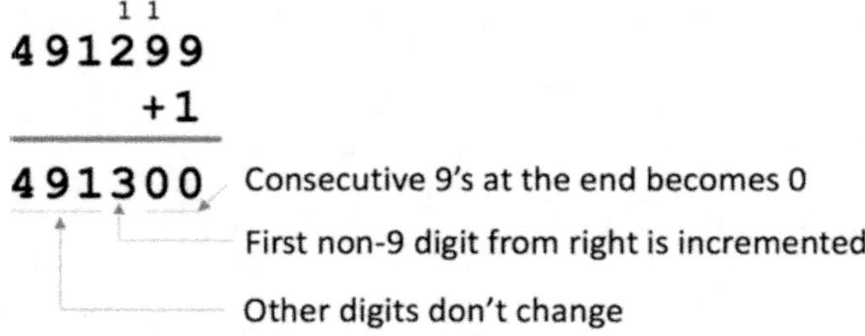

Keep two pointers. One is to reach the end of the list and other will point to the right most digit that is <9. The rightmost digit less than 9 need to be incremented and all trailing nodes with value 9 will be made zeros. If all nodes are 9 (e.x. number `99,999,99999`, etc.), then all nodes will be set to zero and a new node, with value `1`is added at the head. Consider below code

```
Node* increment(Node* head)
{
  // IF NO NODE.
  if(head == NULL){ return newNode(1); }
```

```
  // pre POINTS TO RIGHTMOST NODE WITH VALUE < 9
  Node *prev = head, *cur = head;
  while(cur != NULL){
    cur = cur->next;
    if(cur!= NULL && cur->data != 9)
      prev = cur;
  }

  if(prev == head && prev->data ==9){
    // ALL NODES ARE 9. SET THEM ALL TO 0
    // AND ADD 1 (99+1 = 100)
    prev = newNode(1);
    prev->next = head;
    while(head != NULL){
      head->data = 0;
      head = head->next;
    }
    head = prev;
  }else{
    // ALL NODES AFTER prev ARE 9's.
    // BUT NODE POINTED TO BY prev IS < 9
    prev->data++;   // INCREMENT prev
    prev = prev->next;
    while(prev!= NULL){ // SET NODES AFTER prev TO 0
      prev->data = 0;
      prev = prev->next;
    }
  }
  return head;
}
```

Assignment 3.1: Subtract two numbers from a linked list.

Assignment 3.2: Given a singly linked list of `0`s and `1`s representing a binary number. Write a function that return the decimal equivalent of that number.
`Input List: 0->0->1->1->0->1->1->0Output: 54`
`Input List: 1->1->0Output:6`

Question 3.13: Sort the linked list in which each node has either `0` or `1`.
`Input: 1->0->0->1->0->1->1->0`
`Output: 0->0->0->0->1->1->1->1`

Solution: One of the easiest solution is to count the `0`'s. If there are `k` nodes with zero values, then set first `k` nodes to `0` and the other `n-k` nodes to `1`.

If the change of Node's content is not allowed, we have to do actual sorting. But since there are only two values, we can keep two pointers pointing to the start and the end of the list. Each node is deleted from the list and is inserted at start or end depending on whether its values is 0 or 1 respectively.

```
Node* sortBinaryList(Node* head)
{
  Node* start = head;
  Node* end = start;

  // SETTING end TO FIRST NODE WITH 1
  while(end != NULL && end->data == 0)
    end = end->next;

  if(end== NULL) { return start; } // ALL 0 NODES

  while(end->next != NULL)
  {
    Node* temp = end->next;
    if(temp->data == 0){
      // INSERT temp AT HEAD
      end->next = temp->next;
      temp->next = start;
      start = temp;
    }else{
      end = end->next;
    }
  }
  return start;
}
```

This code takes O(n) time, but it is not doing stable sorting.

Same logic can be used to separate nodes with positive and negative values (or nodes with even and odd values). We can move positive nodes on one side and negative nodes on the other side.

Question 3.14: Sort a linked list where each node stores either 0, 1 or 2.

```
Input: 1->0->2->1->0->1->1->2
Output: 0->0->1->1->1->1->2->2
```

Solution: This question is a natural extension to the previous question. We can keep three pointers `ptrZero`, `ptrOne`, `ptrTwo`, each pointing to the end (or start) of part of sorted list contatining 0's, 1's and 2's respectively.

```
ptrZero      ptrOne          ptrTwo      head
0->0->0->1->1->2->2->2->2->1->0->2->1->1->0
```

Part-1 (0-Nodes) Part-2 (1-Nodes) Part-3 (2-Nodes) Rest of the list (still unsorted)

The list is traversed by using the `head` pointer. Node pointed to by `head` is removed from the list and inserted after either `ptrZero`, `ptrOne` or `ptrTwo`, depending on the value of node. This is also a linear-time solution.

Insertion sort can be used to sort a random list, but that takes more time.

Question 3.15: Given n linked lists where first nodes of every list is connected with each other as shown below.

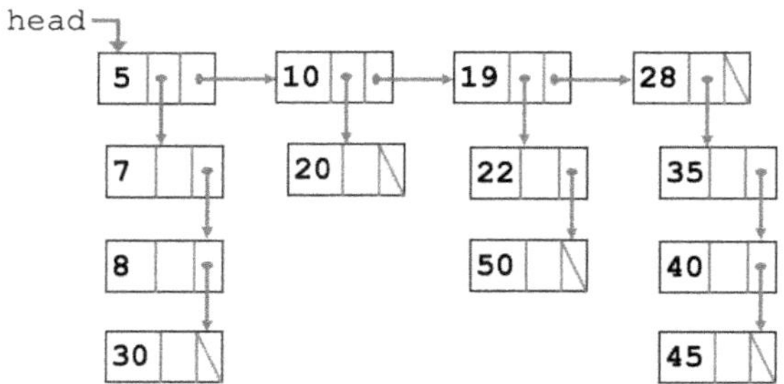

This structure can be formed by having two pointers in each node.

(i) **Right pointer:** This pointer connects the first nodes of each list in a horizontal linked list. For all nodes, other than the first node of each list, this pointer is NULL.

(ii) **Down pointer:** This is like the next pointer which points to the next node of the linked list (below it).

All the linked lists are individually sorted and the top list (formed by using `right` pointers) is also sorted (see the diagram above).

Write a function to merge all the linked list into a single list, in which all nodes are connected by using `down` pointers. The final list should be:
`5->7->8->10->19->20->22->28->30->35->40->45->50`

Solution: The structure of each node is:

```
typedef struct Node
{
  int data;
  struct Node *right;
  struct Node *down;
} Node;
```

Let us write a function to merge two sorted linked list. The function accepts two sorted linked list, merge them on the `down` pointer field and return pointer to the head of new list:

```
Node* merge( Node* a, Node* b ){
  // TERMINATING CONDITIONS
  if(a == NULL){ return b; }
  if(b == NULL){ return a; }
```

```
  Node* head;  // POINTER TO THE HEAD OF MERGED LIST
  if(a->data < b->data){
    head = a;
    head->down = merge(a->down, b);
  }else{
    head = b;
    head->down = merge(a, b->down);
  }
  return head;
}
```

It is a recursive function; you may write (a better) non-recursive function also. The logic to merge all the lists is simple:

```
Node* mergeAll(Node* head){
  // TERMINATING CONDITION
  if(head == NULL || head->right == NULL)
    return head;

  // MERGE ALL THE LISTS AFTER CURRENT LIST
  Node* temp = mergeAll(head->right);

  // NOW WE HAVE ONLY TWO LISTS. MERGE THEM
  return merge(head, temp);
}
```

Case Study: Simulating a linked list in an array

Let us take an array `arr` of `struct Node`, defined as below:

```
struct Node{
  int data;
  int next;
  bool isFree;
} arr[10];
```

Array definition

	data	next	isFree
arr[0]			1
arr[1]			1
arr[2]			1
arr[3]			1
arr[4]			1
arr[5]			1
arr[6]			1
arr[7]			1
arr[8]			1
arr[9]			1

Memory allocated to array

The Memory allocated to `arr` is shown vertically (rather than horizontally) for convenience. Each element of the array has three fields, `data`, `next` and `isFree`. If `arr[i].isFree` is false then `arr[i].data` and `arr[i].next` stores a valid node of linked list.

A Linked list is a dynamic data structure, and a new node is allocated on the heap independently of existing nodes. All nodes are linked with each other by `next` pointers.

We will use array `arr` to simulate Heap memory, index of an array represents the address of memory. `malloc` function allocates memory on the heap and return its address and `free` function receives a memory address and deallocates that memory. Let us define two parallel functions to allocate and free memory from array `arr`.

```
// ALLOCATE ONE ELEMENT OF ARRAY AND RETURN INDEX
int allocateMem()
{
  for(int i=0; i<N; i++)
    if(arr[i].isFree){
      arr[i].isFree = false;
      return i;
    }
  return -1; // NO MEMORY AVAILABLE
}

// SET MEMORY AT INDEX address FREE.
void freeMem(int address){
  arr[address].isFree = true;
}
```

Function `allocateMem` look for the first free memory in the array, allocate that memory (set `isFree` to `false`) and return its address (i.e index). Function `freeMem` accept and address in the form of memory index and set it free.

Let us rewrite the function of Code 3.3 to store linked list in array arr maintaining the dynamic nature of heap.

```
// HELPER FUNCTION. ALLOCATE & INITIALIZE NODE
int newNode(int value){
  int temp = allocateMem();
  if(temp != -1){
    arr[temp].data = value;
    arr[temp].next = -1;
  }
  return temp;
}
```

```
void insertInSortedList(int* hp, int val){
  int temp = newNode(val);
  int head = *hp;

  // UNABLE TO ALLOCATE MEMORY
  if(temp == -1){ return; }

  if(head == -1 || arr[head].data > val){
    // INSERTING AT HEAD
    arr[temp].next = head;
    *hp = temp; // CHANGING head OF main FUNC
  }else{
    // MAKE head POINT NODE AFTER WHICH TO INSERT
    while(arr[head].next != -1 &&
          arr[arr[head].next].data < val)
      head = arr[head].next;

    arr[temp].next = arr[head].next;
    arr[head].next = temp;
  }
}
```

When we insert the following elements in this list:

```
int main(){
  for(int i=0; i<N; i++)
    arr[i].isFree = 1;

  int head = -1; // COMPARE WITH Node* head = NULL
  insertInSortedList(&head, 4);
  insertInSortedList(&head, 8);
  insertInSortedList(&head, 2);
  insertInSortedList(&head, 0);
  insertInSortedList(&head, 5);
  insertInSortedList(&head, 10);
  insertInSortedList(&head, 1);
}
```

The memory will look like below:

	data	next	isFree
arr[0]	4	4	0
arr[1]	8	5	0
arr[2]	2	0	0
arr[3]	0	6	0
arr[4]	5	1	0
arr[5]	10	-1	0
arr[6]	1	2	0
arr[7]	-	-	1
arr[8]	-	-	1
arr[9]	-	-	1

head = 3

The first node of list is at index `head`, `arr[head].data` represents the value of first node and `arr[head].next` holds the index of second node. If we put the nodes in linked form, then they actually represent the following linked list:

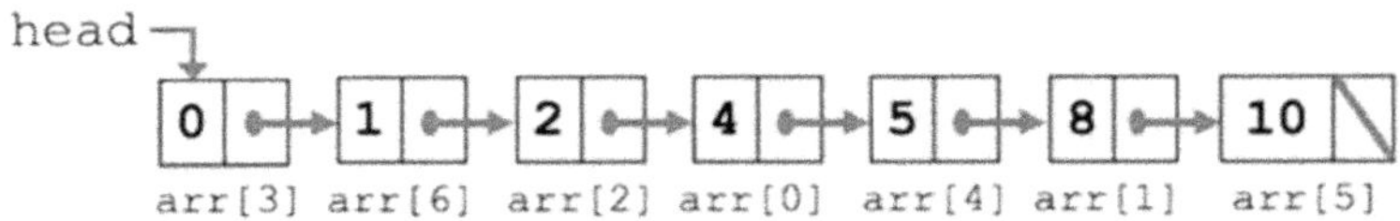

What we have seen for array `arr` is very similar to a linked list, the difference is that here addresses are `int` and not pointers, as they are when memory is on heap.

Doubly linked list

A doubly linked list is a linked list where a node also has pointer to previous node, along with a pointer to the next node. The structure of Node is:

```
struct Node{
  int data;
  struct Node *prev;
  struct Node *next;
};
```

prev data next

The `prev` pointer points to the previous node and `next` pointer points to the next node. Data in this node is integer, but it can be of any type including user-defined data type. A typical doubly linked list looks like:

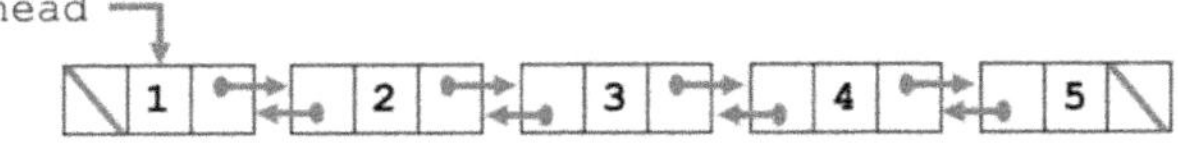

`head` pointer points to the first node of the list. `prev` pointer of the first node and `next` pointer of the last node is `NULL`.

Because there is a previous pointer, we can traverse in any direction. To delete a node from a doubly linked list, we need to set both `prev` and `next` pointers of neighboring nodes. If we want to delete Node with value 4 from the above list, let `ptr` points to it's previous node:

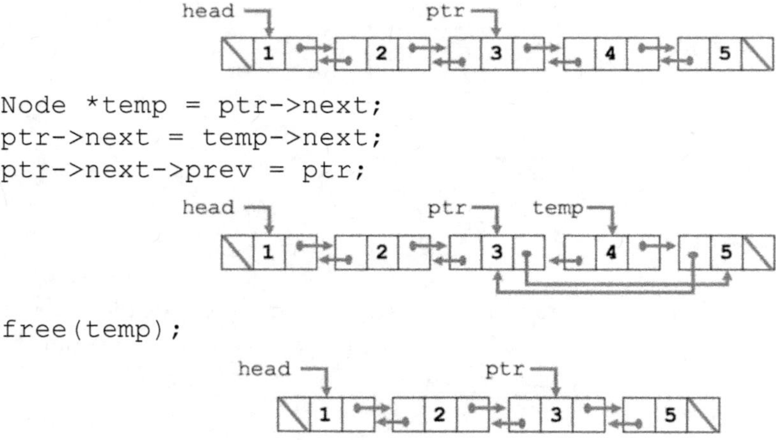


```
Node *temp = ptr->next;
ptr->next = temp->next;
ptr->next->prev = ptr;
```

```
free(temp);
```

If pointer to a node that need to be deleted is given, then also we can delete that node (unlike singly linked list).

Similarly, inserting a new node is also about manipulating the `prev` and `next` pointers of adjacent nodes. Like singly linked list, insertion at head requires to change the `head` pointer.

Chapter 4

Stack

A stack is an ordered linear data structure that follows a "Last-In-First-Out" order for inserting and deleting elements in it. Both insertion and deletion of element happens at the same end. The last element inserted is the first one to be removed, hence the name, `LAST-IN-FIRST-OUT`.

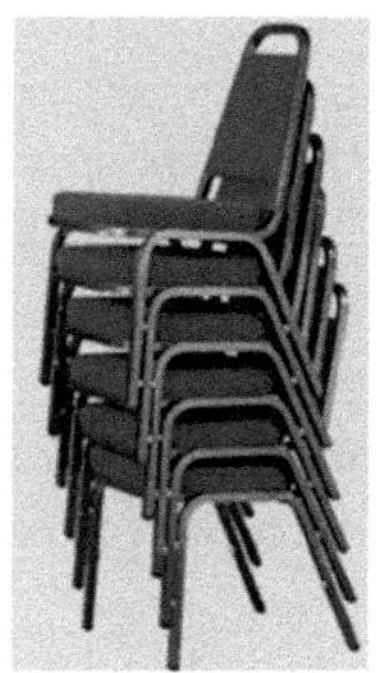

STACK OF CHAIRS

If we want to add another chair to the stack of chairs on the right side, then that chair will be added at the top. If we want to remove a chair, the chair will be removed from the top only. The most recently added chair is the one that will be removed.

Both insertion and deletion happens at the same end, therefore size of stack grows and shrink from one end only. A stack of numbers look like:

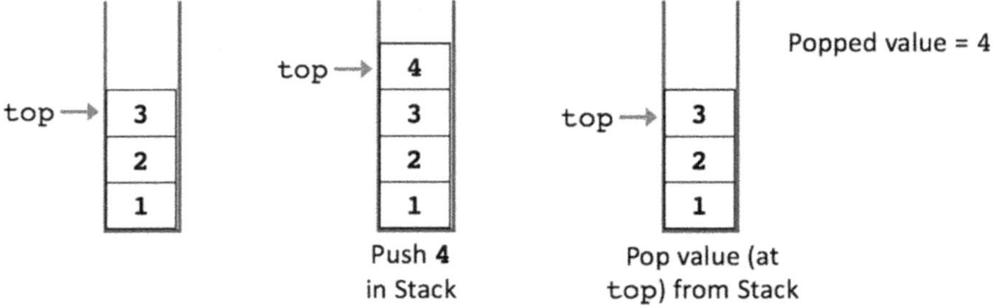

The operation to insert an element in the stack is called **Push,** and operation to deleted an element from the stack is called **Pop**. The most recently pushed element is always at the top of stack. Operation `isEmpty()`, returns `true` if the stack is empty and `false` otherwise. Similarly, other operations can be added to the stack depending on requirements. A very common operation is `peek`, that returns top element without removing it from the stack.

In computer science, a stack data structure is used in many situations. Consider the back button in your browser:

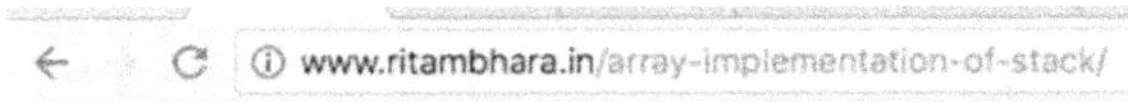

When a link on a webpage is clicked, current page's URL is pushed in a stack and the link URL is opened in the current tab. When back button is pressed, the most recent page URL is popped up from the stack and that page is loaded.

Similarly, in an editor, each operation performed on a document is pushed in a stack. When `UNDO` button is pressed, the most recent operation is popped from the stack and undone.

Even the notifications on a phone are shown in a 'Most-Recent-first' order, so are the call logs. Stack comes naturally and should be most intuitive choice of data structure when elements are in such an order.

Important thing to note is that Stack does not have its own underlining implementation. We may choose to store the elements in any data structure, as long as we can maintain the order of insertion and deletion in a LIFO order.

If we choose to store elements in an Array, it is simpler and easier to code, but there is a limit on the total number of elements. Linked list can expand dynamically but it involves pointer management. Below class in C++ declares a Stack that use array to store the values:

```
#define MAX_LEN 10
class Stack
{
  int top;// HOLDS INDEX OF TOP ELEMENT
  int a[MAX_LEN];   // HOLDS THE STACK ELEMENTS
public:
  Stack():top(-1){} // DEFAULT CONSTRUCTOR. top = -1
  void push(int x);
  int pop();
  bool isEmpty();
};
```

Elements of stack are inserted in the array and stack grows rightward in the array.

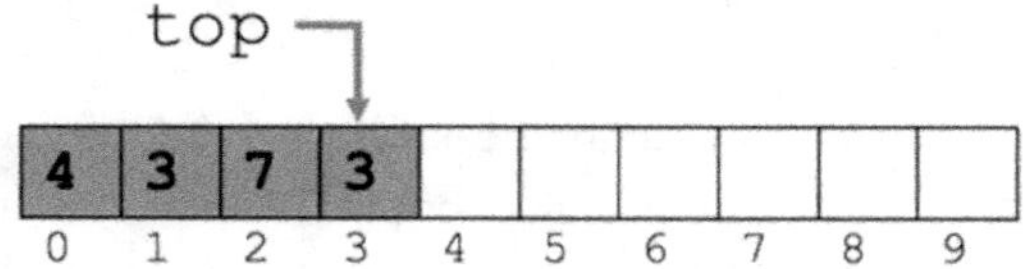

At any point, `top` holds the index of most recently added element in the stack. A push operation increment `top` and put the element at new `top` index in the array, similarly a pop operation return the element at index `top` in the array and then decrement `top`.

Push operation should also handle the situation when the array is full and pop should handle the situation when there is no element in the array. Below is the code of these two operations..

```
void Stack::push(int x){
  if(top< MAX_LEN)
    a[++top] = x;
  else
```

```
    cout<<"STACK OVERFLOW"; // STACK FULL
}

int Stack::pop(){
  if(isEmpty()){               // STACK EMPTY
    cout<<"STACK EMPTY";
    return -1;
  }
  else
    return a[top--];
}
bool Stack::isEmpty(){
  return (top == -1);
}
```

A better implementation is to throw a custom exception from `push` (`pop`) function when the stack is full (empty). We are printing error to avoid clutter. `pop` function returns -1 when the stack is empty which means that `-1` cannot be a valid entry in the Stack.

If we are not throwing an exception, `push` function should return something to tell its calling function whether or not it was able to insert element in the stack. `printf` is not a good way to do it, especially in production, because most often it ends up printing in logs.

Both `push` and `pop` operations take constant time. More functions can be added to the `Stack` class depending on requirements.

If we choose to store elements in a linked list, push operation can be implemented as Insert-Element-At-Head and pop operation can be implemented as Delete-First-Node of a linked list.

Adding a new node at the head of a list takes constant time, and, removing the first node of a linked list also takes constant time.

Structure of a linked list Node is similar to Code 3.1. Let us add constructors to the Node structure definition to make node creation easier (in C++ `struct` can have functions).

```
struct Node{
  int data;
  Node* next;
  // CONSTRUCTORS
  Node():data(0), next(NULL){}
  Node(int v):data(v), next(NULL){}
};
```

Now we can initialize the node while allocating it on heap:

```
Node* temp = new Node(5);
```

The `data` and `next` fields of node pointed to by `temp` are 5 and `NULL` respectively. First node of the list holds the top element and `head` of the list is our `top` pointer. The Stack class with linked list implementation looks like:

```
class Stack
{
  Node* top;
public:
  Stack():top(NULL){} //DEFAULT CONSTRUCTOR.top=NULL

  void push(int x);
  int pop();
  bool isEmpty();
};
```

The `push` operation now does not need to bother about the overflow condition and just insert new node as the first node:

```
void Stack::push(int x){
  Node* temp = new Node(x);
  temp->next = top;
  top = temp;
}
```

The `pop` operation deletes the first node and return its value. It also need to take care of the stack-empty condition:

```
int Stack::pop(){
  if(isEmpty()){// STACK EMPTY
    cout<<"STACK EMPTY";
    return -1;
  }
  else{
    int value = top->data;   // VALUE OF FIRST NODE
    Node* temp = top; // temp POINTS TO FIRST NODE
    top = top->next;  // MAKE top POINT NEXT NODE
    delete temp;      // DEALLOCATE FIRST NODE
    return value;
  }
}
```

As expected, the two functions take constant time. It may be a good choice, because it allows dynamic growing and shrinking of the stack.

Using stack to Simulate Recursion

Because of its Last-In-First-Out order, elements are deleted in reverse order of their insertions. A stack data structure can therefore, be used to reverse the order.

Let us consider an example of printing an array in reverse order (Question 2.1). In the given recursive solution, the function first print rest of the array in reverse order and then (after printing rest of the array) print the first element. If there are `n` elements, then there are `n` instances of function stack frames (activation records) in the memory Stack. We can use the logic of recursion without actually calling the function recursively by keeping an extra stack as shown below.

```
void printReverse(int *arr, int n){
  Stack s;

  for(int i=0; i<n; i++)
    s.push(arr[i]);

  while(!s.isEmpty())
    cout<<s.pop()<<" ";
}
```

This logic can be used to convert any recursive implementation to non-recursive. Take an extra stack and push contents of stack frame of a function into that stack and simulate the recursion. We know the function to print pre-order traversal of a binary tree

```
void preOrder(Node*r)
{
  if(r==NULL){ return; } // TERMINATING CONDITION

  cout<<r->data<<" ";
  preOrder(r->left);
  preOrder(r->right);
}
```

We first print the data at `root` and then perform the preorder traversal of its left subtree, followed by preorder traversal of its right subtree. After printing the root, we call the same function recursively for left sub-tree and when recursion returns back, call same function for right sub-tree.

Above logic of preorder traversal can be simulated with a stack as shown below:

```
void preOrderStack(Node*r){
  if (r == NULL) return;

  stack<Node*> s;
  s.push(r);

  while(!s.empty())
  {
    Node* temp = s.pop();
    cout << temp->data << " ";

    if(temp->right)
      s.push(temp->right);
```

```
    if(temp->left)
      s.push(temp->left);
  }
}
```

Code 4.1

Asymptotic time taken by both the functions is `O(n)`, but if we compare the absolute time take, `preOrderStack` takes less time.

Stack class in Java Library

Java library has `Stack` class in `java.util` package. It extends `Vector` class and provide with five operations that allow a vector to be treated as a stack. Along with `push`, `pop` and `isEmpty` operations, `peek` method is provided that returns the top element without poping it from stack. Method `search` is also there that enables searching for an element in the stack.

Below piece of code demonstrates the use of `Stack` class in Java.

```
public static void main(String[] args)
{
  Stack<Integer> s = new Stack<Integer>();

  // PUSHING IN STACK
  for(int i=0; i<5; i++)
    s.push(i);

  // SEARCHING IN STACK
  int pos = s.search(2);
  if(pos != -1)
    System.out.println("2 FOUND AT POS "+pos);
  else
    System.out.println("2 NOT FOUND");

  // POPING ALL VALUES
  System.out.print("POPPED VALUES : ");
  while(!s.isEmpty()){
    System.out.print(" " + s.pop());
  }
}
```

The output of above code is

```
2 FOUND AT POS 3
POPPED VALUES:  4 3 2 1 0
```

Similarly, there is a `stack` class in C++ STL. We saw it working in Code 4.1.

Practice Questions

Question 4.1: Given an array, implement two stacks in that array such that its memory is used in an optimal manner.

Solution: One way is to divide the array into two equal parts and give one part to each stack.

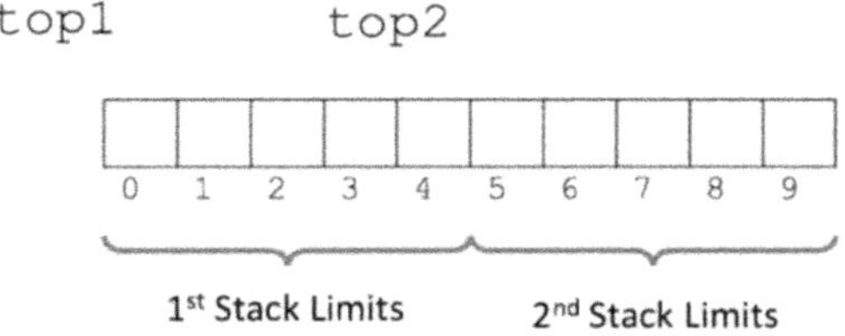

But, it is not the optimal use of memory because of fixed boundaries. If the second stack is full, we cannot push an element in second stacks even if the first stack is empty. So, we may end up using half of the total available space.

A better implementation is to have the two Stacks grow in opposite direction toward each other.

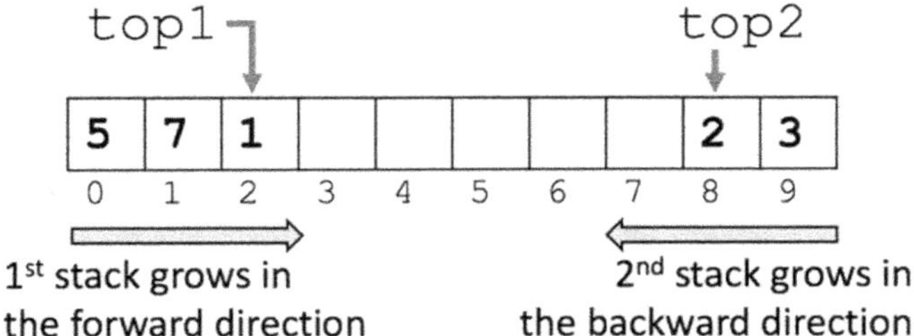

The first Stack (`top1`) will grow in the forward direction and second stack (`top2`) grows in the backward direction. If array size is `10`, then initially,

```
top1 = -1
top2 = 10
```

This indicates that the two stacks are empty. With this implementation, the user may insert `1` element in the first stack and 9 elements in the second or 5 in first and 5 in second or in any other ratio. None of the two stacks is full until there is space left in the array. If the user inserts in only one stack, then that stack may take the entire array space. Hence, memory usage is optimized.

The Empty Condition for two stacks are as follows:

```
// EMPTY CONDITION FOR forward STACK
if (top1 == -1)
    ... Forward Stack is Empty

// EMPTY CONDITION FOR backward STACK
```

```
if(top2 == 10)
    .. Backward Stack is Empty
```

None of the two stacks will be full unless all the 10 memory locations in the array are used. FULL condition for both stacks are:

```
// FULL CONDITION FOR BOTH THE STACKS
if (top1 == top2-1)
    ... Both the stacks are full
```

When an element in forward stack is pushed, `top1` is incremented, and when element in backward stack is pushed, `top2` is decremented.

```
#define FORWARD 1
#define BACKWARD 2

int arr[100];           // COMMON ARRAY
int top1 = -1;          // top OF Stack-1
int top2 = 100; "//" top OF Stack-2

// POP ELEMENT FROM stackNo
int pop(int stackNo){
  if(stackNo == FORWARD)
    return (top1 == -1) ? -1 : arr[top1--];
  else
    return (top2 == 100) ? -1 : arr[top2++];
}

// PUSH value INTO stackNo
bool push(int stackNo, int value)
{
  if(top1 == top2-1){ return false; } // BOTH FULL

  if(stackNo == FORWARD)
    arr[++top1] = value; // PUSHING IN Stack-1
  else
    arr[--top2] = value;  // PUSHING IN Stack-2
}
```

I am sure you can write the others functions like `isEmpty()` for each stack.

Question 4.2: Print next greater element for each element in a given array of numbers. If a greater number does not exist on the right side of an element, print `-1`.

```
Input Array: {5, 9, 1, 15} Output: 9, 15, 15, -1
Input Array: {20, 17, 15, 1} Output: -1, -1, -1, -1
```

Solution: The brute-force way is to use two loops. Outer loop picks all elements one by one and inner loop looks for the first greater element on its right side. If a greater element is found, it is printed as next greater element, otherwise `-1` is printed for the picked number.

```
void printNextGreater(int *arr, int n){
  for(int i=0; i<n; i++){
    int j=i+1;
    for( ; j<n; j++){
      if(arr[i] < arr[j]){
        printf("%d ", arr[j]);
        break;
      }
    }
    if(j>=n)
      printf("-1 ");
  }
}
```

This method takes `O(n²)` time in the worst case. There is a better algorithm that takes linear time by using a Stack. The algorithm is as follows

1. Push first element into Stack.
2. Pick other elements one by one (call it, current element):
   a) If stack is not empty, keep popping from the stack while the popped element is smaller than current element. Current element is the next greater element for all of these popped elements.
   b) Push current element in Stack.
3. After the loop in step 2 is over, pop all the elements from the stack and print -1 as next greater element for all of them.

```
void printNGE(int *arr, int n)
{
  int i = 0;
  Stack s;
  int element, next;

  s.push(arr[0]);  // PUSH FIRST ELEMENT

  for (i=1; i<n; i++){
    next = arr[i];

    if(!s.isEmpty()){
      element = s.pop();
      while (element < next) {
```

```
        cout<<"\n "<<element<<" --> "<<next;
        if(s.isEmpty() == true)
          break;
        element = s.pop();
      }
      if (element > next)
        s.push(element);
    }
    s.push(next);
  }

  while (!s.isEmpty()){  // REMAINING ELEMENTS
    element = s.pop();
    next = -1;
    cout<<"\n "<<element<<" --> "<<next;
  }
}
```

This algorithm takes `O(n)` time.

Question 4.3: A valid mathematical expression can also have duplicate parenthesis as shown below:

```
((a+b))
(((a+(b)))*(c+d))
(a+b())
```

All of these expressions are valid, but darker parenthesis are duplicate and can be removed without affecting the value of overall expression. Write code to find if given expression has duplicate parenthesis.

Solution: For simplicity, let us assume that given expression does not have any white spaces and operands are single characters. Below is the algorithm:

```
For each character (traversing left to right)
• If (current character == ')'),
    pop an element from Stack, if it is open parenthe-
    sis '(', we found a duplicate parenthesis. Else,
    pop all characters till we find an open parenthe-
    sis.
• else
    push current character into the stack.
```

Below is C++ implementation of above logic.

```
bool findDuplicateparenthesis(char* str){
  MyStack stack;

  // TRAVERSE THE EXPRESSION
```

```
  for(char ch; ch != '\0'; str++){
    ch = *str;

    if (ch == ')'){
      char top = stack.pop();

      // IF IMMEDIATE pop IS OPEN PARENTHESIS '(',
      if (top == '(')
        return true;
      else
        while (top != '(')
          top = stack.pop();
    }
    else
      stack.push(ch);
  }
  return false;    // NO DUPLICATES
}
```

This function takes `O(n)` time and `O(n)` extra memory.

Question 4.4: Design a Stack with one extra operation (along with regular `push` and `pop`) called `getminimum()`. This function should return minimum element in the stack at that time without deleting any element.

Solution: One way of solving this problem is to use one extra stack in addition to the main stack that we have. Let us call the main stack `S`, and the new stack `MinStack`.

Modify the Push operation of the original Stack, S

```
If (MinStack's top element < value being inserted)
   Insert current element in the MinStack also
```

Also modify the Pop operation of the original Stack, S.

```
If (MinStack's top element <= S's top element)
   pop from MinStack and ignore that element
```

`getMinimum` function returns (and not pop) top element from `MinStack`. At any point, the Stack and `MinStack` looks as shown below:

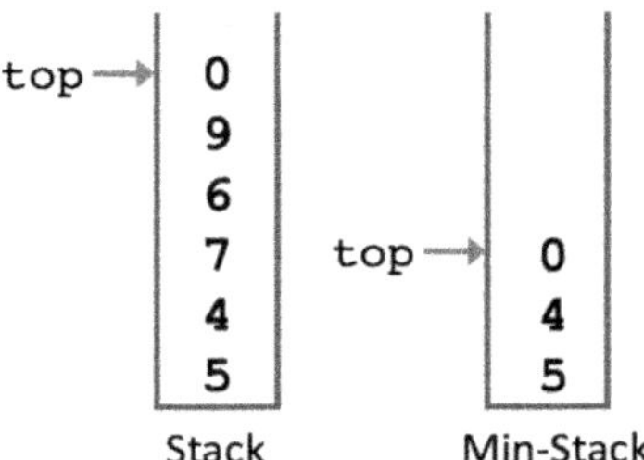

If we pop 0 from the Stack, 0 is also poped from MinStack and then getminimum will return 4.

Method-2: Using Min Pointer

Let us see this method with the linked list implementation of Stack. Modify the Node structure to hold a pointer to the next minimum as shown below,

```
struct Node{
  int data;
  Node* next;
  Node* nextMin; // POINTS TO NEXT MINIMUM

  // Default Constructor. Set the pointers to NULL
  Node(int d): data(d), next(NULL), nextMin(NULL){}
};
```

If you don't know C++ and wonder, how a function comes in the definition of a struct, please ignore it and just focus on the logic below.

Add one more pointer to the Stack class, in addition to the top pointer, call it min; min points to the current minimum element in the Stack, and is updated after each push and pop operation. The Stack class looks like below:

```
class MinStack{
private:
  Node* top;
  Node* min;
public:
  MinStack():top(NULL),min(NULL){}

  void push(int d);
  int pop();
  int getMinimum();
};
```

There are three functions, push, pop and getMinimum. Let us modify the original push and pop operation as below:

Push Operation:

```
IF stack is Empty
   Push the element and make both min & top point to it.
ELSE
   Push element in stack.
   IF(New Elemenet is less than element at min)
      top->nextMin = min
      min = top;
```

Pop Operation:

```
IF top element is also Minimum element
  min = min->nextMin
Pop the element at top
```

`getMinimum` operation return the element pointed to by `min` pointer. Below figure shows how the state of stack changes with operations:

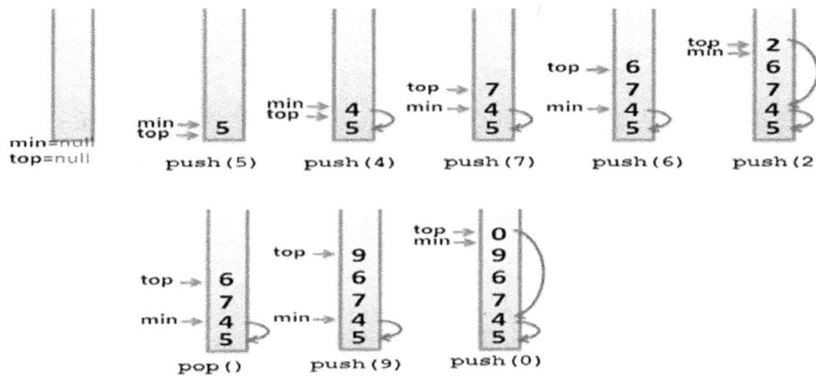

Let us write the definition of `MyStack` class functions

```
void MinStack::push(int d){
  Node* t = new Node(d);          // CREATE NEW NODE
  t->next = top; top = t;         // INSERT AT HEAD

  if(min == NULL || min->data > top->data){
    top->nextMin = min;
    min = top;
  }
}

int MinStack::pop(){
  if(top == NULL){ return -1; }

  if(min == top){ min = min->nextMin; }

  int value = top->data;
  // DELETING FIRST NODE
  Node* temp = top; top = top->next; delete temp;
  return value;
}

int MinStack::getMinimum(){
  if(min == NULL)
```

```
    return -1;
  else
    return min->data;
}
```

Question 4.5: Given a string having only '}' and '{' representing brackets in a mathematical expression. The brackets may not be balanced. You can toggle the bracket, changing '}' to '{' or '{' to '}'. Find minimum number of toggle operations needed to make the expression balanced. For example, consider the following input strings and corresponding output:

```
char str[]="{{{{";       // Output: 2
char str[]="{{{{}}";     // Output: 1
char str[]="}{{}}{{{";   // Output: 3
char str[]="}}}";        // Output: IMPOSSIBLE TO BALANCE
```

Solution: The solution does not exist if the number of brackets are odd.

At each position, there are two possibilities – an open bracket or a closed bracket. Fix an open bracket at index-0 and call the function recursively for string starting from index-1, then fix a closed bracket at index-0 and call the function recursively for index-1. Repeat the same for all the positions and keep track of minimum changes required when the expression is balanced.

This is a very time consuming logic and will take exponential time, `O(2^n)` to be precise.

A better solution, is to first remove all balanced parts of the expression. For example, if the given expression is "}{{}}{{" then the highlighted part of the expression is balanced, "}{**{}}**{{". If we remove this part, then we are left with the expression "}{{{". After removing the balanced part, the expression left is of the form, "}{..{" , i.e the expression contains zero or more closing brackets followed by zero or more open brackets.

Let us say that there are `m` closing brackets followed by `n` open brackets. In this case, we just need $\lceil m/2 \rceil + \lceil n/2 \rceil$ toggles. The expression, "}{{{" requires 1+2 = 3 toggles. Let us write the code for this logic

```
int minToggles(char* exp)
{
  if(strlen(exp)%2 == 1)
    return -1;    // IMPOSSIBLE

  stack<char> s;   // USING STACK OF CHAR
  for (int i=0; exp[i] != '\0'; i++)
  {
    if (exp[i]=='}' && !s.empty())
    {
      if (s.top()=='{')
```

```
            s.pop();
        else
            s.push(exp[i]);
    }
    else
        s.push(exp[i]);
}

int red_len = s.size();

int n = 0;
while (!s.empty() && s.top() == '{'){
    s.pop();
    n++;
}

return (red_len/2 + n%2);
}
```

To remove the balanced part of the parenthesis, we use a stack as shown in the code above.

Using Stack to evaluate expressions

Infix, Prefix and Postfix notations are three different (but equivalent) ways to write a mathematical expression. The '**In**', '**Pre**' and '**Post**' in notation's name represent the relative position of an operator (w.r.t operands) in the expression:

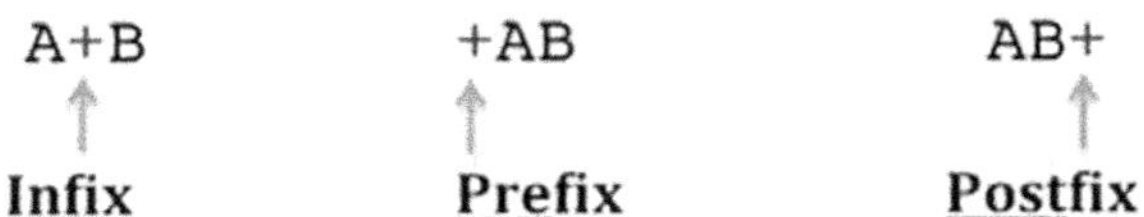

Infix Notation is the usual way of writing a mathematical expression where operator appears between two operands (in case of binary operators). It is ambiguous and requires knowledge of operator hierarchy for its evaluation. In below expression:

`A + B * C`

We must know that we have to perform multiplication (`B*C`) before addition. The expression is evaluated as `A+(B*C)` and is different from `(A+B)*C`. Parentheses can be used to override operator hierarchy.

Note that this is different from associativity or order of evaluation of operands.

Prefix Notation has operator written before the operands. This is also called Polish Notation. Hence, `A+B` is written as `+AB`.

The expression `A+B*C` is an Infix expression and equivalent prefix expression is

+A*BC. Conversion from infix to prefix is done as below:

Because multiplication has higher precedence, it is considered first. The two operands of multiplication are B and C. Therefore, B*C becomes *BC. After this, the two operands of plus operator are A and *BC. Note that *BC is considered as single operand by plus operator, and now (A)+(*BC) is converted to prefix.

Postfix Notation, also called the reverse-polish notation has operator at the end of expression. A+B in postfix is AB+. The expression A+B*C in reverse polish notions is written as ABC*+. The conversion from infix to postfix is:

Following table shows the prefix and postfix forms of some infix expressions

InFix	PreFix	PostFix
A+B	+AB	AB+
A+B)*C	*+ABC	AB+C*
A*(B+C)-D/E	-*A+BC/DE	ABC+*DE/-

Converting from InFix to PostFix

To solve this problem, we use a Stack. Let the Infix expression be given as String, and postfix expression will go in another string. Initially the Stack is empty and postfix expression string is also empty. Following is the algorithm to convert an infix expression to postfix expression:

1. Read the tokens (in this case characters) from the infix string one at a time from left to right
2. If the token is an operand, add it to the Postfix string.
3. If the token is a Left parenthesis, Push it on to the Stack
4. If the token is a Right parenthesis:
   a. Repeatedly Pop the stack and add the poped out token in to postfix string until a left parenthesis is encountered.
   b. Remove left parenthesis from the stack and ignore it

5. If the token is an operator
    a. If the stack is empty push the operator into the stack
    b. If the stack is not empty compare the precedence of the operator with the element on the top of the stack.
        i. If the operator in the stack has higher precedence, then Pop the stack and add the poped element to the string S. else Push the token into the stack.
        ii. Repeat this step until the stack is not empty or the operator in the stack has higher precedence than the current token.
6. Repeat this step till all the characters are read.
7. After all the characters are read and the stack is not empty then Pop the stack and add the tokens to the string S.
8. Return the Postfix string S.

Let us dry-run our algorithm for the following infix expression:

Infix Notation: A*(B+C)-D/E **Postfix Notation:** ... **top** = NULL **STACK**	**Infix Notation:** *(B+C)-D/E **Postfix Notation:** A **top** = NULL **STACK**	**Infix Notation:** (B+C)-D/E **Postfix Notation:** A **top** → * **STACK**
STEP-1	**STEP-2**	**STEP-3**
Infix Notation: B+C)-D/E **Postfix Notation:** A **top** → (* **STACK**	**Infix Notation:** +C)-D/E **Postfix Notation:** A B **top** → (* **STACK**	**Infix Notation:** C)-D/E **Postfix Notation:** A B **top** → + (* **STACK**
STEP-4	**STEP-5**	**STEP-6**
Infix Notation:)-D/E **Postfix Notation:** A B C **top** → + (* **STACK**	**Infix Notation:** -D/E **Postfix Notation:** A B C + **top** → * **STACK**	**Infix Notation:** D/E **Postfix Notation:** A B C + * **top** → - **STACK**
STEP-7	**STEP-8**	**STEP-9**

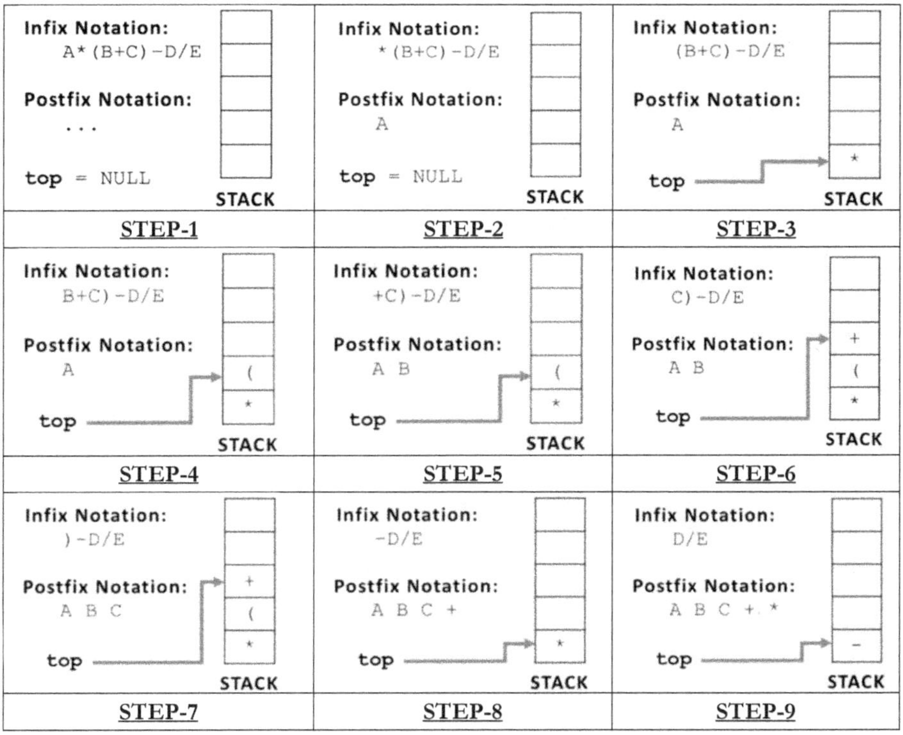

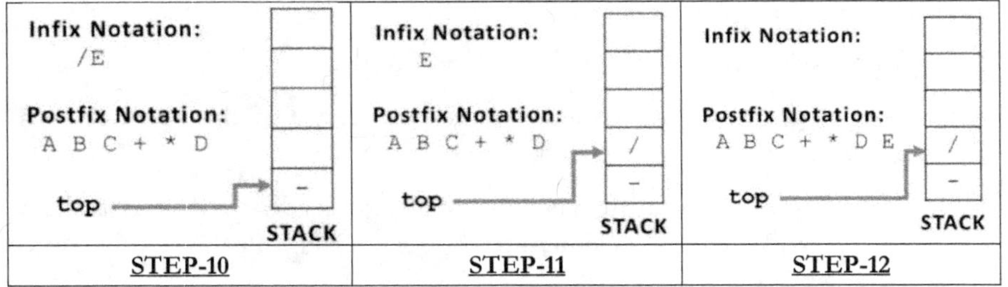

When the entire Infix string is considered, pop out everything from the stack and append it to the postfix expression. The final postfix expression is:

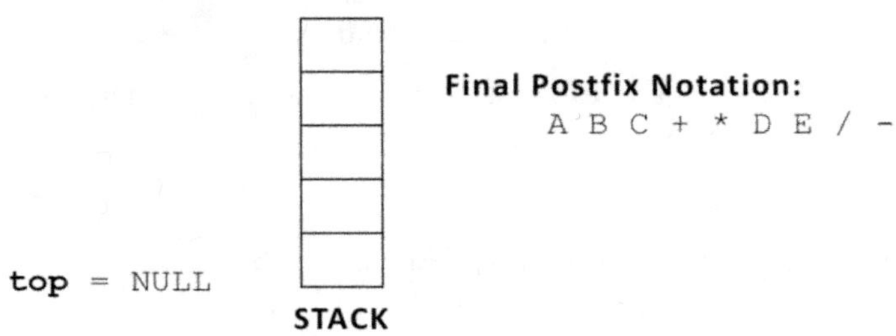

For simplicity of implementation, let us assume that expression has only three things:

1. **Operators:** Single character representing the operator.
2. **Operands**: All operands are single character.
3. **Parenthesis**: Expression has only parenthesis, '(' and ')'.

we are using two helper function, one to check if the given character is operand or not and second to return the precedence of the given operator.

```
bool isOperand(char ch){
  return (ch>='a' && ch<='z')||(ch>='A' && ch<='Z');
}
int precedence(char ch){
  switch (ch){
    case '+':
    case '-': return 1;

    case '*':
    case '/': return 2;

    case '^': return 3;   // EXPONENTIAL OPERATOR
  }
  return -1; // INVALID OPERATOR
}
```

Below function does the actual conversion, and prints the postfix expression:

```
// RETURNS 1 IF exp IS INVALID
int infixToPostfix(char* exp){
  int i, k;
  Stack stack;

  for (i = 0, k = -1; exp[i]; ++i)
    if (isOperand(exp[i]))          // OPERAND
      exp[++k] = exp[i];
    else if (exp[i] == '(')         // OPEN PARENTHESIS.
      stack.push(exp[i]);
    else if (exp[i] == ')'){        // CLOSE PARENTHESIS.
      while(!stack.isEmpty()&&stack.peek()!='(')
        exp[++k] = stack.pop();
      if (!stack.isEmpty() && stack.peek() != '(')
        return 1;
      else
        stack.pop();
    }else{                          // OPERATOR
      while( !stack.isEmpty() &&
              ( precedence(exp[i]) <=
                precedence(stack.peek()))
        exp[++k] = stack.pop();
      stack.push(expression[i]);
    }
  }

  // REMAINING ELEMENTS ARE ALL OPERATORS
  while (!stack.isEmpty())
    exp[++k] = stack.pop();

  exp[++k] = '\0';
  cout<<exp;

  return SUCCESS;
}
```

Evaluating a postfix expression

Note that there is no need for any brackets in either postfix or prefix expression. These notations represent an expression in an unambiguous way without the need for any parenthesis. The logic to evaluate a postfix expression is simpler than infix expression because you do not have to worry about handling the brackets. Below algorithm evaluates an expression given in the postfix notation:

```
Step-1:FOR each token in the expression
  IF it is an operand,
     Push it in the Stack.
  ELSE, if it is operator
     Pop operands from stack and perform this operation
     on those operands and push the result back in the
     stack.
Step-2: After operating on each token of expression, there
     will be just one value left in the stack. This is the
     final value of the expression, just pop it out.
```

Let us see the above algorithm for the following postfix expression (keeping all numbers as single digit for simplicity).

`4 2 3 + * 6 -`

The above postfix expression represents the following infix expression:

`4 * (2 + 3) - 6`

Step-1: The first token encountered in given expression while reading it from left to right is 4. 4 is an operand, push it in the stack.

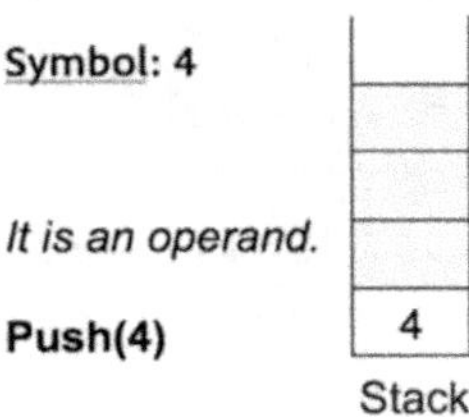

Step-2: Next symbol is 2, it is also an operand, push it in the stack.

Symbol: 2

It is an operand.

Push(2)

Stack
2
4

Step-3: Next symbol is 3, it is also an operand, push it in the stack.

Symbol: 3

It is an operand.

Push(3)

Stack
3
2
4

Step-4: Next symbol is the operator, plus (+). When an operator is encountered, pop the number of operands that this operator need (plus is a binary operator and need two operands). Perform the operations on these poped values and push the result back into the stack.

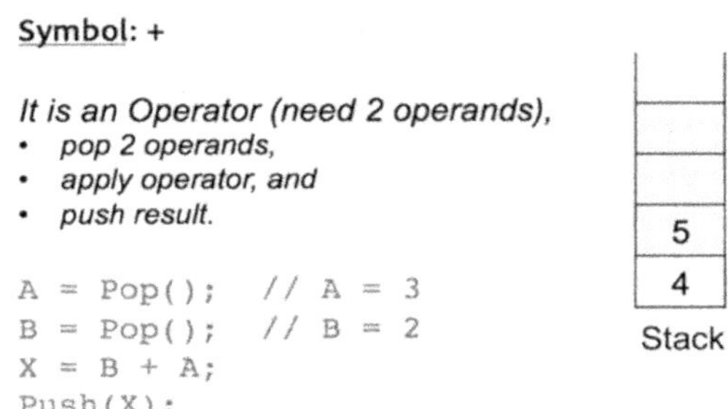

Symbol: +
It is an Operator (need 2 operands),
pop 2 operands,
apply operator, and
push result.
A = Pop(); // A = 3
B = Pop(); // B = 2
X = B + A;
Push(X);
5
4
Stack

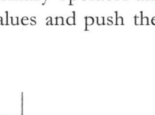

Step-5: Next token is an operator (multiply). Perform the same steps.

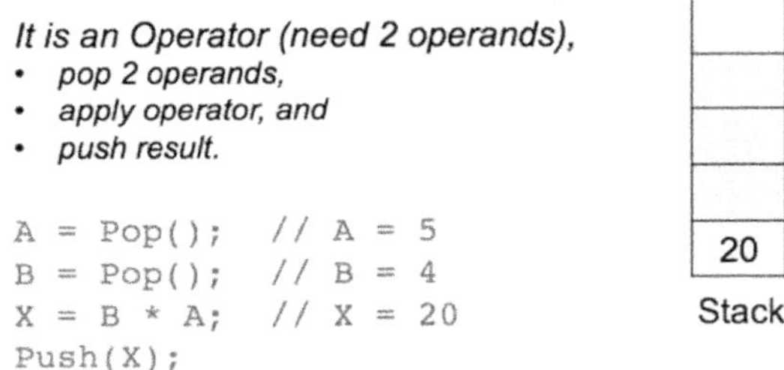

Symbol: *
It is an Operator (need 2 operands),
pop 2 operands,
apply operator, and
push result.
A = Pop(); // A = 5
B = Pop(); // B = 4
X = B * A; // X = 20
Push(X);
20
Stack

Step-6: Next Symbol is an operator, 6, push it in the stack.

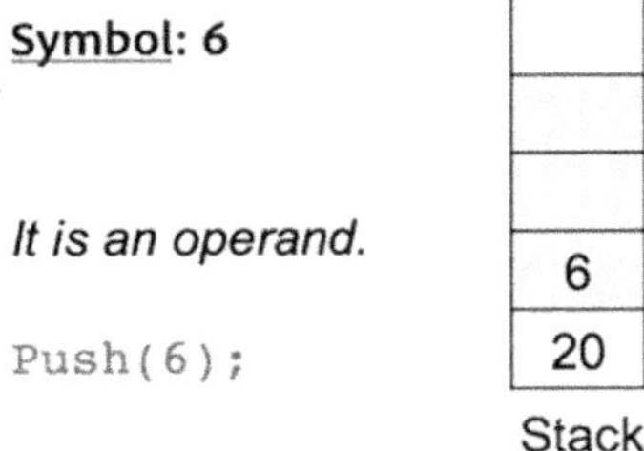

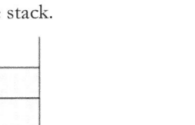

Step-7: Next symbol is an operator, minus. Pop two operands, perform the subtraction, and push the result back in the stack.

Symbol: -
It is an Operator (need 2 operands),
• pop 2 operands,
• apply operator, and
• push result.
A = Pop(); // A = 6
B = Pop(); // B = 20
X = B - A; // X = 14
Push(X);
14
Stack

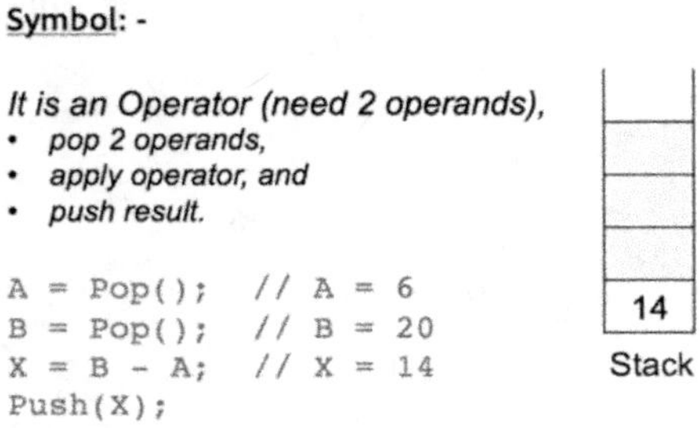

Step-8: Now the expression is complete, and hence there should be only one value in the stack which is the final value of the expression. Pop the final result out. The result of expression is: `14`.

Function `evaluatePostfix` evaluates the postfix expression and returns the result:

```
bool isNumericDigit(char ch){
  return (ch>='0' && ch<='9');
}

int evaluatePostfix(char* exp){
  Stack stack;

  for (int i = 0; exp[i] != '\0'; ++i){
    if (isNumericDigit(exp[i])) // OPERAND
      stack.push(exp[i] - '0');
    else{
      int a = stack.pop();
      int b = stack.pop();
      switch (exp[i]){
        case '+': stack.push(b + a); break;
        case '-': stack.push(b - a); break;
        case '*': stack.push(b * a); break;
        case '/': stack.push(b/a); break;
      }
    }
  }
  return stack.pop();
}
```

Since stack operations take constant time, the overall time taken to evaluate this expression is linear.

Function call Stack in the memory

The program that we write in C language becomes a binary executable after compilation and linking. When this executable binary is actually executing, it is called a Process. Operating system allocates memory to a process. This memory is used by the process while executing.

Area of the memory where a process is loaded is called process address space. Figure 4.1 shows the broad layout of the process address space (it is independent of platform, actual layout may differ for operating system and for program). The process address space has following segments:

- Code segment (or Text segment)
- Data segment
- Stack segment
- Heap segment

Code Segment is the read-only area where actual machine language instructions of the executable go. Size of this segment is fixed at load time.

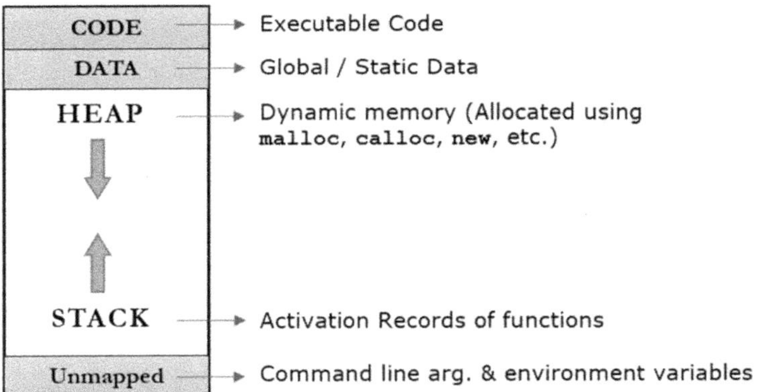

Figure 4.1: Broad layout of address space

Data Segment is the area where memory to global and static variables are allocated. Variables allocated memory in this area has a lifetime equal to that of the program. In C language, these variables are initialized with zero, if no default value is provided in the program. Size of this segment is also fixed for the lifetime of a program execution.

When we allocate memory dynamically (using malloc, calloc, realloc functions in C language) then memory is allocated in the **Heap segment**. Allocating and deallocating memory in this segment is the responsibility of the developer. The memory allocated on heap does not have a name and is accessed by using pointers and references. A bad handling of heap memory may result in memory leaks.

Stack segment contains Activation Records (also called Stack Frames) of all the active functions. An active function is a function that is currently under the call. Consider the below code:

```
int main(){
  fun1();
}
void fun1(){
  fun2();
}
void fun2(){
}
void fun3(){
  // NEVER CALLED
}
```

When `main` is called, it is the only active function. Then `main` calls `fun1`. At this point `fun1` is executing but both `main` and `fun1` are active. When `fun1` calls `fun2`, then execution controls go in `fun2`, but `main`, `fun1` and `fun2` are all active and has their activation records in Stack.

When function `fun2` returns, activation record of `fun2` is poped from the Stack and execution is back in `fun1`. At this point, `main` and `fun1` are active and has their activation records in the memory Stack.

`fun3`, is never active because it is never called and hence its activation record never gets created on the Stack.

✓ When a function is called, its Activation Record is created and pushed on the top of stack.

✓ When a function returns, the corresponding Activation Record is poped from the Stack.

✓ Size of Stack keeps changing while the program is executing because the number of active functions keeps changing.

✓ Non-static local variables of a function are allocated memory in the Activation Record of that function when the function is active.

✓ Variables allocated memory on Stack are not initialized by default. If the initial value of a variable is not given, then it is not initialized and its value is garbage (this is different from variables allocated memory in Data Segment).

✓ Activation record also contains other information required in function execution.

✓ Stack Pointer (SP) keeps track of the top of the Stack.

✓ The function whose Activation Record is at the top of the Stack is the one that is executing currently. A function whose activation record is in Stack, but not at the top is active but not yet executing.

✓ If a function is recursive then multiple activation records of that function may be present on the Stack.

After compilation and linking, the executable code (in machine language) gets generated. The first thing that happens when this executable code is executed is that it is loaded in the memory. Loading has following steps:

✓ **Code goes in the code area.** Code is in the form of binary machine instructions and Instruction Pointer (IP) holds the address of current instruction being executed.

✓ **Global and static variables are allocated memory in the data area.** Data area has two sub-parts, Initialized and Un-Initialized data area, If the initial value of a variable is given by us, it gets allocated in the Initialized data area, else memory to the variable is allocated in the un-initialized data area and it is initialized with zero.

✓ **Global and static variables are initialized.** If we have given the initial value explicitly, then variables are initialized with that value otherwise they are initialized with zeros of their data types.

```
int x = 5; // initialized with 5
int y;     // initialized with 0
```

After these steps, we say that the program is loaded. Once the loading is complete, the `main` function is called and actual execution of the program begins. Read the entire program given in below carefully:

```
// Go in data area at load time. Initialized with 0.
int total;

/** Code (machine instructions) of function goes in
 * code area. When this function is called, then
 * Activation Record of the function is created on
 * Stack.
 */
int square(int x){
  // x goes in Activation Record of this function
  // when it is called.
  return x*x;
}

/** Code of function goes in the code area. When this
 * function is called (at run-time), its AR gets
 * created on Stack and memory to non-static local
```

```
 * variables (x and y) is allocated in that AR.
 * count, being a static variable, is allocated in
 * data area at load time.
 */
int squareOfSum(int x, int y){
  static int count = 0; // Load-time var
  printf("Fun called %d times", ++count);
  return square(x+y);
}

/** Code goes in code area. When main is called, its
 * activation record gets created on Stack and memory
 * to non-static local variables (a and b) is
 * allocated in that Activation Record.
 */
int main(){
  int a=4, b=2;
  total = squareOfSum(a, b);
  printf("Square of Sum = %d",total);
}
```

This program computes `(a+b)`^{2} and print the result. To keep it simple, we are using the hard-coded values 4 and 2 for `a` and `b` respectively. The function `squareOfSum` also keeps a count of how many times it is called in a static variable `count`, and print the count every time it is called.

This may not be the best implementation, but it serves our purpose. Read the code again, especially the comments before each function and make sure that you understand everything.

After compilation and linking the executable of the program is created and when this executable is run, the first thing that happens is that this executable is loaded in the memory (RAM). At this point, the `main` function is not yet called and the memory looks like below:

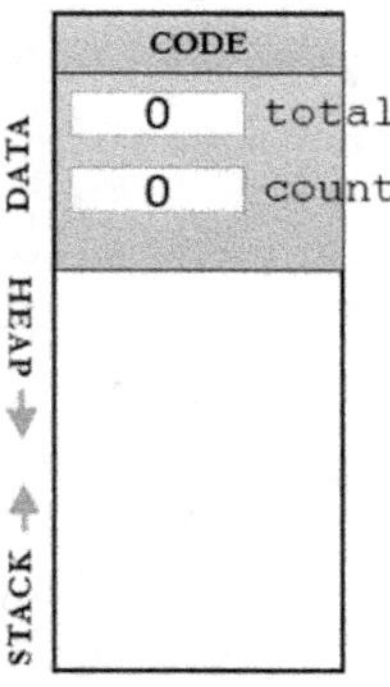

After loading is complete, `main` function is called. Whenever a function is called, its Activation Record is created and pushed in the Stack. The AR has:

- ✓ Local (non-static) variables of a function (`a` and `b` for `main`).
- ✓ Other things stored in the Activation Record.

In the diagrams, we are only showing non-static local variables in Activation Records. After `main` is called, the memory looks like below:

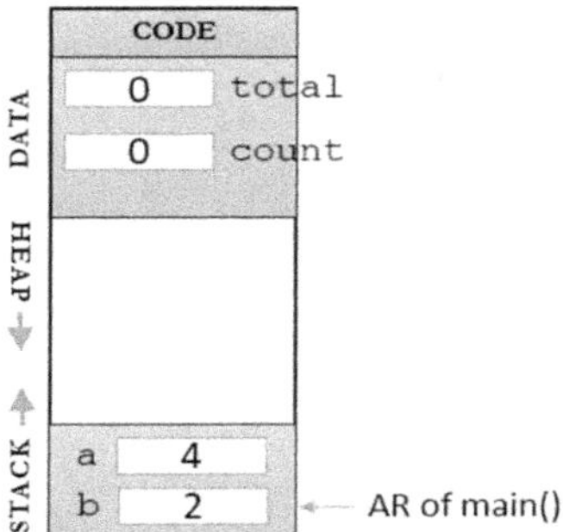

At any time, the point of execution (Instruction Pointer) is in the function whose AR is at the top of Stack. Let us understand what all happens internally when a function is called by another function.

When a function is called:

1. State (register values, Instruction Pointer value, etc.) of calling function is saved in the memory.
2. Activation record of the called function is created. Local variables of called function are allocated memory inside the AR.
3. Instruction pointer (IP register) moves to the first executable instruction of called function.
4. Execution of the called function begins.

Similarly, when the called function returns back (to the calling function), following work is done:

1. Return value of the function is stored in some register.
2. AR of called function is popped from the memory (Stack size is reduced and freed memory gets added to the free pool, which can be used by either the stack or heap).
3. State of the calling function is restored back to what it was before the function call (Point-1 in function call process above).

4. Instruction pointer moves back to the instruction where it was before calling the function and execution begins from the point at which it was paused.
5. Value returned from the called function is replaced at the point of call in the calling function.

```
Clearly, a function call is a lot of overhead both in
terms of time and memory.
```

One of the reasons behind the popularity of macros in C language (even after all the evil that they bring along) is this overhead in function call. Another was the type independence that macros bring.

Some compilers optimize the performance by replacing function call with the entire code of the function during compilation, hence avoiding the actual function call overheads. This is called in-line expansion. For example, the compiler may just put the entire code of function `square` inside `squareOfSum` and remove the function call all together as shown below.

```
int squareOfSum(int x, int y){
  static int count = 0; // Load-time var
  printf("Fun called %d times", ++count);
  return (x+y) * (x+y);
}
```

Recursive functions are very difficult to expand in line because the compiler may not know the depth of function call at compile time.

Let us see how memory looks like if we miss the terminating condition of a recursion. Below Code is an example of infinite recursion:

```
int main(){
  int x = 0;
  x++;
  if(x<5){
    printf("Hello");
    main();
  }
}
```

When the program is executed after compilation, it is first loaded in the memory and then the `main` function is called. At this point, Stack has only one activation record of function `main`.

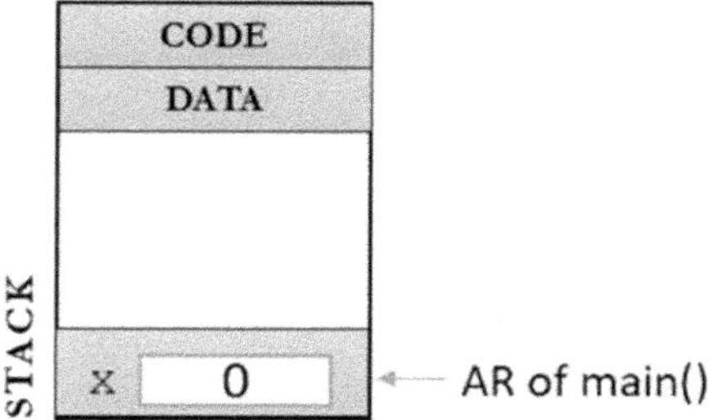

Initial value of x is 0. After increment x become 1, since x<5, the condition is true and main is called again. A new AR for this newly called main is created on the Stack and this AR also has local variable x that is different from variable x in AR of previous call.

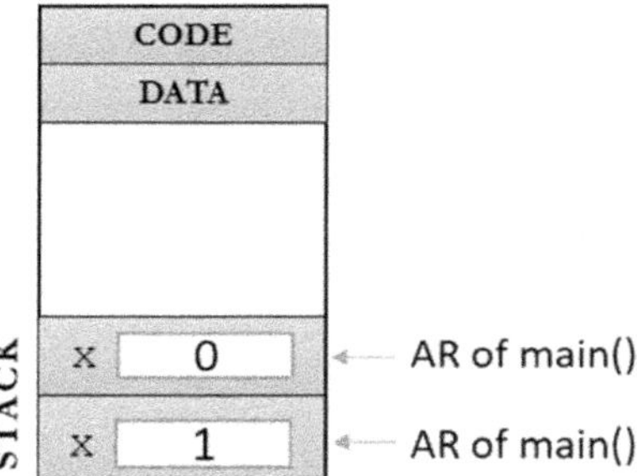

Value of this new x is again 0, and main is called again. Every time main is called, the value of x in the new activation record is 0and it will call main again. This becomes a an unstoppable chain.

The problem is that, every instance of main is actually using a different x (from their own instance of AR).

It will continue to print "Hello", until a point when no space is left in the Stack to create new AR. At this point main cannot be called further and the program will crash.

Recursive v/s Non-Recursive inside memory

Consider below recursive function, that return sum of first n natural numbers.

```
int sum(int n){
  if( n== 1)
    return 1;
  else
    return n + sum(n-1);
}
```

When this function is called for n=3, as sum(3); It will call sum(2); which will in-turn call sum(1);

At this point, the memory stack will have three instances of activation records of function `sum`, each having a local variable `n`, as shown below:

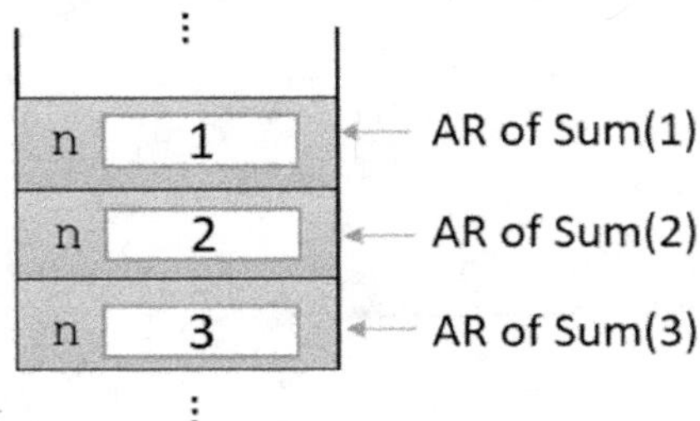

Consider the solution version of this problem

```
int sum(int n){
   int s = 0;
   for(int i=1; i<=n; i++)
      s += i;
   return s;
}
```

In this case, there is only one function call to `sum(3)` and local variables `n`, `i` and `s` on the Activation Record (AR) of function `sum` looks like:

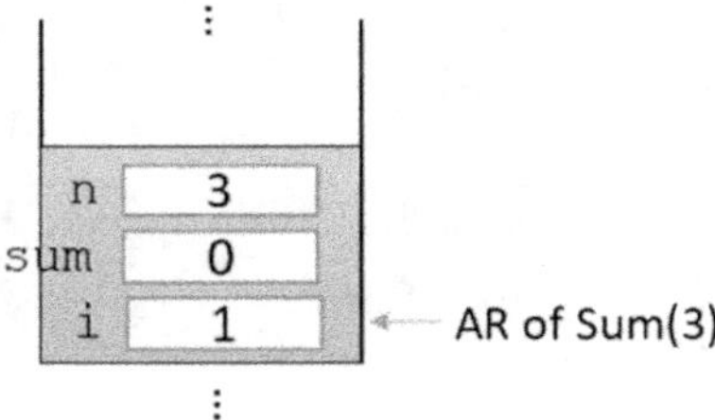

In recursive version, one activation record is created for each value of `n`. If `n=1000` then `1000` ARs are created. Therefore, the extra memory used is `O(n)`. The Table below gives a comparison of asymptotic running time and memory for recursive and non-recursive sum functions.

	Recursive	Non-Recursive
Time	O(n)	O(n)
Memory	O(n)	O(1)

The asymptotic time may be same for both, but actual time taken by recursive version is more than the iterative version.

```
Recursion is a huge overhead. Both in terms of memory and
execution time.
```

The examples of recursion seen till now are simple linear recursions. One of the major problems with recursive function comes when recursive calls start overlapping at sub-problem level. Overlapping subproblems is discussed in detail in our book "**Dynamic Programming for Coding Interview**".

Having discussed all this, the important point is that Stack frames of active functions are stored in a Stack data structure. Whenever we use the term, "memory stack', we are actually referring to this stack.

Question 4.6: How will you check if the memory stack on a machine is growing in the forward or backward direction?

Solution: Stack growing forward actually means that when a function is called, then the address of the activation record of called function is more than the address of activation record of calling function. We can check this by comparing the address of local variables in the activation records of two functions

```
bool isForwardTempFun(char* prevFunVar)
{
  char thisFun;
  return (&thisFun > prevFunVar);
}
bool isGrowingForward()
{
  char temp;
  return isForwardTempFun(&temp);
}
```

Question 4.7: Let us re-visit Question 3.4, to check if a linked list is a palindrome or not. Following linked lists are palindromes

```
2->3->4-> 4 ->3->2
M->A->L->A->Y->A->L->A->M
```

But following lists are not palindromes

```
2 -> 3 -> 4 -> 5 -> 4 -> 6 -> 2
M -> O -> K -> S -> H -> A
```

Solution: We may use Stack to solve this problem using the following algorithm:

1. **STEP-1:** Traverse the given list and push each node to the stack.
2. **STEP-2:** Traverse the list again, and for each node, do the following:
   a. Pop an element from the Stack.
   b. If content of current node is not equal to the one poped from Stack, return `false`.

3. **STEP-3:** If all nodes are traversed, return true.

This code takes O(n) time and O(n) extra memory.

Don't fall into the trap

If the same question is asked for an array, to check if an array is a palindrome or not. The solution is much simpler and takes constant memory because in array we can traverse backward.

```
bool isPalindrome(char* arr, int n)
{
  for(int i=0; i<n/2; i++)
    if(arr[i] != arr[n-1-i])
      return false;
  return true;
}
```

Chapter 5

Queue

A Queue is also a linear data structure much like a stack, that talks about the order in which elements are inserted and removed from the collection and not about how they are stored internally.

The only difference is that the order of insertion and deletion follows First-In-First-Out order, much like the real queues.

All the insertions happen at one end, say, rear-end, and all the deletions happen at another end, say, front-end. This is unlike stack data structure where all operations were happening at only one end.

Stack has one pointer, `top`. But queue need two pointers one each, for front and rear end. Let us call them `front` and `rear`. There are two major operations:

1. **Enqueue** (Insert): Insert an element at rear end and make the `rear` pointer point to this newly added element.
2. **Dequeue** (Remove): Removes the element at front and adjust the `front` pointer to point to the next element.

There can be other operations like `isEmpty` or `peek` which are similar to corresponding operations on stack.

Next, we discuss two major implementations of a queue data structure. First, when elements are kept in a linked list and second when elements are kept in an Array.

Linked list implementation

In a queue, operations are performed at both the ends, unlike stack, where all operations happen at only one end. Therefore, we need pointers, one for each end of the list. Let us defined the Queue class in C++.

```
class Queue
```

```
{
  Node *front;   // POINT TO FIRST NODE
  Node *rear;    // POINT TO LAST NODE
public:
  void enqueue(int x);
  int dequeue();
  bool isEmpty();
  void printQueue();
public:
  Queue():front(NULL), rear(NULL){}// CONSTRUCTOR
  ~Queue();   // DESTRUCTOR
};
```

Initially, both `front` and `rear` pointers are `NULL`. These two pointers are maintained in a way that `front` always points to the first node and `rear` always points to the last node

Function `enqueue` inserts at the rear end of the list and `dequeue` delete the first node of the list. `isEmpty` can just check one of the two pointers. Below are the implementations of these three functions:

```
void Queue::enqueue(int x){
  Node* temp = new Node(x);

  if(front == NULL)
    rear = front = temp;      // IF FIRST NODE
  else{
    rear->next = temp;        // INSERT AT END
    rear = temp;
  }
}
int Queue::dequeue(){
  if(isEmpty()){ return -1; } // QUEUE EMPTY

  int value = front->data;
  Node* temp = front;
  front = front->next;
  delete temp;
  return value;
}
bool Queue::isEmpty(){
  return (front == NULL);
}
```

All of these functions take constant time.

Array implementation of Queue

If we keep the queue elements in an array and use similar logic taking two integers to hold front and back indices, then the queue functions will look like below:

```
class Queue{
  int arr[QUEUE_LEN];  // ARRAY TO HOLD QUEUE
  int front;
  int rear;

public:
  void enqueue(int x);   // INSERT IN QUEUE
  int dequeue();         // REMOVE FROM QUEUE
  bool isEmpty();

  // CONSTRUCTOR & DESTRUCTOR
  Queue():front(-1),rear(0){}
  ~Queue();
};
```

Let us say that `front` always holds the index of the first element (the element to be deleted) and `rear` always hold the index where new values will come. Initially `front` is `-1`, and `rear` is `0`. The check for an empty queue is

```
bool Queue::isEmpty(){
  return (front == -1);
}
```

If we use a logic parallel to linked list, then `enqueue` operation insert value at `arr[rear]`, and `dequeuer` returns `arr[front]`. Both the functions also keep a check of boundary conditions when queue is full or empty. Below are these two functions:

```
void Queue::enqueue(int x){
  if(rear == QUEUE_LEN){    // OVERFLOW
    cout<<"OVERFLOW"<<endl;
    return;
  }

  arr[rear++] = x;
  if(front == -1){front = 0; } // IF EMPTY EARLIER
}

int Queue::dequeue(){
  if(isEmpty()){ // EMPTY
    cout<<"UNDERFLOW"<<endl;
    return -1;
  }
  int value = arr[front++];
```

```
    if(front >= rear){ // IF QUEUE BECOMES EMPTY
        front = -1; rear = 0;
    }

    return value;
}
```

But this logic has a flaw! Even after element at a position is removed, that position cannot be reused to store newly inserted values (unless the entire queue becomes empty and `front` and `rear` are reset).

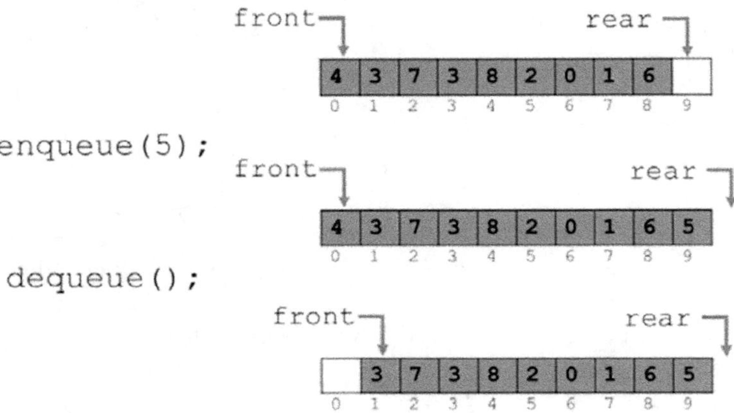

We have an empty space in the queue, but cannot insert any value. One solution is to fix this problem is to shift elements. Let us see it with an example. Initially, when the Queue is empty, `front` is `-1` and `rear` is `0`.

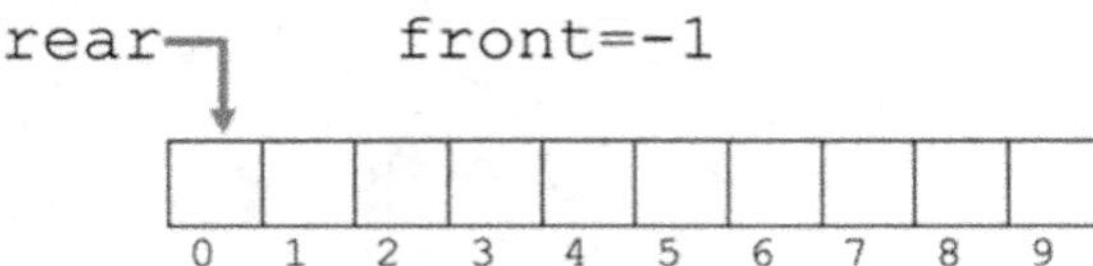

After inserting one element in the queue (say 2), the situation looks like below:

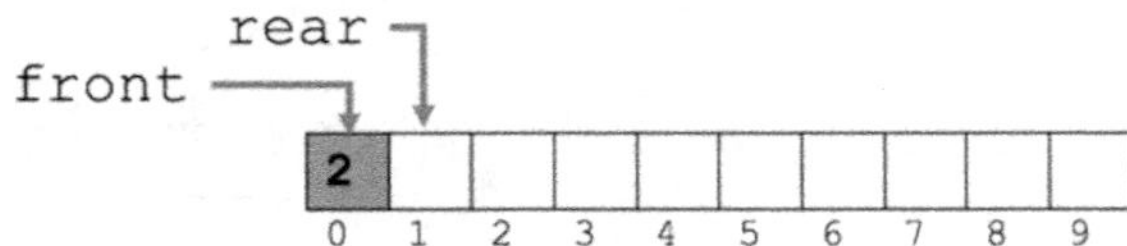

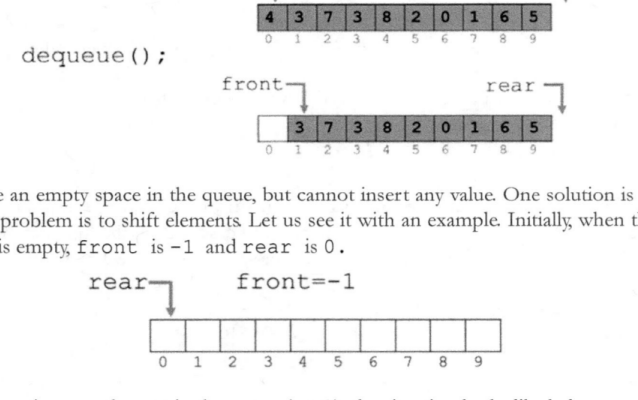

After inserting an element, the `rear` pointer moves forward and point to the next empty position. After some insertions, the queue will look like below:

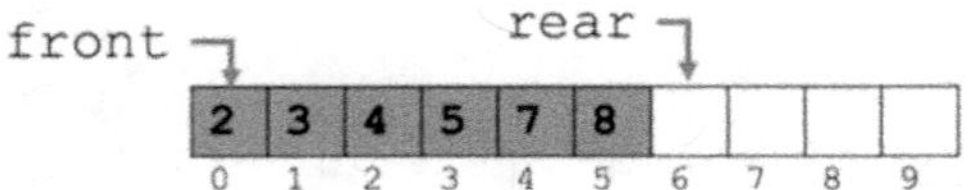

When an element is to be removed, the element is removed from `front` and all elements are shifted toward left so that there is no empty gap in the queue.

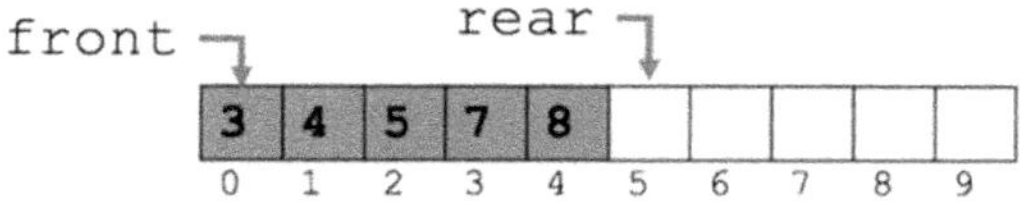

The first position of queue is never empty unless the entire queue is empty. The queue is empty when (`rear==0`) and queue is full when (`rear==n`).

In this implementation, we do not need `front` pointer at all because front element is always at index 0. Please note that size of the queue is fixed.

```
class QueueOnArray{
public:
  // CONSTRUCTOR
  QueueOnArray(int n): rear(0), size(n){
    arr = new int[size];
  }
  // DESTRUCTOR
  ~QueueOnArray(){
    delete[] arr;
  }

  void enqueue(int data);
  int  dequeue();
  bool isEmpty();

private:
  int rear;
  int size;
  int *arr; // Array to hold the queue
};
```

Code for construction and destructor is given in the class definition itself. Below is the code of other functions:

```
void QueueOnArray::enqueue(int data){
  if(rear == size){ throw; } // QUEUE FULL
  arr[rear++] = data;
}

int QueueOnArray::dequeue(){
  if(rear == 0){ return -1; } // QUEUE EMPTY

  int value = arr[0];

  // SHIFT ALL ELEMENTS.
```

```
  for(int i=0; i<rear-1; i++)
    arr[i] = arr[i+1];
  rear--;

  return value;
}
```

This implementation is correct, but, the time complexity of `dequeue` operation is now `O(n)`. Ideally both the operations should take constant time. So, this is not the best implementation.

Queue in a Circular Array

In this case also, there are two pointers `front` and `rear`. Initially, both are pointing to the 0th position

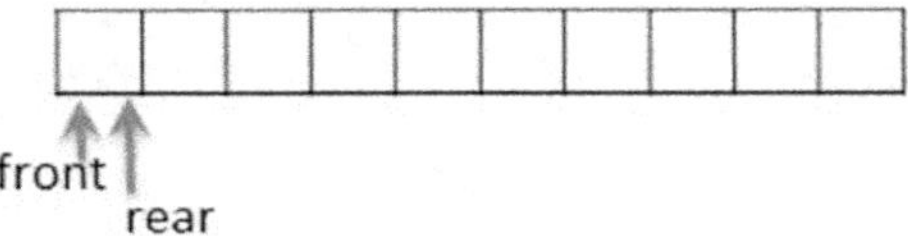

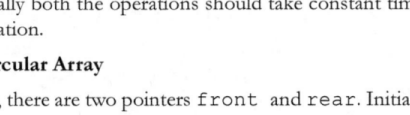

This can be used as a condition to check if queue is empty:

```
bool isEmpty(){
  return (rear == 0);
}
```

When an element is inserted in the Queue, `front` pointer remains as it is, and the element is inserted at the `rear` position and `rear` is incremented as

```
rear = (rear + 1) % SIZE
```

The state of Queue after inserting elements 2, 3, 4 and `6` is:

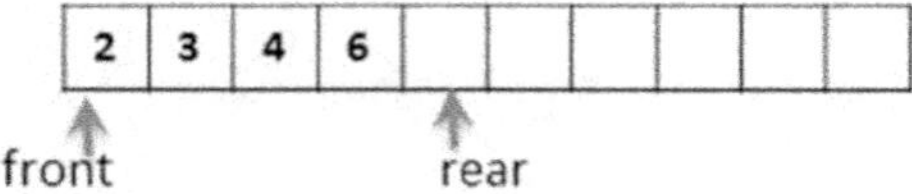

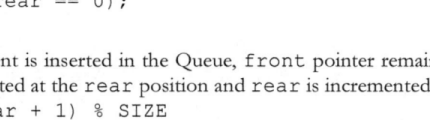

The element is dequeued from the front position and after deleting, `front` is incremented by using the same logic:

```
front = (front + 1) % SIZE
```

After removing the front element, i.e 2, the queue looks like:

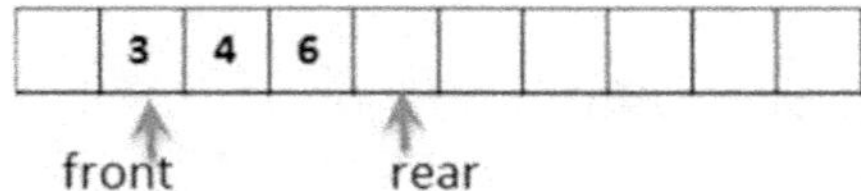

After few insertions and deletions, when `rear` reach the end, the queue may look like:

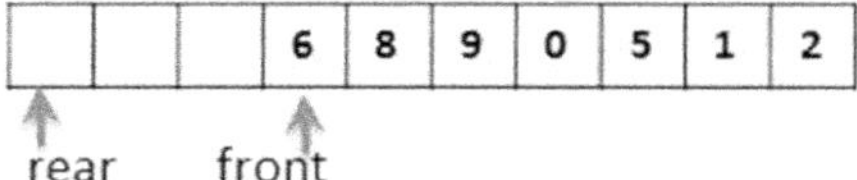

If we insert three more elements, then queue will look like:

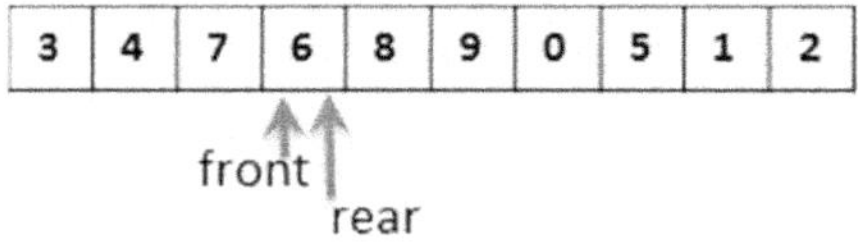

With this logic, when the queue is full, then also `front` and `rear` will become equal. We need to define a way to distinguish between whether Queue is empty or full. We can have a flag (say, `isEmpty)` to indicate whether the queue is full or empty. The initial state is:

```
front = 0;
rear = 0;
isEmpty = true;
```

Below is the implementation of queue:

```
class QueueOnArray{
public:
  QueueOnArray():rear(0),front(0),isEmpty(true){}
  ~QueueOnArray();

  void enqueue(int data);
  int dequeue();
  bool isEmpty();

private:
  int rear;
  int front;
  bool isEmpty;
  int arr[N];      // ARRAY TO HOLD QUEUE.
};
```

The initial condition is very clear in the constructor. Other functions are shown below (notice the difference in the `printQueue` function).

```
void QueueOnArray::enqueue(int data){
  if(rear==front && !isEmpty){ return; } //QUEUE FULL

  arr[rear] = data;
  rear = (rear + 1) % N;
```

```
  isEmpty = false;
}

int QueueOnArray::dequeue(){
  if(rear==front && isEmpty){ return -1; } //Q. EMPTY

  int value = arr[front];
  front = (front+1) % N;   // N IS SIZE OF ARRAY

  // IF QUEUE BECOMES EMPTY (AFTER DELETION)
  if(front == rear){ isEmpty = true; }

  return value;
}

bool QueueOnArray::isEmpty(){
  return isEmpty;
}

void QueueOnArray::printQueue(){
  cout<<"\n QUEUE IS : ";

  if(isEmpty){ return; }
  int i = front;
  do{
    cout<<" "<<arr[i];
    i = (i+1)%N;
  }while(i != rear);

  cout<<endl;
}
```

This code takes constant time in both enqueue and dequeue operations. We are using three variables, `front`, `rear` and `isEmpty` flag. We can manage with two variables instead of three.

1. `front`: pointing at the first element of the queue.
2. `count`: holding the number of elements in the queue.

Initially, `count` is zero. When an element is inserted in the queue, `count` is incremented and when an element is removed from the queue it is decremented.

Queue is empty when `count` is zero and queue is full when `count` is equal to length of array. `front` is incremented like `front=(front+1)%size` only.

We may also define the logic by using just `front` and `rear` variables, without any flag.

Double-Ended Queue

A double-ended queue, also called a dequeue; is a queue in which insertion and deletion can happen from either of the two ends. There are four operations that affect the size of the queue:

- ✓ Insert at front-end
- ✓ Insert at rear-end
- ✓ Delete from front-end
- ✓ Delete from rear-end

The implementations of queue can be easily extended for these operations.

Practice Questions

Question 5.1: Implement a queue using two stacks.

The question here is that given two Stacks, say `S1` and `S2`, and no other data structure, simulate the functionality of a Queue. Data will reside in these two stacks (instead of an array or linked list) and the order of insert and delete should be FIFO (First-In-First-Out). The class definition for this Queue is:

```
class SpecialQueue{
private:
  Stack s1;
  Stack S2;

public:
  void enqueue(int data);
  int dequeue();
  bool isEmpty();
};
```

You may also define some helper functions, constructor(s) and destructor. They are not included here to keep it small and to the point.

Solution: This question is a typical example of **abstraction**. User is given an interface to access the Queue (enqueue-dequeue) and is not concerned about the underlining data structure used to store actual data of queue.

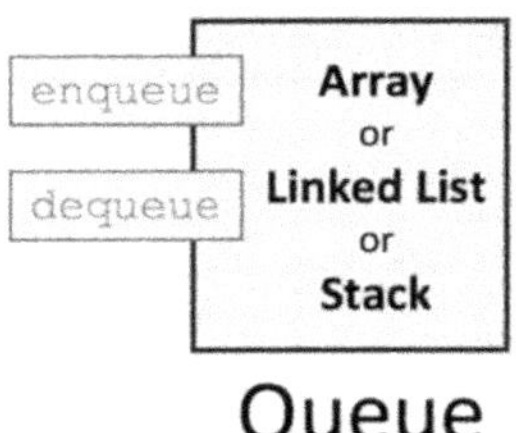

Queue

In this case, we insert elements in one stack and remove them from another stack. Let us call these two Stacks `InputStack` and `OutputStack`. Element is always inserted (pushed) in `InputStack`, and removed(poped) from `OutputStack`. If a stack is empty, then all elements are poped and pushed into that stack. This way, the order of elements in `InputStack` and `OutputStack` is always opposite. If an element is inserted first, then it remains at the bottom of `InputStack`, but it will come at the top of `OutputStack`. Hence, the first element to be removed is the one that was inserted first (i.e FIFO).

Queue operations are all constant time operations, but in this case, since elements are moved from one stack to another stack, at least one of the two operations will be `O(n)`.

Depending on our requirements we can opt to make either enqueue faster or dequeue faster, as discussed below:

Method-1: Enqueue takes constant time, Dequeue is O(n)

```
void enqueue(int x):
  1. Push x to InputStack.

int dequeue():
  1. IF (both stacks are empty) THEN error.
  2. IF (OutputStack is empty)
       Pop each element of InputStack
       and push it into OutputStack.
  3. Pop element from OutputStack and return.
```

Method-2: Enqueue is O(n), Dequeue takes constant time

```
void enqueue(int x):
  1. Pop all elements of OutputStack
     and push them into InputStack.
  2. Push x into InputStack.
  3. Pop all elements of InputStack
     and push them into OutputStack.

int dequeue():
  1. IF (OutputStack is empty) THEN error.
  2. Pop element from OutputStack and return.
```

Method-3: Mixed Approach

```
void enqueue(int x):
  1. IF (OutputStack IS Non-Empty)
       Pop all elements from OutputStack
       and push them into InputStack.
  2. Push x into InputStack.

int dequeue():
  1. IF (InputStack is Non-Empty)
       Pop all elements from InputStack
       and push them into OutputStack.
  2. IF (OutputStack is Empty) Return ERROR
     ELSE  Pop and return from OutputStack.
```

In this approach, we are not moving the data unless required.

If there are n-consecutive Enqueue (or n-consecutive Dequeue) operations, then the first one takes `O(n)` time, but rest `(n-1)` takes constant time. The worst case is when we have alternate enqueue and dequeue operations. Below is the code for Method-1:

```
class SpecialQueue{
  private:
    Stack inputStack;
    Stack outputStack;
  public:
    void enqueue(int data);
    int dequeue();
    bool isEmpty();
};

void SpecialQueue::enqueue(int x){
  inputStack.push(x);
}

int SpecialQueue::dequeue(){
  if(!outputStack.isEmpty())
    return outputStack.pop();

  if(!inputStack.isEmpty()){
    while(!inputStack.isEmpty())
      outputStack.push(inputStack.pop());
    return outputStack.pop();
  }

  // ERROR case. Queue is Empty.
```

```
  return -1;
}

bool SpecialQueue::isEmpty(){
  return ( inputStack.isEmpty() &&
           outputStack.isEmpty());
}
```

Question 5.2: Implement a stack by using two queues.

Solution: This question is similar to the previous one. Use two Queues, push operation of stack will insert element in one Queue and Pop operation will remove element from another Queue.

In an ideal implementation of a stack, both push and pop operations are constant time operations. But, that is when the Stack is implemented by using either array or linked list. If Stack is implemented by using two Queues, then both the operations cannot be constant time operations. At least one of the operations has to be expensive.

The class definition of such a stack looks like the below `SpecialStack`:

```
class SpecialStack{
  public:
    void push (int data);
    int pop();
  private:
    Queue queue1;
    Queue queue2;
};
```

In real life, such a class should be a template because we do not want our Stack or Queue to be restricted to store only `int` values. But we will stick to this definition only for the sake of simplicity.

Like the previous question, there can be two methods. Below is the code of both the methods.

Method-1: Push is constant time, but Pop operation takes O(n) time

```
void SpecialStack::push(int data){
  queue1.enqueue(data)
}

int SpecialStack::pop(){
  int returnValue =-1;   // INDICATE EMPTY STACK.

  while(!queue1.isEmpty()){
    returnValue = queue1.dequeue();

    if(queue1.isEmpty())
```

```
      break;
    else
      queue2.enqueue(returnValue);
  }

  // SWAP queue1 AND queue2. IF SWAPPING IS NOT
  // POSSIBLE, MOVE ALL ELEMENTS FROM queue2 TO
  // queue1.
  Node * temp = queue1;
  queue1 = queue2;
  queue2 = temp;

  return returnValue;
}
```

Method-2: Pop is constant time but Push operation is `O(n)` time

```
void SpecialStack::push(int data){
  queue2.enqueue(data);

  while(!queue1.isEmpty())
    queue2.enqueue(queue1.dequeue());

  Node * temp = queue1; // SWAPPING
  queue1 = queue2;
  queue2 = temp;
}

int SpecialStack::pop(){
  return queue.dequeue();
}
```

Chapter 6

Binary Trees

All Data structures discussed till now are linear in nature. A Binary tree is a hierarchical data structure where each node is either `NULL` or has two children, left sub-tree and right sub-tree, and both of them are also binary trees.

Node structure of a Binary tree is very similar to that of a linked list, with one extra pointer along with data field. Each node has two pointers, to store addresses of left and right child.

```
struct Node{
  int data;
  struct Node *left;
  struct Node *right;
};
```

left	data	right

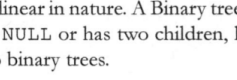

Structure of Node of a Binary tree

Figure 6.1, below shows a sample binary tree based on above structure of Node.

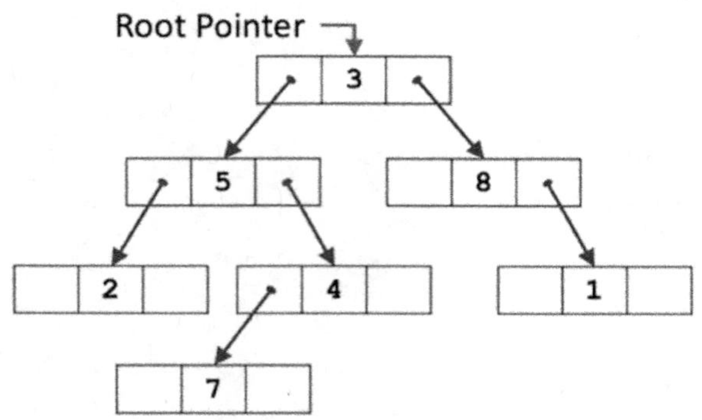

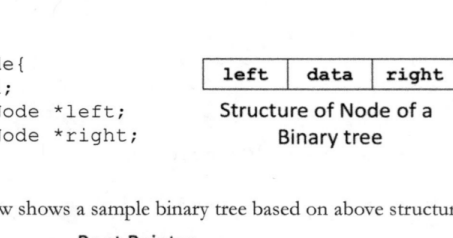

Figure 6.1: Binary tree

There is one node at the top of the hierarchy, called the **root node**. We keep a pointer to the root node (the way we keep a pointer to the head of linked list). All other nodes are either in left or right sub-tree of the root node and can be reached via this pointer. Below are some common terminologies used for a binary tree (Figure 6.1):

- **Root:** Top node of tree. (Node with value 3)
- **Child:** Node whose address is stored in either left or right field. (5 and 8 are child nodes of 3)

- **Parent:** Node that is pointing to the current node. (5 is the parent of 2 and 4)
- **Siblings:** Nodes with the same parent. (5 and 8 are siblings)
- **Cousin:** Nodes at the same level with different parents. (4 and 1 are cousin nodes)
- **Leaf:** Node without any children. (2, 7 and 1 are leaf nodes)
- **Non-Leaf:** All nodes other than leaf nodes (3, 5, 8 and 4 are non-leaf nodes)

Edges and Path in a binary tree

We already know what a node is. An Edge is a line that connects two nodes. It shows the connection between two nodes. In binary trees the connection is one way, i.e we can only move from a parent node to its child node, but in Graphs, we may have bi-directional (called undirected) edges also.

```
As a general Rule: If a connected and acyclic Graph has
N vertices, then it will have exactly N-1 edges.
```

If there are `N` nodes in a binary tree, then there will be exactly `(N-1)` edges irrespective of the structure of the tree.

A path in a binary tree is a sequence of nodes where each node is connected with the next node directly with an edge. One of the paths in the below binary tree is `A->C->D->E`, as shown in the diagram.

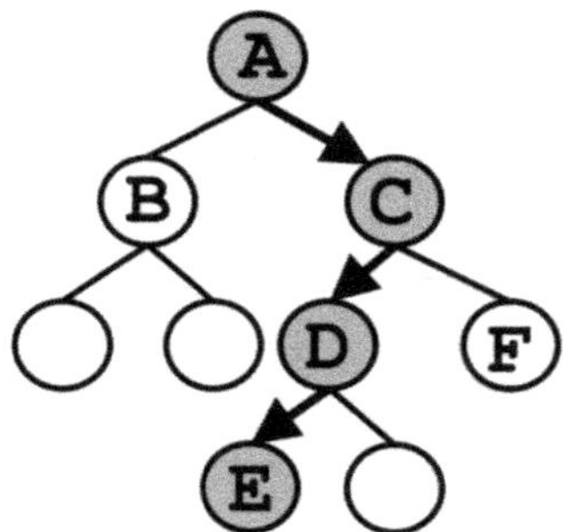

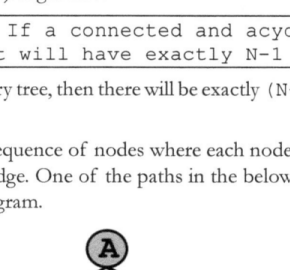

The direction of the path in a binary tree is always downwards. The length of a path is equal to the number of edges in that path.

Level of each node and Height & Depth of a binary tree

Level of root node is zero. A node at a distance `k` from the root is at level `k`. The following diagram shows the level of nodes for the given binary tree:

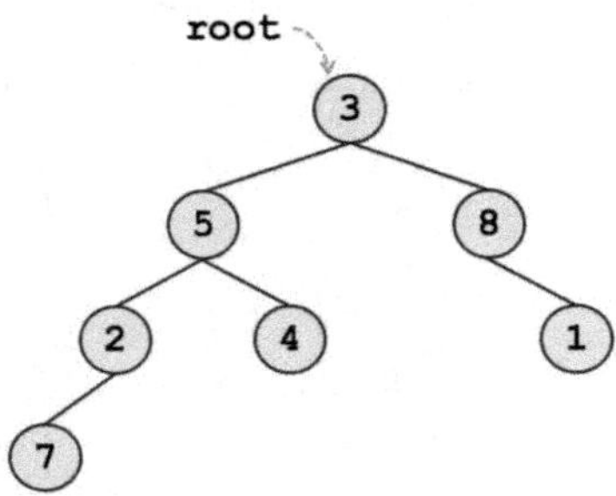

The height of a tree is equal to the maximum level of that tree. If there is only one node in the tree, then height of the tree is zero. Depth of the tree is same as height of the tree.

Below function returns the height of a binary tree:

```
int height(Node* r)
{
  if(r==NULL || (r->left == NULL && r->right == NULL))
    return 0;
  return max(height(r->left), height(r->right))+1;
}
```

This function takes O(n) time because it is visiting each node only once.

Later in this book, we also discuss special types of Binary trees like Complete Binary tree and Almost complete binary tree.

Traversals of a binary tree

A Linear structure can only be traversed in forward or backward direction, but a hierarchical data structure like a binary tree can be traversed in many ways. Most common traversals are given below (along with example of tree shown in Figure 6.1)

Pre Order: `<3, 5, 2, 4, 7, 8, 1>`

1. Print data at root.
2. Traverse left sub tree in Pre Order traversal.
3. Traverse right sub tree in Pre Order traversal.

In Order: `<2, 5, 7, 4, 3, 8, 1>`

1. Traverse left sub tree in In Order traversal.
2. Print data at root.
3. Traverse right sub tree in In Order traversal.

Post Order: `<2, 7, 4, 5, 1, 8, 3>`

1. Traverse left sub tree in Post Order traversal.
2. Traverse right sub tree in Post Order traversal.
3. Print data at root.

Level Order: `<3, 5, 8, 2, 4, 1, 7>`

1. Print all nodes at each level starting from level-0 till last level from left-to-right.

Code for first three traversals is a direct translation of the corresponding algorithm. Function `inOrder` in Code 6.1 below prints the tree in inorder traversal.

```
void inOrder(Node *root)
{
  if(root == NULL) { return; } // TERMINATING COND.

  inOrder(root->left);
  printf("%d", root->data);
  inOrder(root->right);
}
```

Code 6.1

Code 6.1 takes `O(n)` time because each node is visited only once and time taken at each node is constant.

INTERVIEW TIP

In real life, time taken by a function is calculated more intuitively then mathematically. I have not seen anyone solving equations to come up with time complexity of a function, especially during interviews. Below pointers may help in calculating time taken by a code:

1. *See how many times each element is visited and what is the work done at each element.*
2. *See how recursion is unfolding. In Code 6.1, each instance of the function is calling itself twice if the node is not-NULL. If there are N nodes, function is called 2*N times (Time complexity is still O(n)). This gives us an idea to optimize the code and remove unnecessary function calls. See the code below and compare it with Code 6.1.*

```
void inOrder(Node *root)
{
  if(root == NULL) { return; }

  if(root->left != NULL)
    preOrder(root->left);

  printf("%d", root->data);

  if(root->right != NULL)
    preOrder(root->right);
}
```

In any binary tree, number of NULL pointers is more than number of NON-NULL pointers. Number of function calls in the above code is almost half of Code 6.1 even when both takes asymptotic linear time.

3. *In recursion, try to draw function call tree. It helps in analyzing both time and memory taken by the code.*

Constructing tree from traversals

If we are given the preorder and inorder traversal of a Binary Tree, then we can construct the original tree from these traversals. Let us take an example,

```
InOrder: 1 4 5 8 10 30 40
PreOrder: 10 5 4 1 8 30 40
```

In the preorder traversal root node is always traversed first. Hence, `10` is the root of the given Binary Tree.

In the inorder traversal, root node is traversed after traversing the entire left subtree and before starting traversing the right subtree. Because `10` is the root, we can say, all elements on the left side of `10` in the inorder traversal are in the left sub-tree, and all elements on the right side of `10` are in the right subtree. After the first step, we are able to infer, that the binary tree looks like below

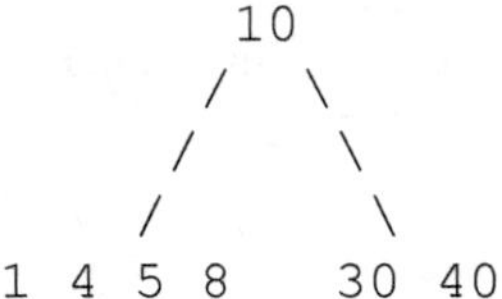

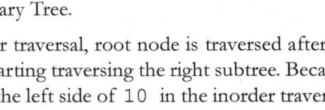

The root node is 10, left subtree has four nodes with values `1,4,5` and `8` and right subtree has two nodes with values `30` and `40`. We do not (yet) know the structure of left and right subtrees, but, if we recursively follow the above steps for left and right subtrees, we can construct the entire tree.

In the right-subtree, we have `30` and `40`. In the preorder traversal, `30` comes first, so `30` is at the root of right subtree. `40` can be the right-child or left child of `30`, to find whether it is the right child or the left child we have to look into the inorder traversal and see the relative positioning of `30` and `40`.

In the inorder traversal, `40` comes after `30`, therefore `40` is right child of `30`. The right sub-tree of `10` is now completely constructred:

```
          10
         /  \
        /    30
       /       \
  1 4 5 8       40
```

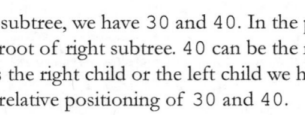

Similarly, the left subtree can be constructed. The final tree is:

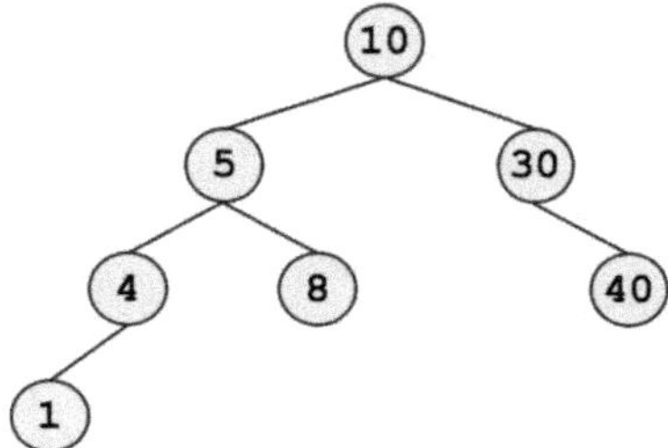

Similarly, we can construct the original tree when post-order and in-order traversals are given. The logic remains similar to the above (when in-order and pre-order is given) except, in post-order traversal root is the last element.

But, can we construct a tree if only pre-order and post-order traversal of the tree is given?

No !

From both the traversals (pre & post) we can find the root of binary tree, but we will not be able to separate nodes belonging to the left and right subtree.

Level Order traversal

If we apply BFS graph algorithm to a binary tree, taking root as the source node, then we actually traverse the binary tree in level order.

Take a queue, where each element of the queue is `Node*`. The following code, traverse the tree in level order traversal:

```
void printLevelOrder(Node* r){
  if(r == NULL) { return; }

  Queue q;
  q.enqueue(r);

  while(!q.isEmpty()){
    Node* temp = q.dequeue();
    cout<<temp->data<<" ";
    if(temp->left){  q.enqueue(temp->left); }
    if(temp->right){ q.enqueue(temp->right); }
  }
}
```

Insert root of the tree in the queue and then repeat the following two steps until the queue becomes empty:

1. Remove the front element from queue and print its value.
2. Insert the left and right child of newly removed Node if they exist.

If the given binary tree is

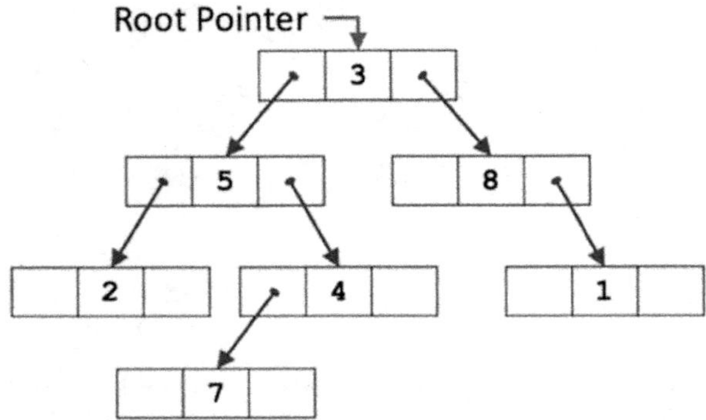

Then the root is initially inserted in the queue. The queue contains pointer to the root node. In the below figures, we show content of node in the queue rather than its address. Initially the queue has only root.

In the `while` loop, we remove the node, print its value and then insert the non-null children in the queue. After the first iteration, root is printed and nodes 5 and 8 are inserted in the queue

Queue 5 8

Removed Elements = 3

Then, 5 is removed and printed and its children are inserted in the queue.

Queue 8 2 4

Removed Elements = 3 5

After this 8 is removed and its child is inserted in the queue.

Queue 2 4 1

Removed Elements = 3 5 8

Following the same approach, elements get printed in level order traversal. It does not use recursion, unlike the other three traversals discussed earlier. But, it uses `O(n)` extra memory for Queue.

The output of level order traversal is: `3 5 8 2 4 1 7.`

What if we want to print nodes of each level in a separate line? i.e for the given tree, the output should be:

```
3
5 8
2 4 1
7
```

One solution to this problem is to first compute the height of tree and then for each level, print all nodes in that level from left to right.

A better solution, however, is to add a delimiter after each level in the queue while doing a level order traversal. This delimiter can be a `NULL`. Initially, root and `NULL` are inserted in the queue (there is only one node in the top level):

```
q.enqueue(r);
q.enqueue(NULL);
```

In the `while` loop when we remove value from the queue. If the element removed from the queue is `NULL`, it indicates the end of current level, print the new line character. At this point, queue contains only nodes from the next level, and all the nodes of the next level are in the queue. Insert a `NULL` back in the queue, indicating the end of next level inside the queue.

```
while(!q.isEmpty())
{
  Node* temp = q.dequeue();
  if(temp == NULL)          // END OF LEVEL
  {
    if(!q.isEmpty())
      q.enqueue(temp);
    cout<<endl;
    continue;
  }
  cout<<temp->data<<" ";
  if(temp->left){  q.enqueue(temp->left); }
  if(temp->right){ q.enqueue(temp->right); }
}
```

Code 6.2

Please note that we are inserting `NULL` back in the queue only when the queue is not empty. This check is important, else we will be in an infinite loop.

Another way of solving this is to store the level number of each node in the queue along with their address. The fundamental element of queue, in this case, is the below structure:

```
struct QueueElement{
  int level;
  Node* node;
};
```

The level for root is zero and each time we insert a new value in the queue, its level will be one more than the level of current node (its parent). This algorithm can also be used to find the distance of a node from the root.

If the level of node removed from the queue is not same as level of the previous node, then it also inserts a new-line character before printing the data.

INTERVIEW TIP

In almost all interview questions related to Binary trees, we end up traversing the tree in any one of these traversals, and may be performing some operation at the root. For example, if we want to count the number of leaf nodes, then traverse the tree and increment the count if current node is leaf node. Another way of looking at the solution is:

Number of leaf nodes in tree = Number of leaf nodes in the left subtree
+
Number of leaf nodes in the left subtree

The code for above logic is:

```
int numOfLeafNodes(Node* r){
  if(r == NULL) { return 0; }

  // IF CURRENT NODE IS LEAF NODE
  if(r->left == NULL && r->right == NULL)
    return 1;

  return ( numOfLeafNodes(r->left) +
           numOfLeafNodes(r->right) );
}
```

We are actually traversing the tree in preorder traversal and at each node, we are checking if it is a leaf node or not.

We can very well use other traversals to count the number of leaf nodes. The below code use level-order traversal to count the number of leaf nodes in the tree.

```
int numOfLeafNodes(Node* r){
  if(r == NULL) { return 0; }

  Queue q;
  q.enqueue(r);
  int cnt = 0; // TO COUNT LEAF NODES

  while(!q.isEmpty()){
    Node* temp = q.dequeue();
    cout<<temp->data<<" ";
    // IF LEAF NODE
    if(r->left == NULL && r->right == NULL)
      cnt++;

    if(temp->left){ q.enqueue(temp->left); }
    if(temp->right){ q.enqueue(temp->right); }
  }
}
```

Practice Questions using pre, in or post order traversal

Some questions require us to process the root node first and then the left and right subtree depending on the results of processing of root node. This order of accessing nodes is same as preorder traversal of a binary tree. The code in such a situation is similar to the preorder traversal. Consider the following question.

Question 6.1: Given a binary tree and a number `x`. Search for `x` in the binary tree.

Solution: We know, linear search traverses a collection in sequential order and while traversing, compares the current element with the value being searched. If the two are equal, we have found the element, else we continue searching with the next element. If end of the collection is reached without hitting a success, we return a failure.

Linear search in binary trees traverse all node one-by-one in some deterministic sequential order and compares the value of each node with the value being searched till a match is found or no element is left in the tree.

This deterministic sequential order can be of any of the tree traversals discussed earlier.

Position of an element does not make much sense because of hierarchical nature of the tree. Below function traverse the given binary tree in preorder and returns `true` if the value is found at either the root, left subtree or right subtree.

```
int linearSearch(Node* r, char x){
  if(r == NULL){ return 0; } // NOT FOUND

  if(r->data == x){ return 1; } // SUCCESS

  return ( linearSearch(r->left, x) ||
           linearSearch (r->right, x) );
}
```

This code takes `O(n)` time, same as traversal. We can very well traverse the tree in level order traversal and for each node, check if the value of that node is equal to `x`.

Note that this search is linear in nature. We are visiting each node of the tree to check if that node has the value we are looking for. Linear search takes `O(n)` time because of its nature of sequential traversal. We know that binary search can be used to reduce the search time, but for binary search to be applicable in an array, it has to be sorted. Similarly, Binary search can be used in a binary tree, but to use Binary search, nodes of the tree has to be in a particular order. Such an ordered tree is called **Binary Search Tree**; it is discussed in detail in the next chapter.

Question 6.2: Write code to check if two trees are identical or not. Two trees are identical when they have same data and exact same structure.

Solution: The solution to this problem is to traverse both the trees simultaneously and check the data of corresponding nodes in the two trees.

```
bool isIdentical(Node* r1, Node* r2){
  if (r1 == NULL && r2 == NULL){ return true; }
  if (r1 == NULL || r2 == NULL){ return false; }

  if(r1->data != r2->data) { return false; }

  return (isIdentical(r1->left, r2->left) &&
          isIdentical(r1->right, r2->right));
}
```

Question 6.3: Given a Binary Tree and a number `x`, write a function to see if there exist a root to leaf path in the tree in which sum of nodes is equal to `x`. For Example, if the given binary tree is:

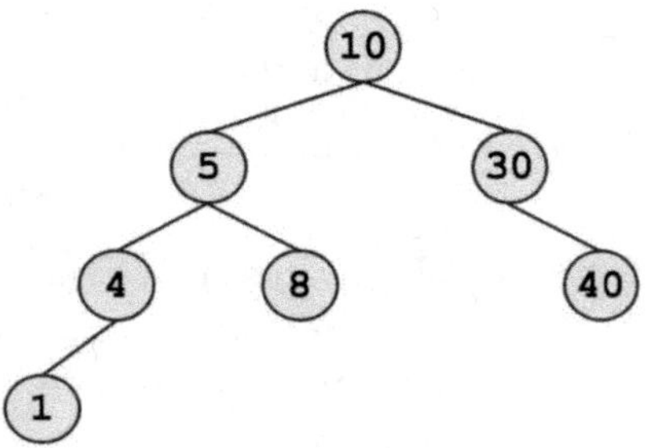

And `x = 23`, then your function should return `true`, because there exist a path from root to leaf `10->5->8` where sum of nodes in the path is equal to `23`. But if `x=15`, then the function should return `false`, because there is no path from root to leaf with sum of nodes `15`. Note that we consider the sum of nodes in the entire path and not the partial path.

Solution: There are only three root to leaf paths in the tree:

```
10 - 5 - 4 - 1
10 - 5 - 8
10 - 30 - 40
```

If `x=20`, then the problem, "find a path in the tree **(with root at 10), whose sum is 20**" can be seen as:

- Find a path in the tree with **root at 5**, whose **sum is 10.**
- Find a path in the tree with **root at 30**, whose **sum is 10.**

This is the recursive definition. Return `true` if any of the above recursive call returns `true`, `false` otherwise.

```
bool hasPathSum(node* r, int x)
{
  // TERMINATING CONDITIONS
```

```
  if(x == 0){ return r ? false : true; }
  if(r == NULL) { return false; }
  if ( NULL == r->left && NULL == r->right )
    return !(x - r->data);

  return ( hasPathSum(r->left, x - r->data) ||
           hasPathSum(r->right, x - r->data) );
}
```

Some questions require us to process the left and right sub tree and then use the information we receive from left and right sub trees to compute the information required at the root. Consider following examples:

Question 6.4: Given a binary tree with integer values at each node. Find sum of all the nodes in the tree.

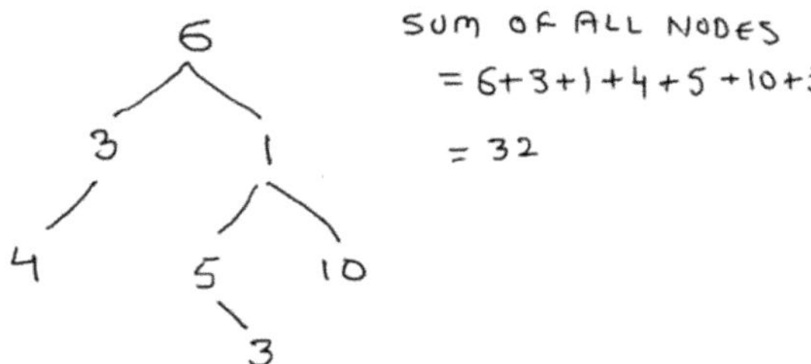

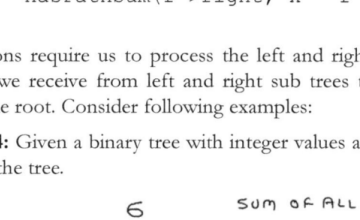

Solution: Signature of the function is:
`intsumOfNodes(Node* root);`

It receives pointer to the root, and return the sum of all the nodes in the tree.

Like other Binary tree questions, this one also use recursion and define the solution in terms of left and right subtrees. The function can be defined as:

```
Sum of Nodes = Sum of nodes in left subtree +
               Sum of nodes in right subtree +
               Data of root node
```

Code of function after putting the terminating condition is:

```
int sumOfNodes(Node* root){
  if(root == NULL)     // TERMINATING CONDITION
    return 0;

  return sumOfNodes(root->left) +
         sumOfNodes(root->right) +
         root->data;
}
```

This code is also traversing the tree.

Question 6.5: All the nodes in a binary tree are unique. Write code to print all ancestors of a given node in the tree.

Solution: The function receives a pointer to the root of tree and a number `x`. If a node with value `x` exists in the tree, then it should print all ancestors of that node in the tree. If no such node is present, nothing should get printed.

For example, for the below binary tree, if `x` is `4`, the function should print `6` and `3` in any order.

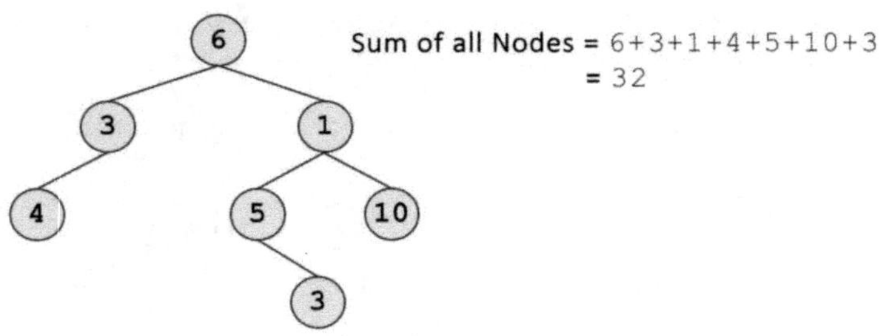

Conside below signature of the function:

```
bool printAncestors(Node *root, int x);
```

If `x` exists in the subtree rooted at `root`, function returns `true`, otherwise if `x` is not present in subtree rooted at `root`, function returns `false`. At the root we need to check for following three things:

1. If data at root is equal to `x`. Do not traverse further in that sub-tree and return `true` (all nodes are unique).
2. If `x` is present in either the left or right subtree, current node (pointed to by `root`) is among the ancestors. Print `root->data` and `true`.
3. Else, If `x` is not present in either the left or right subtree. Current node is not among the ancestors, do not print and return `false`.

Below is the code for above logic.

```
bool printAncestors(Node *r, int x)
{
  if(r == NULL){ return false; }

  // IF r IS THE NODE WITH x
  if(r->data == x){ return true; }

  // IF x IS IN LEFT OR RIGHT SUBTREE,
  // THEN CURRENT NODE IS IT'S ANCESTOR
  if( printAncestors(r->left, x) ||
      printAncestors(r->right, x))
  {
```

```
        cout<<r->data<<" ";
        return true;
    }else{
        return false;
    }
}
```

Question 6.6: Write a function that accepts pointer to the root of a binary tree and delete the entire tree.

Solution: Sometimes, developers make a mistake of deallocating the root before deallocating the left and right subtree. If we deallocate the root, we loose the addresses of left and right child and they all becomes memory leak.

The way to delete all the nodes, is to start deleting the tree from the leaf nodes, and deallocate the root only when both left and right subtrees are completely deallocated.

```
void deleteTree(Node* r)
{
  if(r == NULL){ return; }

  deleteTree(r->left);   // DELETE LEFT SUBTREE
  deleteTree(r->right);  // DELETE RIGHT SUBTREE
  free(r);
}
```

After calling this function, the calling-function should set its root pointer to `NULL`, because it is no more pointing to an allocated memory. This function is not taking the responsibility of setting the root pointer in the calling-function to `NULL`.

Question 6.7: Write code to evaluate an expression tree.

Solution: Expression tree is a binary tree that represents a unique mathematical expression. Each non-leaf node of the binary tree stores an operator and each leaf node stores an operand. Below table shows the expression tree corresponding to the given expression:

Expression	**Expression Tree**
`2+3`	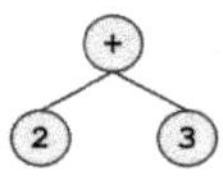
`(2+3)*5`	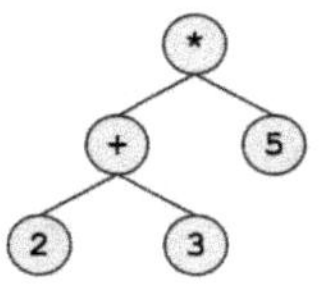

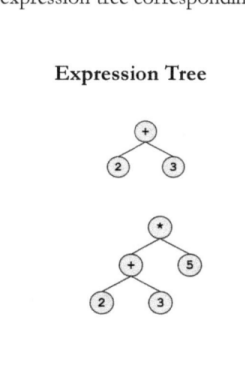

The algorithm to evaluate and expression tree is as follows (assuming only binary operators):

```
Int Evaluate(Node* r)
{
  IF(r != NULL)
  {
    IF (r->data IS OPERAND)
      RETURN r->data
    ELSE              // IF r->data IS OPERATOR
    {
      x = Evaluate(r->left)
      y = Evaluate(r->right)
      APPLY OPERATION r->data ON x AND y AND
                          RETURN THE VALUE
    }
  }
}
```

For simplicity, let us assume that all operands are single digit integers, and there are only four operators: add, subtract, multiply and divide. This is to keep a uniform structure of node that can be used for both types of nodes.

```
struct Node{
  char data;
  Node* left;
  Node* right;
};
```

Also, let us add two helper functions, first one to check if the given character is an operator or an operand and second function to apply the operation on two values (we are considering only binary operators)

```
bool isOperator(char ch){
  return (ch=='+' || ch=='-' || ch=='*' || ch=='/');
}
```

```
int applyOperator(int x, int y, char op){
  switch(op){
    case '+':
      return (x+y);
    case '-':
      return (x-y);
    case '*':
      return (x*y);
    case '/':
      return (x/y);
    default:
      return 0; // UNREACHABLE CODE
  }
}
```

The logic to evaluate the function is now simple:

```
int evaluate(Node *r)
{
  if(r == NULL){ return 0; }

  if(isOperator(r->data))
    return applyOperator( evaluate(r->left),
                          evaluate(r->right),
                          r->data );
  else
    return r->data - '0';
}
```

This code takes `O(n)` time because we are visiting each node only once. Note that we are storing integers as characters inside a node to keep things simple, you may want to have two different kind of nodes one each for operand and operator. Both of these nodes can extend a common 'Node' class.

Practice Questions that track the level of nodes

There are some questions that require us to track the level of the node or its distance from the root node. In such cases, it is usually a good idea to pass the level of node as an extra parameter. The initial value of this parameter is 0 (root node is at level zero).

Question 6.8: Given a binary tree and a number, `x`, return the level at which node with value `x` is present in the tree.

Solution: Traverse the binary tree in preorder, and if data of the current node is equal to `x`, return the level of current node.

To accomplish this, we need to track the level of current node. This can be done by passing the current-node level as a parameter. In the below code, `curLevel` represent level of the node pointed to by `r`.

```
// level IS THE LEVEL OF CURRENT NODE
int getLevel(Node* r, char x, int curLevel)
{
  if(r == NULL){ return -1; }

  if(r->data == x){ return curLevel; }

  int level = getLevel(r->left, x, curLevel+1);

  if(level != -1) // x PRESENT IN LEFT SUBTREE
    return level;

  return getLevel(r->right, x, curLevel+1);
}
```

Root node is at level zero, this function should be called as `getLevel(root, data, 0);`

Each time we move down, either the left subtree or right subtree, we increment the `curLevel`.

Question 6.9: Given a number `k` and a pointer to the root of a binary tree, print all nodes at distance `k` from the root. For example, in the below tree the nodes at distance 2 from the root are 4, 5, and 10.

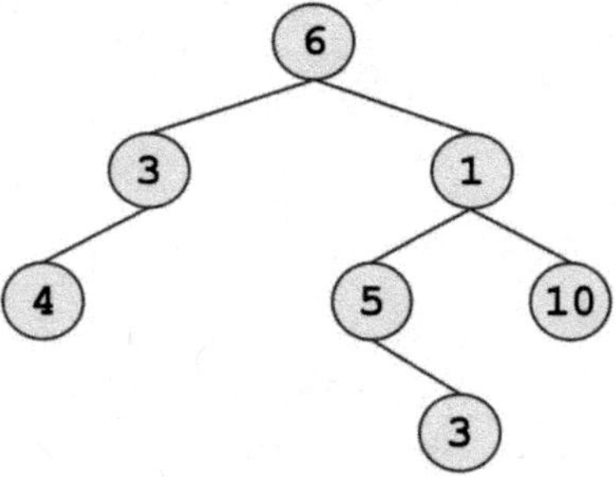

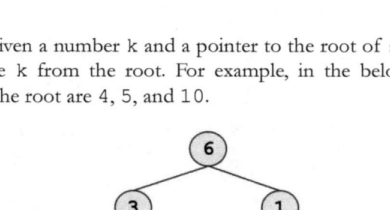

Solution: The Nodes at distance `k` from root are at distance `k-1` from both left and right child of the root. When `k` becomes equal to zero, then that node should be a part of the output. If root becomes `NULL` before k becomes zero, it indicates that there are no nodes at a distance `k` in this path.

```
void printNodesAtDistance(Node* r, int k)
{
  if(r == NULL){ return; } // TERMINATING CONDITION

  if(k == 0)
    cout<<r->data<<" ";

  printNodesAtDistance(r->left, k-1);
  printNodesAtDistance(r->right, k-1);
}
```

Question 6.10: Given a binary tree write a function to check if all leaf nodes of the tree are at the same level. For example, in all the below binary trees, all leaf nodes are at same level:

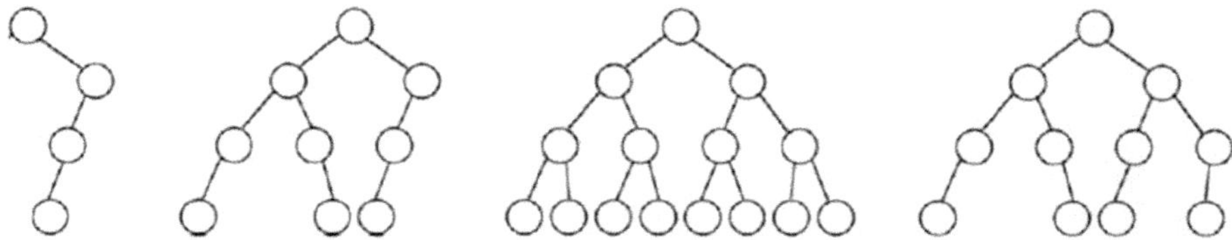

But in the below trees, all the leaf nodes are not at the same level:

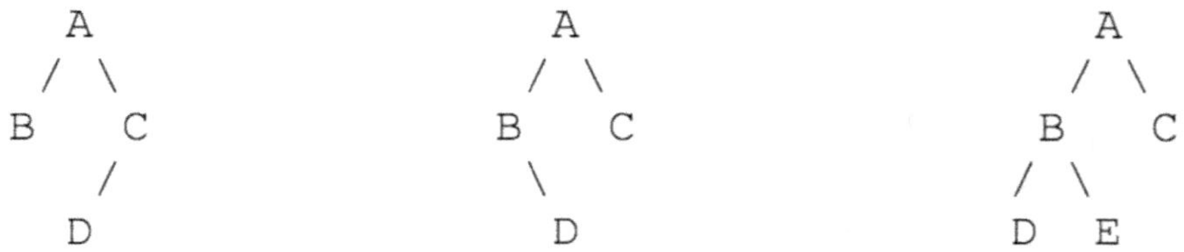

Solution: The question can be rephrased as, "check if all leaf nodes are at the same distance from the root". This can be checked by making small changes in the level order traversal algorithm.

In the level order traversal, if all nodes after the first leaf node are at the same level, then return `true`, else `false`.

Another method is to keep a reference variable, say `leafLvl`, initialized to `-1` (which is same in all the recursive functions), and keep track of the level of current node. Set the reference variable with the level of node when the first leaf node is encountered while traversing the tree.

After that, when a leaf node is found again, its level is checked against `leafLvl`. If they are different, then level of new leaf node is different from level of leaf node found earlier.

```
bool sameLeafLvls(Node* r, int curLvl, int& leafLvl)
{
  // IF LEAF NODE.
  if(r->left == NULL && r->right == NULL){
    if(leafLvl == -1){ // FIRST LEAF
      leafLvl = curLvl;
      return true;
    }
    else if(curLvl == leafLvl) // NOT FIRST LEAF
      return true;
    else
```

```
        return false;
    }

    // NOT LEAF NODE
    bool lRes = true;
    bool rRes = true;
    if(r->left)
      lRes = sameLeafLvls(r->left, curLvl+1, leafLvl);
    if(r->right)
      rRes = sameLeafLvls(r->right, curLvl+1, leafLvl);

    return (lRes && rRes);
}
```

The function should be called like below:

```
int x = -1;
bool res = sameLeafLvls(r, 0, x);
```

If `res` is `true`, all leaf nodes are at the same level, and `x` holds that level.

Assignment 6.1: Given a binary tree with nodes holding integer data. Find the difference in sum of all the nodes at odd levels and sum of all nodes at even levels.

Question 6.11: A new field is added to the structure of `Node` of a binary tree:

```
struct Node
{
  char data;
  Node* left;
  Node* right;
  Node* next;
}
```

`next` field of each node stores `NULL`. For each node, make the `next` pointer of the node point to the next node in its level. The `next` field of the right most nodes should be `NULL` for each level.

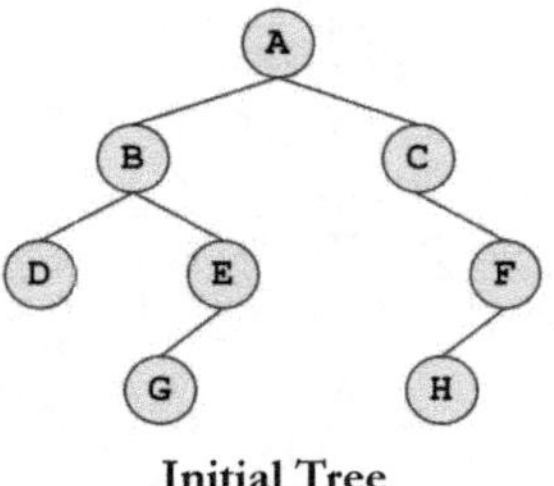

Initial Tree

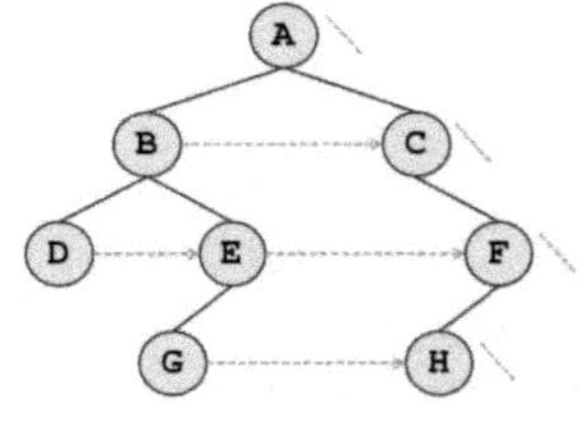

Final Tree

Solution: In simple words we want to arrange nodes at each level like a linked list (thru `next` pointers).

The level order traversal can be easily extended to connect nodes at the same level. In Code 6.2 we saw how to print nodes at each level in a separate line by putting a dilimiter in the queue at the end of each level.

In the same code, keep a pointer to the previous node and set the `next` field of previous node to point to the current node and do a special handling for the first node of each level. Below is the C++ code. It uses the queue class available in STL

```
void fixNextPts(Node *r)
{
  queue <Node*> q;
  q.push(r);
  q.push(NULL);
  Node* prev = NULL;
  while(!q.empty())
  {
    Node* temp = q.front(); // DO NOT REMOVE ELEMENT
    q.pop();                // JUST REMOVE FIRST ELEMENT
    if(prev != NULL)
      prev->next = temp;
    prev = temp;
    if(temp == NULL)        // END OF LEVEL
    {
      if(!q.empty())
        q.push(temp);
    }
    else
    {
      if(temp->left){ q.push(temp->left); }
      if(temp->right){ q.push(temp->right); }
    }
  }
}
```

This code is taking `O(n)` memory for queue.

The problem can also be solved with constant extra space. If `next` pointers of all the nodes in level-k are already set, then these nodes of level-k can be traversed linearly from left-to-right like a linked list using `next` pointers.

While linearly traversing the nodes of level-k, we can set the next pointers for nodes of level-(k+1). Then we can traverse the nodes of level-(k+1) and set the next pointers for level-(k+2) and so on.

```
// HELPER FUN TO FIND NEXT RIGHT NODE
Node* getNextRightNode(Node *r)
{
  r = r->next;

  while (r != NULL)
  {
    if (r->left != NULL)
      return r->left;
    if (r->right != NULL)
      return r->right;
    r = r->next;
  }
  return NULL;
}

void fixNextPts(Node* r)
{
  if(r == NULL) { return; }

  // FIX ALL NODES AT NEXT LEVEL
  while(r != NULL)
  {
    // TRAVERSE CURRENT LEVEL USING next POINTERS
    for(Node* temp=r; temp!=NULL; temp=temp->next)
    {
      if(temp->left != NULL)
      {
        if(temp->right)
          temp->left->next = temp->right;
        else
          temp->left->next = getNextRightNode(temp);
      }

      if(temp->right != NULL)
        temp->right->next = getNextRightNode(temp);
    }

    // SET r TO THE FIRST NODE OF NEXT LEVEL
    if(r->left)
      r = r->left;
    else if (r->right)
      r = r->right;
```

```
      else
        r = getNextRightNode(r);
    }
}
```

This is an iterative function that takes O(n) time and constant extra space.

Views of a binary tree

A person viewing a binary tree from left side may see different nodes than a person who see the same binary tree from right side. For example, in the below tree, if nodes visible from the right side are printed, then shaded nodes will be in the output:

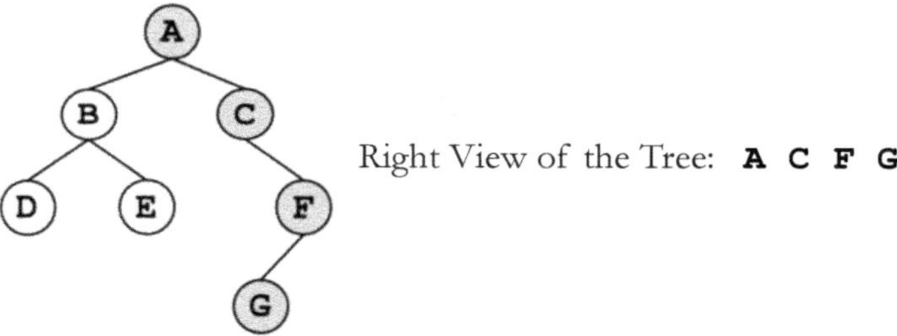

But in the same tree, if we print nodes visible from the left side, then output will be different:

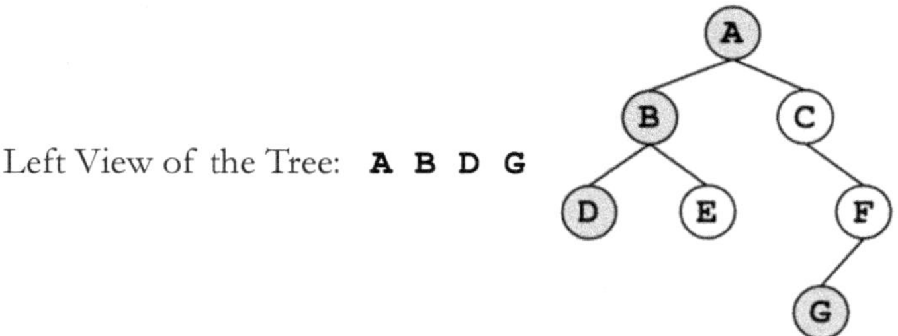

While viewing the tree from left side, Node `B` hide the node with value `C`, therefore `C` is not visible.

Similarly, if we view the tree from top, then `A` will hide Node `E`, because both the nodes are at the same vertical level. The top view of the above tree will print: `D, B, A, C, F`.

The bottom-view of a binary tree can also be defined on the same lines.

Let us first discuss how to print the left view. Left view actually prints the left-most node of each level in the tree.

When a tree is traversed in preorder traversal the left-most node in a particular level is visited before visiting other nodes of that level.

The solution to this problem is simple, we traverse the given tree in pre-order traversal and keep a check on whether or not the left-most node in the current level is printed. If no nodes for the current level is printed, print the current node, else continue to traverse the tree in pre-order.

We need two variables in this case:

1. A variable for the current level of node.
2. A reference variable for the maximum level up to which the left-most nodes are printed.

Also note that the left-most node of level `k` will always be encountered before we move to any node in level `(k+1)`. Below is the C language code that prints the left-view of a tree:

```
void printLeftView(Node *r, int level, int* maxLevel)
{
  if(r == NULL){ return; }

  if(level > *maxLevel)
  {
    printf("%c ", r->data);
    *maxLevel = level;
  }

  printLeftView(r->left, level+1, maxLevel);
  printLeftView(r->right, level+1, maxLevel);
}
```

Note that `maxLevel` is a reference variable. This function should be called as below:

```
  int maxLevel = -1;
  printLeftView(r, 0, &maxLevel);
```

To print the right-view of the tree, just change the order in which function is called for the subtrees (call for right subtree first and then call for left subtree).

For printing the top-view of the tree, we need to consider the vertical order of the nodes in a tree. Vertical order of the following tree is shown as:

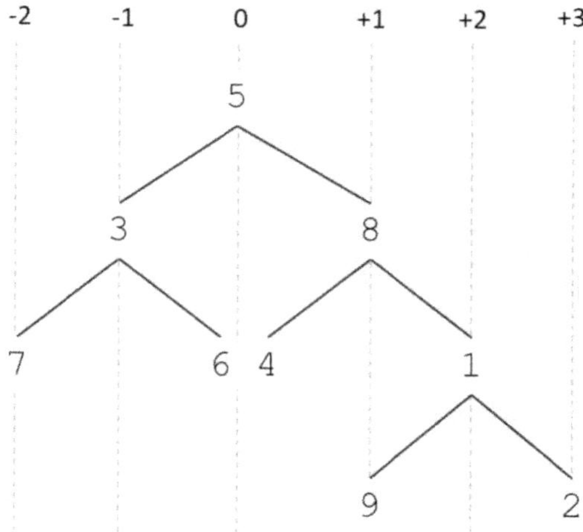

Let us define the concept of horizontal distance (HD) of a node from root. HD of the root node is zero. When we move to the right child, increment the horizontal distance, and when we move to the left child decrement the horizontal distance.

The HD of node 3 is -1 and HD of 8 and 9 is +1. The HD of 5, 6 and 4 is 0 which is same as root. It means that in the top view, root node hides both 6 and 4 and in the bottom-view 4 will hide the other two (let us say that node in right subtree hides node in the left subtree at the same vertical level).

Nodes with the same HD are on the same line (vertical level). To print the top view, print the first node encountered for any vertical level while traversing the tree in level order traversal (or preorder traversal).

To achieve this, we can use a hash, when a node is first found for a particular vertical level, we store the value in the hash (with key = vertical level). If the vertical level of the current node is already present in the hash, just skip the node.

A similar logic can be used to print the bottom-view of a binary tree. The difference in logic for bottom-view is to traverse the tree in level-order and update the hash when a new value for a particular HD is found. This way the hash will hold the last node for each vertical level.

Other problems

Question 6.12: Given the ancestor matrix of a Binary tree, write an algorithm to construct the corresponding tree. The ancestor matrix below represents tree on the right side

	1	4	5	8	10	30	40
1	0	1	1	0	1	0	0
4	0	0	1	0	1	0	0
5	0	0	0	0	1	0	0
8	0	0	1	0	1	0	0
10	0	0	0	0	0	0	0
30	0	0	0	0	1	0	0
40	0	0	0	0	1	1	0

Ancestor Matrix

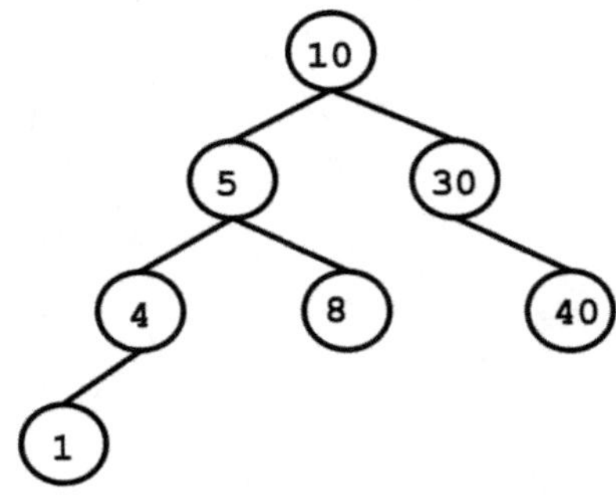

Binary Tree

The order of rows in the above matrix is not defined. It is shown in the ascending order because the data of nodes is numeric.

The value at `a[i][j]` is `1` iff Node is represented by `j`th column is the ancestor of node represented by `i`th row.

We are given the ancestor matrix and are supposed to construct the binary tree from that matrix.

Note: *Since we don't have information about whether a child is a left child or a right child, the tree that gets constructed is un-ordered Binary tree (more than one ordered binary trees can be constructed from the same Ancestor matrix).*

Solution: The row representing the root node has all zeros because there is no ancestor to the root node.

Let us use a queue with `enqueue`, `dequeue` and `isEmpty` operations. Let us add one column in the matrix to store sum of all the elements in that row and follow the following algorithm:

	1	4	5	8	10	30	40	Sum
1	0	1	1	0	1	0	0	3
4	0	0	1	0	1	0	0	2
5	0	0	0	0	1	0	0	1
8	0	0	1	0	1	0	0	2
10	0	0	0	0	0	0	0	0
30	0	0	0	0	1	0	0	1
40	0	0	0	0	1	1	0	2

1. Find the row with `sum = 0`. This is the root node of the tree.
2. Insert this node in the queue.
   `Q.enqueue(r);`
3. WHILE`(!Q.isempty)`
   a. `temp = Q.dequeue();`
   b. Remove both **row & column** of the `temp` node and update the `sum` column accordingly.
   c. Look for all the rows for which `Sum[i] == 0`
   d. Add them as children to node `temp`.
   e. Insert them at the end of the queue.

The algorithm will proceed as shown in the below diagram:

	1	4	5	8	10	30	40	Sum
1	0	1	1	0	1	0	0	3
4	0	0	1	0	1	0	0	2
5	0	0	0	0	1	0	0	1
8	0	0	1	0	1	0	0	2
10	0	0	0	0	0	0	0	0
30	0	0	0	0	1	0	0	1
40	0	0	0	0	1	1	0	2

	1	4	5	8	30	40	Sum
1	0	1	1	0	0	0	2
4	0	0	1	0	0	0	1
5	0	0	0	0	0	0	0
8	0	0	1	0	0	0	1
30	0	0	0	0	0	0	0
40	0	0	0	0	1	0	1

Tree

10

Now we have the root node (10), all the Nodes whose sum are zero will be children of this node. Hence the tree will look like:

```
  10
 /  \
5   30
```

10 will be dequeed from the queue and 5 & 30 are also inserted in the Queue.

Next element in the Queue (to be removed from Queue) will be 5. Remove the corresponding rows & columns and is sum value becomes zero corresponding to some nodes then insert them as child nodes of 5.

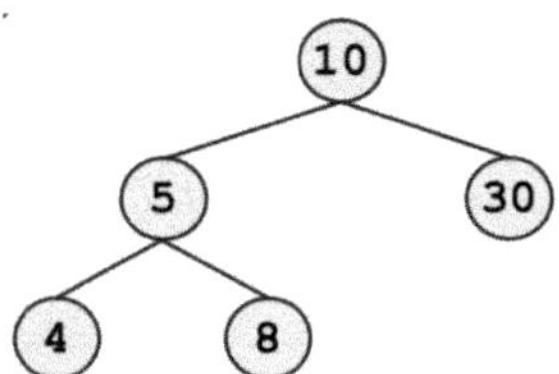

Also insert 4 and 8 in the Queue (and remove them from Matrix). Repeat the same for 30. Then repeat the same for 4, 8 and other nodes.

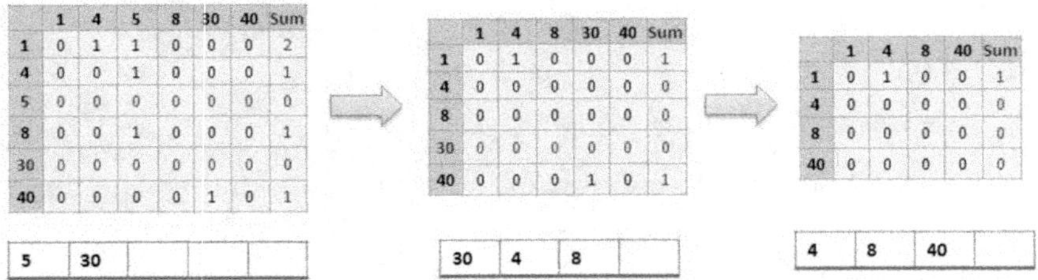

	1	4	5	8	30	40	Sum
1	0	1	1	0	0	0	2
4	0	0	1	0	0	0	1
5	0	0	0	0	0	0	0
8	0	0	1	0	0	0	1
30	0	0	0	0	0	0	0
40	0	0	0	0	1	0	1

5	30			

	1	4	8	30	40	Sum
1	0	1	0	0	0	1
4	0	0	0	0	0	0
8	0	0	0	0	0	0
30	0	0	0	0	0	0
40	0	0	0	1	0	1

30	4	8	

	1	4	8	40	Sum
1	0	1	0	0	1
4	0	0	0	0	0
8	0	0	0	0	0
40	0	0	0	0	0

4	8	40	

Let us try to write the code now. We are using the same logic as above, just that since the physical removal of rows from an array is costly, we just mark the rows as removed.

Data structure used

We use the following data structures in the code:

1. A 2-dimensional array representing the ancestor matrix. This matrix has one extra column for the sum.
2. A one-dimensional Array representing the original values (which are the actual numbers that will be inserted in the tree). This may be a character array if tree nodes are character values.
3. A struct Node, representing the structure of the Node in a Binary tree.
4. Some constant values

```
// CONSTANT VALUES USED
#define N 7
#define REMOVED -1
#define NO_VALUE_FOUND -2

struct Node{ // TREE NODE
    int data;
    Node* left;
    Node* right;
};

// NODE VALUES IN THE TREE
int nodeName[N]  = { 1, 4, 5, 8,10,30,40};

// ANCESTOR MATRIX - LAST COLUMN FOR SUM
int ancestorMat[N][N+1]= {{ 0, 1, 1, 0, 1, 0, 0, 0 },
                          { 0, 0, 1, 0, 1, 0, 0, 0 },
                          { 0, 0, 0, 0, 1, 0, 0, 0 },
```

```
                         { 0, 0, 1, 0, 1, 0, 0, 0 },
                         { 0, 0, 0, 0, 0, 0, 0, 0 },
                         { 0, 0, 0, 0, 1, 0, 0, 0 },
                         { 0, 0, 0, 0, 1, 1, 0, 0 }};
```

Let's look at some of the helper functions. Below function sets up the initial sum values for each row of the matrix. The `sum` column represents the total number of ancestors of that node.

```
// CALCULATE SUM FIRST TIME
void calculateInitialSumAndRemoveRoot(){
  for(int i=0; i<N; i++){
    ancestorMat[i][N] = 0;
    for(int j=0; j<N; j++)
      ancestorMat[i][N] += ancestorMat[i][j];
  }
}
```

Above function is to be called only once before the main logic to set the last column of the matrix correct. Now the matrix is:

```
0 1 1 0 1 0 0 3
0 0 1 0 1 0 0 2
0 0 0 0 1 0 0 1
0 0 1 0 1 0 0 2
0 0 0 0 0 0 0 0
0 0 0 0 1 0 0 1
0 0 0 0 1 1 0 2
```

At this point remove the row with zero `sum` value. We will just mark the row as removed. The below function look for the first row, whose sum is zero and mark it removed and return the value for that row

```
int findAndRemoveFristZeroElement(){
  for(int i=0; i<N; i++){
    if(ancestorMat[i][N] == 0){
      ancestorMat[i][N] = REMOVED;
      return nodeName[i];
    }
  }
  return NO_VALUE_FOUND; // NO MORE NODE
}
```

We need a function that will decrement the sum values from the matrix. When a row is removed, sum values of all its children is decremented

```
void decrementParentCountForNode(int value){
  for(int j=0; j<N; j++){
    if(nodeName[j] == value){
```

```
      for(int i=0; i<N; i++){
        if(ancestorMat[i][j] == 1)
          ancestorMat[i][N]--;
      }
      return;
    }
  }
}
```

Now, let us come to the main logic, function `convertMatToTree` below find the root element, remove it, create the root node of the tree and then call the recursive function to build rest of the tree.

```
Node* convertMatToTree()
{
  // COMPUTE sum OF ALL ROWS
  calculateInitialSumAndRemoveRoot();

  // FIND THE root NODE AND REMOVE IT FROM THE MATRIX
  int rootValue = findAndRemoveFristZeroElement();

  // IF THERE IS NO ROOT
  if(rootValue == NO_VALUE_FOUND){ return NULL; }

  // MAKE A TREE
  Node* root = new Node(rootValue);

  // CREATE REST OF THE TREE AND SET IT TO THE
  // LEFT AND RIGHT CHILD OF root
  convertMatToTreerecursive(root);

  return root;
}
```

Now, implement the function that does the work recursively:

```
void convertMatToTreerecursive(Node* root)
{
  if(root == NULL){ return; }

  // DECREASE sum OF CHILDREN OF root.
  decrementParentCountForNode(root->data);

  // FIND FIRST CHILD AND SET IT TO LEFT
  int value = findAndRemoveFristZeroElement();
  if(value != NO_VALUE_FOUND)
    root->left = new Node(value);
```

```
  // FIND SECOND CHILD AND SET IT TO RIGHT
  value = findAndRemoveFristZeroElement();
  if(value != NO_VALUE_FOUND)
    root->right = new Node(value);

  // CREATE LEFT TREE
  if(root->left != NULL)
    convertMatToTreerecursive(root->left);

  // CREATE RIGHT TREE
  if(root->right != NULL)
    convertMatToTreerecursive(root->right);
}
```

Question 6.13: The structure of a Node of a doubly linked list is same as the structure of Node of a binary tree.

```
struct Node
{
  int data;
  Node* previous;
  Node* next;
}
```

Node of doubly linked list

```
struct Node
{
  int data;
  Node* left;
  Node* right;
}
```

Node of Binary tree

If we treat `left` pointer as `previous` and `right` pointer as `next`, we can use the Node of a Binary Tree to represent Node of a Doubly linked list or vice-versa.

Given a binary tree, modify the pointers in each node of the tree so that they represent a doubly linked list. Order of the nodes in the doubly linked list should be same as that in the in-order traversal of the binary tree.

If the Input Binary tree is:

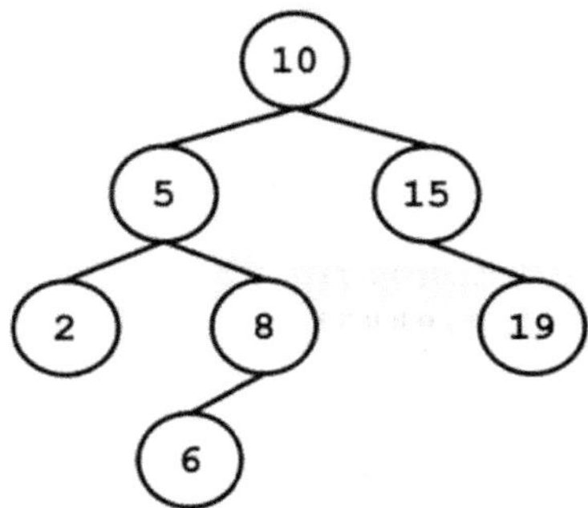

then your function should modify pointers in each node in a way that they form the doubly linked list and return the head of that list.

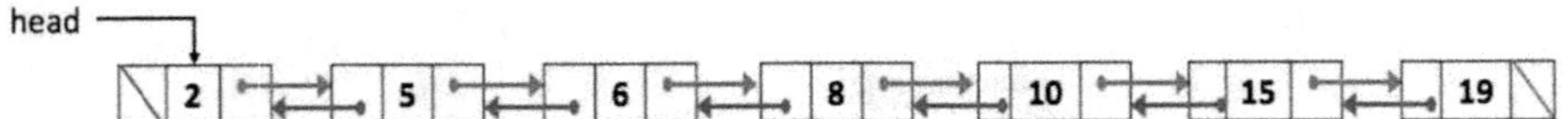

Solution: The solution to this problem, like many others in Binary Trees, uses recursion. Let the signature of the function be:

`Node* convertToDLL(Node *r, Node** tail)`

This function receives pointer to root of the tree. Convert it into a DLL and return the head of the list. It also set the address of last node in `tail` pointer.

We call this function recursively for left and right subtree of the Binary Tree and get the following four pointers:

1. Head of DLL after left subtree is converted to list.
2. Tail of DLL after left subtree is converted to list.
3. Head of DLL after right subtree is converted to list.
4. Tail of DLL after right subtree is converted to list.

Once we have this information, we just need to combine the left and right DLL by inserting root node between them:

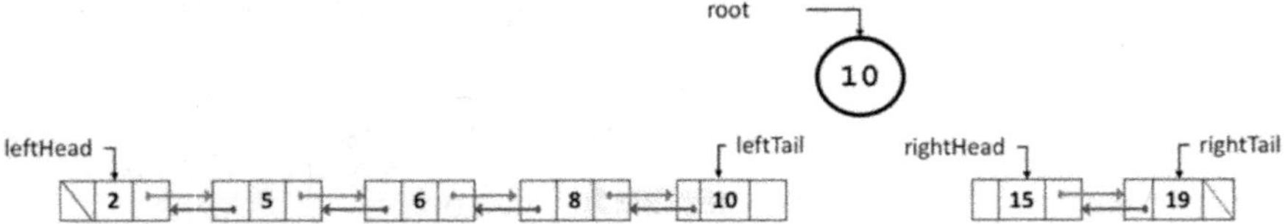

The recursion terminates when we are at leaf node or the root is NULL. Below is the complete code for this solution:

```
Node* convertToDLL(Node *r, Node** tail){
  // 0 or 1 nodes.
  if(r==NULL || (r->left==NULL && r->right==NULL)){
    *tail = r;
    return r;
  }

  Node* leftHead; // Head of DLL - Left subtree.
  Node* leftTail; // Tail of DLL - Left subtree.
  Node* rightHead; // Head of DLL - Right subtree.
  Node* rigthTail; // Tail of DLL - Right subtree.

  leftHead = convertToDLL(r->left, &leftTail);
  rightHead = convertToDLL(r->right, &rigthTail);
```

```
  Node* head = r; // HEAD OF THIS TREE.
  r->left = NULL;
  if(leftHead != NULL){
    head = leftHead;
    leftTail->right = r;
    r->left = leftTail;
  }

  if(rightHead != NULL){
    r->right = rightHead;
    rightHead->left = r;
    *tail = rigthTail;
  }else{
    *tail = r;
    r->right = NULL;
  }

  return head;
}
```

Chapter 7

Advanced Binary Tree Concepts

Binary Tree data structure is used in multiple forms. In the next chapter, we discuss some of those forms.

Our focus is on data structures which are binary trees, but have either some restrictions on their structure or the values that they store. Let us start with the concept of binary search tree.

Binary Search Tree

Binary Search Tree (BST) is sorted version of Binary Tree. It is called Binary Search Tree, because, an algorithm similar to binary search can be used to search in it.

A binary tree is called BST if it satisfies the following two condition:

1. Value of root node is greater than value of all nodes in the left sub tree and less than value of all nodes in the right subtree. The data of node in a BST should be comparable.
2. Left subtree and right subtree are themselves BSTs. Figure 7.1 shows a Binary Search Tree. When a BST is balanced, number of nodes in the left and right sub trees are almost equal (Balanced Binary trees are discussed later in this chapter).

All the below trees are Binary search trees.

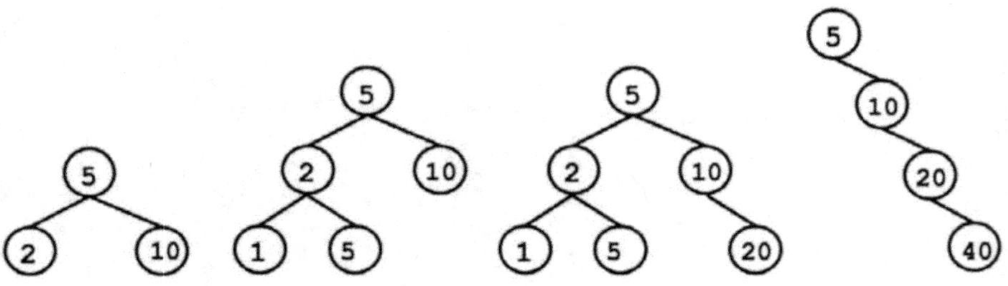

When the node values are equal, they can go on either side of the root. Note that, inorder traversal of a BST is sorted. If a question demands us to print all the nodes of a BST in ascending order, then we can just print the inorder traversal of that Binary search tree.

Search for a value, say `x`, in a BST starts by comparing x with root node. There are 3 possibilities:

1. **Value at root is equal to** `x`**.** Element found, search is successful.

2. **Value at root is greater than** x. All values in the right subtree are greater than the value at root, hence x cannot be present in the right sub-tree. Search the left subtree for x.

3. **Value at root is less than** x. All values in the left subtree are less than value at root and hence x cannot be present in the left sub tree. Discard the left subtree, and search the right subtree for x.

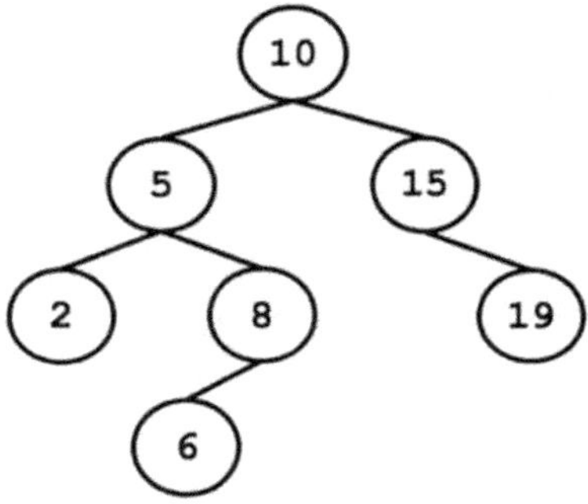

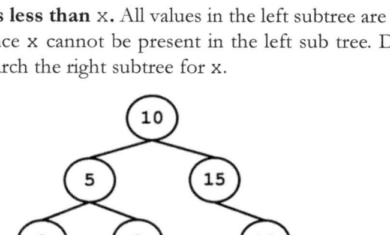

Figure 7.1: Binary Search Tree

Below code shows the function that implements above algorithm. Node represents node of a binary tree

```
bool searchBST(Node *r, int x)
{
  if(r== NULL)
    return false; // ELEMENT NOT FOUND IN THE TREE.
  if(r->data == x)
    return true; // ELEMENT FOUND. SUCCESS.
  else if (r->data > x)
    return searchBST(r->left, x);
  else
    return searchBST(r->right, x);
}
```

If the BST is balanced, number of nodes in the right and left subtrees are (almost) equal and root node is near to the median. In constant time, we discard either the left subtree or the right subtree. This is similar to binary search on a sorted array where we find the middle element of the array and compare it with the value being searched and discard either the left or right half of the array in constant time. Searching in a BST takes `O(lg(n))` time.

But the BST may not necessarily be balanced. If the tree is skewed, then searching in a BST may take `O(n)` time in the worst case. Searching for a larger value (greater than 21) in the following BST takes `O(n)` time. This is no better than linear search. In this case, the binary tree is practically like a linked list.

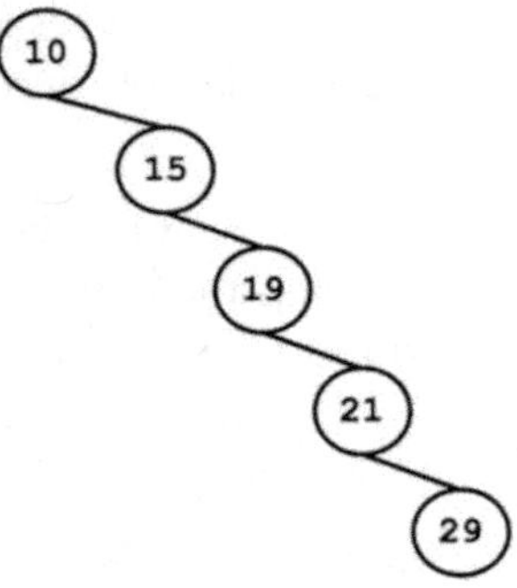

This is the worst-case time. In best case, the search takes constant time, when element is found at the root itself.

Maximum and minimum elements in a BST

Follow right child from the root (without ever visiting the left child of any node), last element gives us the maximum value in the BST irrespective of its structure. If root does not have a right child, then root itself holds the maximum value. The grey nodes in below BSTs holds maximum value in that BST

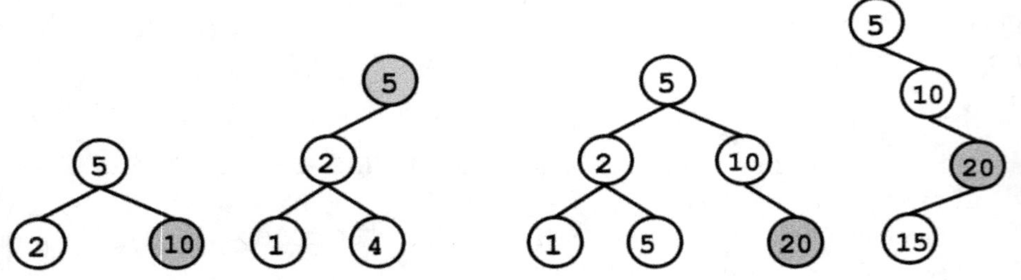

Below function returns the maximum value in the BST:

```
int getMaximum(Node* root){
  // ERROR SITUATION
  if(root == NULL) { return -1;}

  while(root->right != NULL)
    root = root->right;

  return root->data;
}
```

The above function is returns -1 when the tree is empty. In real world, we should throw an exception in such a situation. Similarly, to find the minimum element, follow the left child from the root. If root does not have a left child, then root itself holds the minimum value.

Inserting in a BST

Given a binary search tree, we want to insert a new value in the tree, such that the BST property of the tree is preserved (the tree remains BST even after insertion).

The new value is always inserted as a leaf node. We first locate the position at which the new value needs to be inserted. To find the location of insertion in the tree, follow the search logic discussed in the previous section.

Search for the value that we are inserting in the tree until you hit an empty position (`NULL` node). When the position is found, insert the new node at that position. Let us take an example. If we want to insert `16` in the below tree. We start by searching for 16 from the root. The search for 16 goes as shown below:

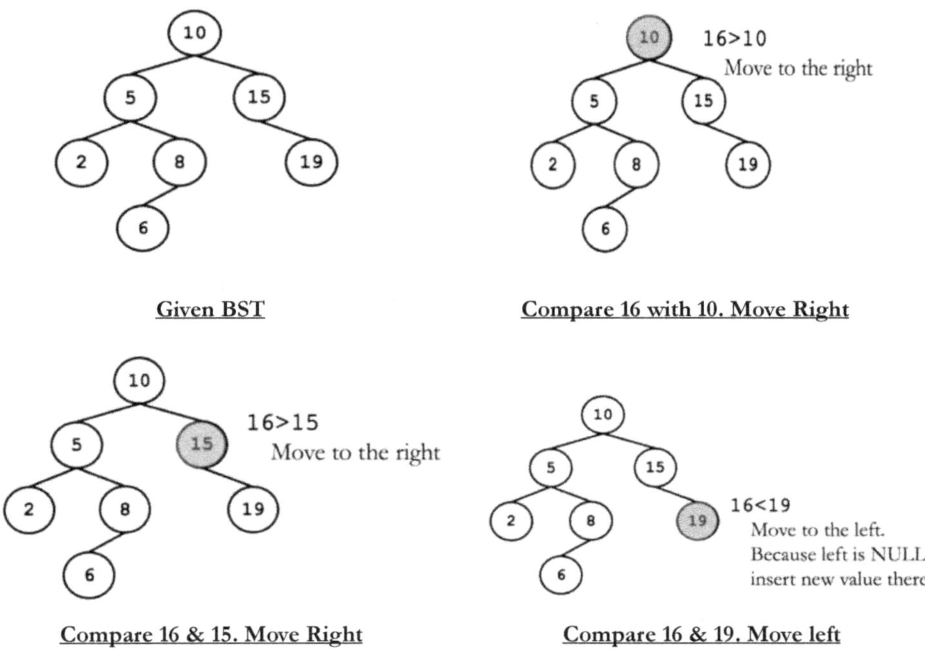

Given BST

Compare 16 with 10. Move Right

Compare 16 & 15. Move Right

Compare 16 & 19. Move left

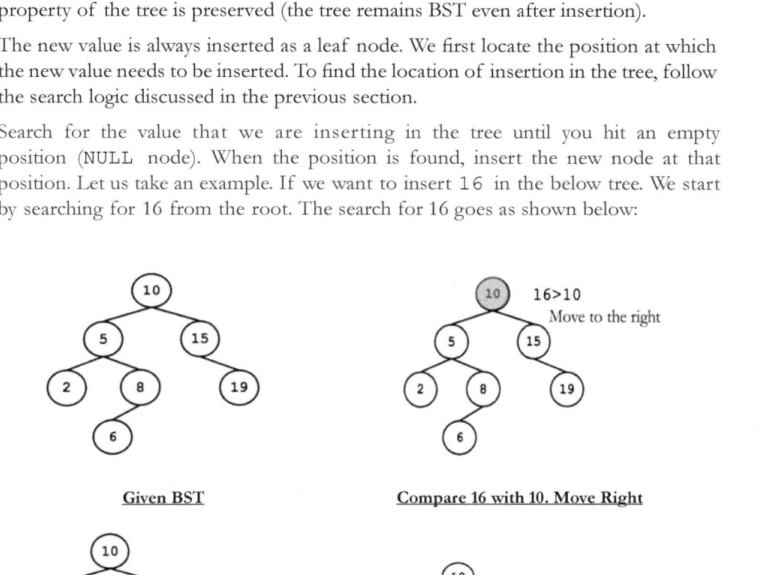

In the end, there is no place to go on the left of 19. The new value, 16 is inserted here

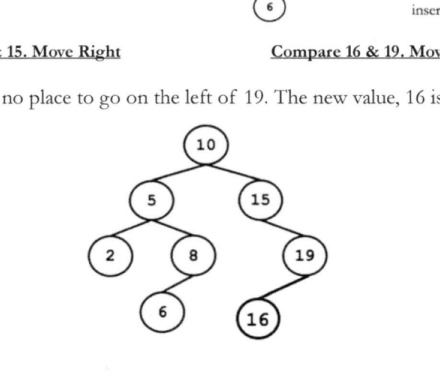

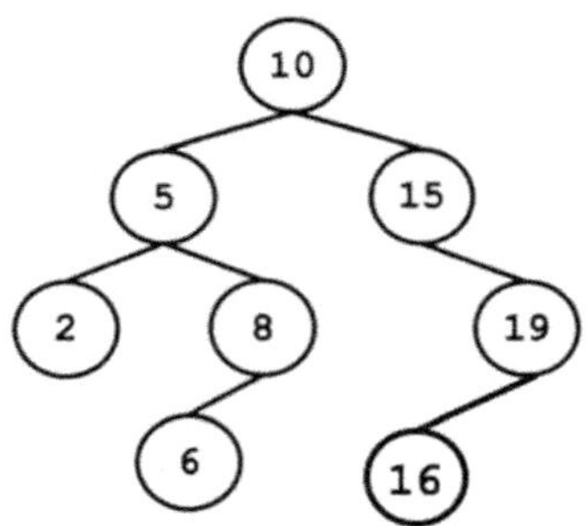

Below is the code for inserting an element in a Binary search tree:

```
Node* insert(Node*r,  int val)
{
  Node* temp = new Node;
  temp->data=val; temp->left=temp->right=NULL;

  // IF INSERTING FIRST NODE. CHANGE ROOT
  if(r == NULL)
    return temp;

  // PRESERVING THE ROOT TO RETURN AT THE END
  Node* origRoot = r;

  while(true)
  {
    if(r->data > val)
    {
      if(r->left == NULL)
      {
        r->left = temp;
        break;
      }
      else
        r = r->left;
    }
    else
    {
      if(r->right == NULL)
      {
        r->right = temp;
        break;
      }
      else
        r = r->right;
    }
  }
  return origRoot;
}
```

The recursive implementation of insert function may be interesting because of the compactness of code;

```
Node* insertRec(Node* r, int val)
{
  if(r == NULL)
  {
```

```
    r = new Node;
    r->data = val;
    r->left = r->right = NULL;
  }
  else if(r->data > val)
    r->left = insertRec(r->left, val);
  else
    r->right = insertRec(r->right, val);

  return r;
}
```

If preorder traversal of a BST is given in an array, then we can construct the original BST by picking one element at a time and inserting it in the BST.

Deleting element from a Binary Search Tree

Follow the search algorithm to reach to the parent node of the node that need to be deleted (deleting the root node can be handled separately).

Deleting a particular node from a BST depends on the number of children that the node has. There are three possible situations:

1. The node to be deleted is a leaf node and has no child.
2. The node to be deleted has only one child (either the left or right, but not both).
3. The node to be deleted has both the children (neither left nor and right child of the node is null).

Let us consider these three situations one-by-one.

When the node to be deleted is a leaf node, just remove that node from the tree, and deallocate it from heap.

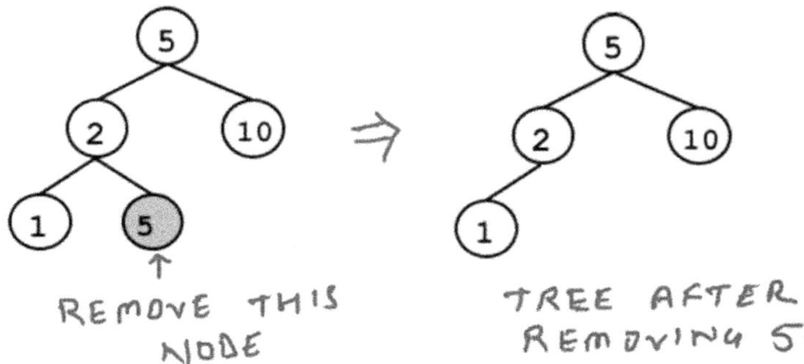

If the node, to be deleted has only one child (it's parent may or may not have both the children), then link its parent node to its only child.

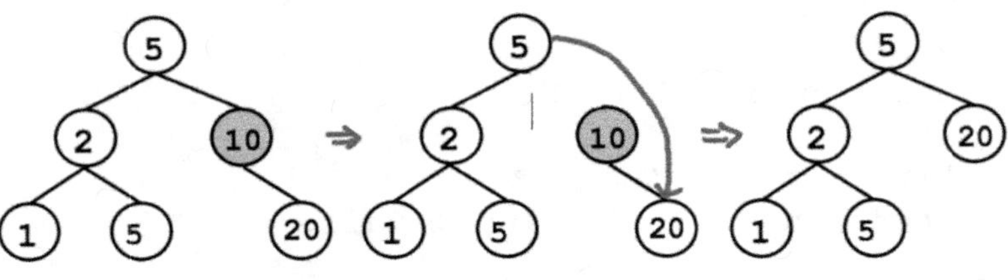

1. Original Tree Node-10 is to be deleted **2. Make parent of 10 Point to its only child** **1. Final Tree (Free Node-10)**

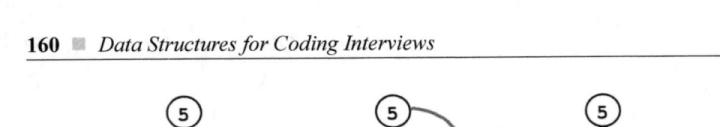

If we want to delete Node with value 10 from the above BST, first link 5 to the only child of 10, i.e 20 and then remove node 10. By attaching the subtree of the deleted node to its parent, we are not violating the BST property of the tree.

Deleting a node with both children can be tricky. We should place some other node in its place that does not violate the BST property of the tree. There are two such nodes that can be put in place of the current node:

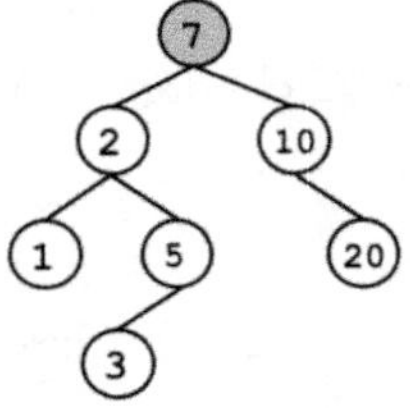

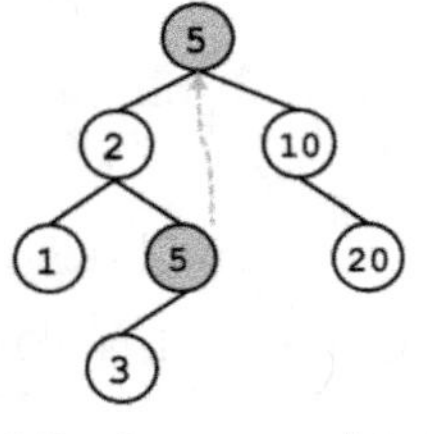

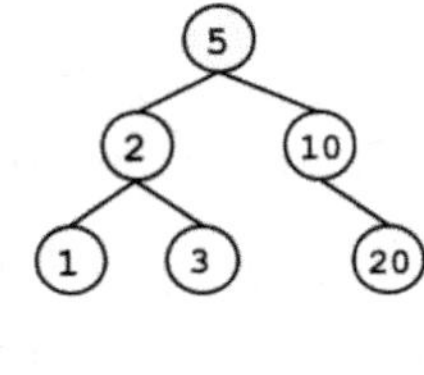

1. Node to be deleted = 7 **2. Put the content of the Inorder predecessor node at the root.** **3. Remove the Inorder predecessor Node**

The node that holds the maximum value in the left subtree, or the node that holds the minimum value in the right subtree. i.e either the inorder predecessor or the inorder successor of the node in the BST.

To delete the node with value 7 (root Node), in the above tree, copy the contents of inorder-predecessor (i.e node 5) or the inorder successor (i.e node 10). Copy the content of that node to the root node and then delete that node.

```
Node* removeNode(Node* r, int val)
{
  if (r == NULL){ return NULL; }

  if(r->data == val)
  {
    // IF EITHER LEAF OR ONLY RIGHT CHILD
    if(r->left == NULL) {
      Node *temp = r->right;
      free(r);
      return temp;
    }else if(r->right == NULL){
```

```
      Node *temp = r->left;
      free(r);
      return temp;
    }

    // HAS BOTH CHILDREN. GET INORDER SUCCESSOR
    // (SMALLEST NODE IN RIGHT SUBTREE)
    Node* temp = r->right;
    while(temp->left != NULL)
      temp = temp->left;

    // COPY CONTENT OF INORDER SUCCESSOR TO NODE
    r->data = temp->data;

    // DELETE THE INORDER SUCCESSOR
    r->right = removeNode(r->right, temp->data);
  }
  else if(val < r->data)
    r->left = removeNode(r->left, val);
  else  // if(val > r->data)
    r->right = removeNode(r->right, val);
  return r;
}
```

Irrespective of how we are deleting the node, the time taken to delete a node from a BST is `O(lg(n))`. We are considering a balanced BST, if the tree is skewed in either side, the worst-case time taken can go up to `O(n)`.

Question 7.1: Given a Binary tree, check if the given tree is BST ?

Solution: Do not fall into the trap, the below code will NOT work:

```
bool isBST(Node* root){
  if(root == NULL)
    return true;

  // DATA AT root VOILATES BST
   if( (root->left && root->data < root->left->data) ||
      (root->right && root->data > root->right->data) )
    return false;

  // BOTH LEFT AND RIGHT SUBTRIES ARE BST
  return (isBST(root->left) && isBST(root->right));
}
```

It returns `true` for the following tree, but the tree is not a BST because 9 is in the right subtree of 15.

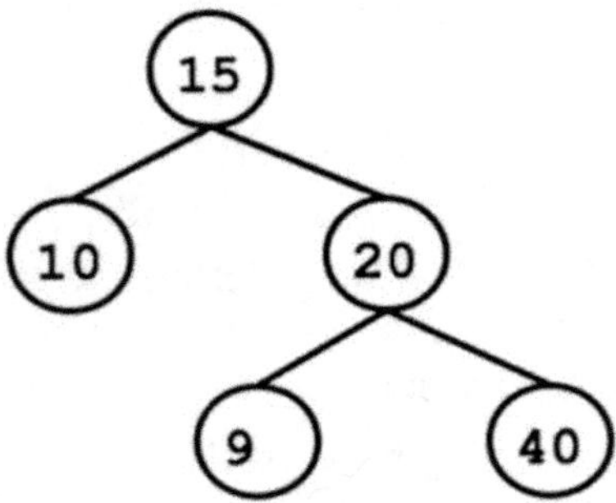

The code does not work because it makes a shallow check and only compares node-value with its direct children. It does not check if a node is in conflict with its grand children or any other node in the subtree.

There can be multiple ways to check if a tree is BST. One way is to check the in-order traversal of the tree

```
IF ( inOrder traversal of tree is sorted in
     ascending order ) THEN
   Tree is BST
ELSE
   Tree is NOT BST
```

Below code populate an array with inorder traversal of the tree.

```
void populateInorder(Node* r, int* arr, int* pos){
  if(r == NULL){ return; }

  populateInorder(r->left, arr, pos);
  arr[*pos] = r->data;
  (*pos)++;
  populateInorder(r->right, arr, pos);
}
```

The function to check if a tree is BST or not is given as below:

```
bool isBST(Node* r){
  if(NULL == r){ return true; }

  // HARD-CODING IT TO 100 FOR SIMPLICITY.
  int arr[1000] = {0};
  int n = 0;            // HOLD NUM OF NODES

  populateInorder(r, arr, &n);

  // CHECK IF ARRAY IS SORTED IN ASCENDING ORDER
  for(int i=0; i < n-1; i++)
```

```
    if(arr[i] > arr[i+1])
      return false;

  return true;
}
```

We may avoid creation of temporary array `arr` because we just need to compare current node with the previous node in the inorder traversal. The value of previous node can be passed to function `populateInorder`. Initial value of previous node can be `INT_MIN`.

```
bool isBSTUtils(Node *p, int& prev) {
  if(p == NULL){ return true; }

  if (isBSTUtils(p->left, prev)){
    if (p->data > prev){
      prev = p->data;
      return isBSTUtils(p->right, prev);
    }else{
      return false;
    }
  }else{
    return false;
  }
}

bool isBST(Node *root){
  int prev = INT_MIN;
  return isBSTUtils(root, prev);
}
```

Another method is to remember the bounds within which a Node's value should exist for the tree to be a BST. For example, if the tree is

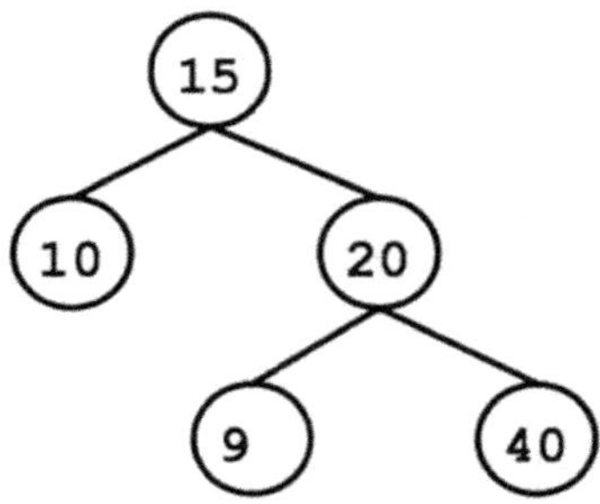

then the bounds for each node is as given below:

Node Value	Lower Bound	Upped Bound	Comments
15	$-\infty$	$+\infty$	Value of root node can be anything
10	$-\infty$	15	Value of this node cannot be >15. Else it will violate the BST property with root.
20	15	$+\infty$	Value of this node cannot be <15.
9	**15**	**20**	**Value in this node should be between 15 and 20**
40	20	$+\infty$	Value of this node must be >20

This tree is not BST, because 9 is not within the bounds. The bounds are passed as parameter to the function.

```
bool isBSTUtils(Node* r, int low, int high){
  if (r == NULL){ return true; }

  if(low > r->data || high < r->data)
    return false;

  // UPDATE low AND high FOR SUBTREES
  return isBSTUtils(r->left, low, r->data) &&
         isBSTUtils(r->right, r->data, high);
}

bool isBST(Node *root){
  return isBSTUtils(root, INT_MIN, INT_MAX);
}
```

Assignment 7.1: Given the preorder traversal of a BST in an array, construct the original tree.

Multi-Way Search Tree

A Multi-way search tree is an extension to the binary search tree. Each node in a binary search tree has one value and two ways to go down the hierarchy (via the left child and right child).

An M-way search tree node has M-1 values and M ways (pointers) to go down.

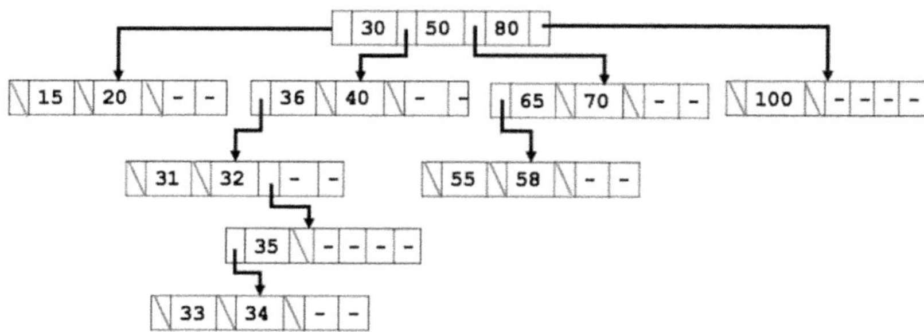

The tree given above is a 4-way tree. Hyphen sign (–) indicate empty places. A slanting line (\) indicate `NULL` value. Each node of 4-way tree can have up to three (sorted) values and four children. If a node has `x` values (`x<=3`), then it has `x+1` children. Structure of such a node is:

```
#define M 4
struct MWayNode
{
  int data[M-1];
  Node* child[M];
}
```

Node values are stored in `data` array and children pointers are stored in `child` array. Subtree to left of `data[i]` is pointed to by `child[i]` and subtree on right of `data[i]` is pointed to by `child[i+1]`.

The algorithm to search in an `M`-way search tree is an obvious generalization of the algorithm to search in a BST. Compare value being searched (say, `x`) with values in the node. Since node values are in sorted order, we may use binary search. But data array of a node is usually so small that even linear search takes constant time.

If `x` is present in the node, value is found. Else, find minimum index `i`, such that `x < data[i]`

if (`i > 0`), `data[i-1]< x <data[i]`. Recursively search in subtree `child[i]`. If `x` is greater than all values of node, search the rightmost subtree of current node recursively (not always `child[M-1]`).

As in BST, if there are no duplicate values, for each node, all elements in entire subtree pointed to by `child[i]` are greater than `data[i-1]` and less than `data[i]`.

Data array of each node is sorted in ascending order. If size of data array increase, height of MST decreases. When nodes are not full, the space utilization is poor. If root is at level zero in a `5`-way tree and tree is complete with each node having `4` values, then there are `15625` nodes with `62500` keys and `78125` subtrees at the 6th level alone, compare it with just `64` nodes at 6th level in a complete binary tree.

Searching for a value in a balanced and complete M-Way tree takes `O(log`$_M$`(n))` time. Structure and depth of tree depends on how the values are inserted. The tree may get skewed if data is inserted linearly, by first populating the root, followed by nodes in next level without re-adjusting the nodes (the way we insert in a normal BST). It wastes a lot of memory.

The 4-way tree below is not balanced. New node gets created at level-2 when there is plenty of room at level-1.

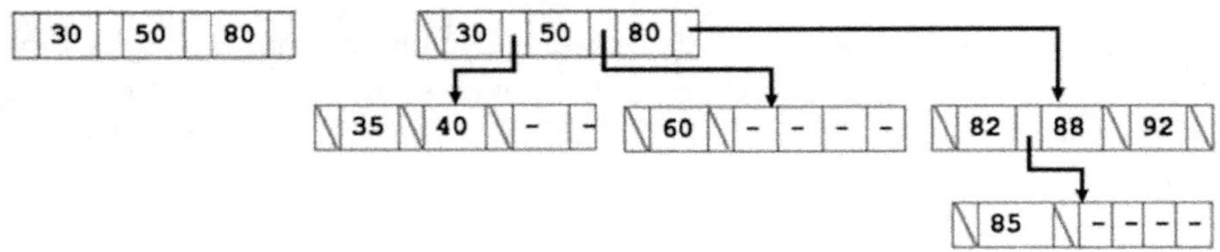

B-Tree is a balanced implementation of M-Way tree where all leaf nodes are at same level and each node, except (possibly) the root has at least M/2 values. The insertion, deletion and other operation of B-Tree are out of scope of this book, but we encourage you to read about it because B-Tree is a good space-utilization (more than 50%) of an M-way tree with all major operations still taking O(log$_m$(n)) time.

Practice questions on BST

Question 7.2: Given values of two nodes in a BST, write code to find the Lowest Common Ancestor (LCA) of these two nodes. For example, in the below BST

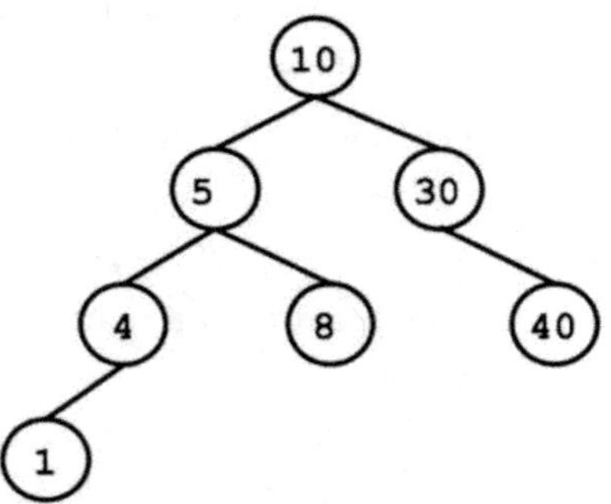

5 is the LCA of 4 and 8. If the two nodes are in the same hierarchy i.e one of the two is an ancestor of the other, then the ancestor node itself is the LCA. For example, LCA of 40 and 10 is 10 itself.

Solution: If given two nodes are not in the same hierarchy, then there is only one common ancestor of these two nodes whose value lies between the two nodes. If the two values are 4 and 8, then 5 is the only node, that is in the hierarchy of both of them and whose value lies between the two.

Lowest Common Ancestor of two nodes is the ancestor to both of these nodes whose value lies between the two nodes. For any two nodes in a BST, there is only one common ancestor whose value is between the two nodes (i.e less than one and greater than the other).

If the two nodes are in a hierarchy i.e. one node is ancestor of the other, then their common ancestor is the node which is above in the hierarchy. For example, LCA of 5 and 1 is 5 itself.

```
Node* lca(Node* r, int val1, int val2)
{
  if (r == NULL) return NULL;

  // IF BOTH val1 & val2 ARE SMALLER THAN ROOT
  // THEN LCA LIES ON THE LEFT SIDE
  if (r->data > val1 && r->data > val2)
    return lca(r->left, val1, val2);

  // IF BOTH val1 & val2 ARE GREATER THAN ROOT,
  // THEN LCA LIES ON THE RIGHT SIDE
  if (r->data < val1 && r->data < val2)
    return lca(r->right, val1, val2);

  // OTHERWISE ROOT IS THE LCA
  return r;
}
```

This problem can be used to find the shortest path between two nodes in a BST, because the shortest path between two nodes in a BST always pass thru the LCA.

Question 7.3: Find the median of a BST. Media is the value that is printed at the middle in the inorder traversal of the tree. If number of nodes in the tree are even, then median is the mean of two middle nodes.

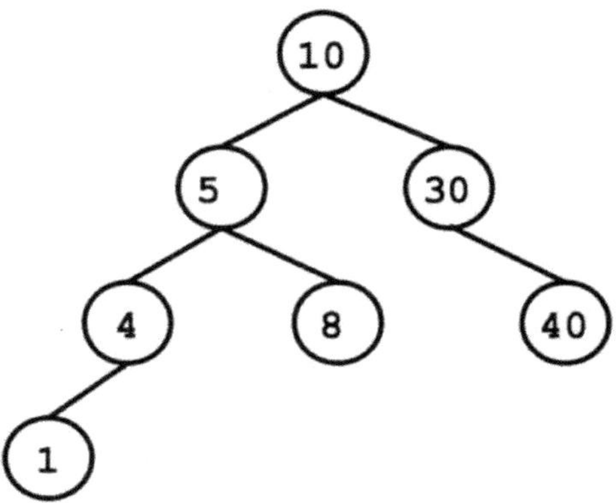

Solution: First count the total number of nodes in the BST. If there are N nodes in the BST, then the $\frac{N}{2}$ th node in the inorder traversal of the tree represents the median.

Question 7.4: Given a Binary Tree, write code to convert it to Binary Search Tree such that the structure of the tree remains the same. For example: Both the below trees

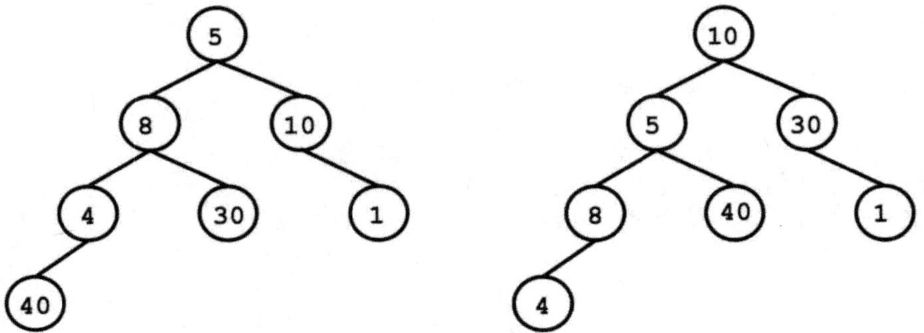

Should generate the below Binary Search Tree.

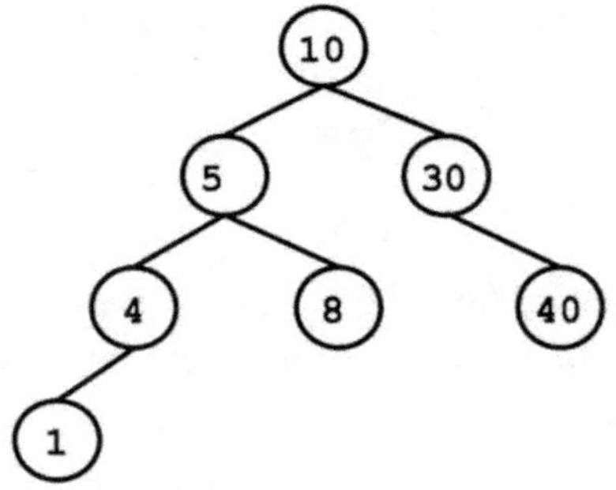

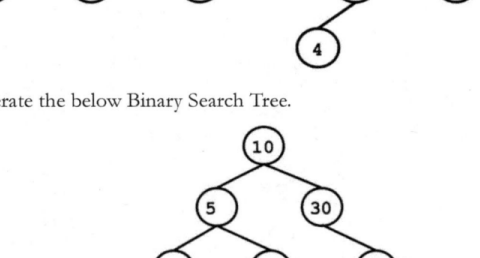

Note that the structure of the tree remains the same. Hence all the three trees above are Similar in terms of structure.

Solution: Let us take an extra array of size n (Number of Nodes) and use the below algorithm:

1. Traverse the tree in inorder and store the values in the array.
2. Sort the Array.
3. Traverse the tree again in inorder and copy corresponding value from the array to the Nodes.

Below is the code for above algorithm:

```
// POPULATE ARRAY arr WITH THE INORDER TRAVERSAL.
void populateArray(Node* r, int* arr, int* pos)
{
  if(r == NULL){ return; }

  populateArray(r->left, arr, pos);
  arr[*pos] = r->data;
  (*pos)++;
  populateArray(r->right, arr, pos);
}

// TRAVERSE THE TREE IN INORDER AND COPY THE
// CORRESPONDING CONTENT FROM THE ARRAY TO THE TREE.
void copyArrayToTree(Node* r, int* arr, int* pos)
{
  if(r == NULL){ return; }

  copyArrayToTree(r->left, arr, pos);
  r->data = arr[*pos];
  (*pos)++;
  copyArrayToTree(r->right, arr, pos);
}

// THIS IS THE MAIN FUNCTION TO BE CALLED FROM OUTSIDE
// NO NEED TO RECEIVE Node** BECAUSE WE ARE NOT
// CHANGING NODE, JUST CHANGING THE CONTENT OF NODES
void converttoBST(Node* r)
{
  if(NULL == r){ return; }

  // HARDCODING FOR SIMPLICITY. DO ONE MORE TRAVERSAL
  // TO FIND NUMBER OF NODES IN THE TREE
  int arr[1000] = {0};
  int n = 0;            // FOR NUMBER OF NODES

  populateArray (r, arr, &n);
  quickSort(arr, 0, n-1);

  n=0;
  copyArrayToTree (r, arr, &n);
}
```

Time taken by the above code is:

- Time to populate Inorder Array = O(n)
- Time to Sort Inorder Array = O(n.lg(n))
- Time to Put values from Sorter array to Tree = O(n)
- **Total Time** = O(n) + O(n.lg(n)) + O(n) = O(n.lg(n))

It also uses O(n) extra space for the array and recursion.

Question 7.5: Given two BSTs consisting of unique positive elements. Check whether they contain same nodes or not. The below two BSTs have different structures, but they hold the same values:

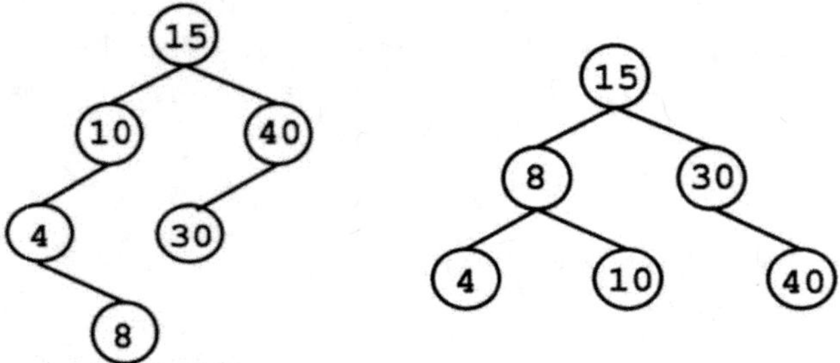

Solution: One of the easiest ways to check this is to put the inorder traversal of two BSTs in two separate arrays. If the two BSTs have same values, then the two arrays have to exactly same.

Instead of storing the values in an array, we may use a hashmap to store the traversal of the first BST. Then traverse the second subtree and check if the current node is present in the hash or not. In this case, we may use any traversal of the tree (and not necessarily the inorder traversal). You may also compare the sum and product of all the nodes in the two BSTs.

Question 7.6: Given the preorder traversal of a BST in an array, check if all nodes of the BST has exactly one child. You may assume that the BST contains unique entries. For example, if the preorder traversal array is
`{20, 11, 14, 18, 16}`

Then the output should be `TRUE`, because the given array represents the following BST in which every node has only child.

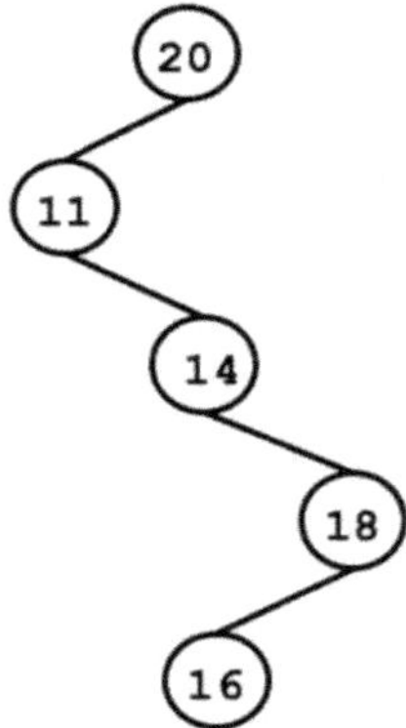

Solution: In preorder traversal, root comes before all other nodes in the tree, therefore `20` is the root node of the given tree. Notice that all the other values in the array are less than `20`. Which means that they are all in the left subtree of the root node and root node has only one child. Similarly, if all the values toward the right of an array element are greater than the element, then all of them belong to the right subtree of that element and there is no left subtree in that node.

In general, if all nodes have only one child in a BST, then all the descendants (every value that comes on the right side) of every node are either smaller or larger than the node.

The brute-force approach is to use two nested loops and for each element in the array check if all the values after it are either less than or greater than the value of that element. This approach will take $O(n^2)$ time in the worst case.

A better solution is to use the fact that in preorder, after traversing the root node, we first traverse the left subtree (where all the nodes are less than root node) and then the right subtree (where all the nodes are greater than the root node). So if a node has both left and right subtree, then the very next value will be from the left subtree and the last value in the array will be from its right subtree (or the right subtree of its ancestor). The check for given problem can be as simple as:

For each node, check the next value and last value in the array. If both are either less than the current node, or greater than the current node, then continue, else, return `FALSE`.

```
bool eachNodeSingleChild(int* arr, int n){

  for(int i=0; i<n-1; i++){
    if( ((arr[i]-arr[i+1]) * (arr[i]-arr[n-1])) < 0)
      return false;
  }
  return true;
}
```

This logic takes linear time and constant extra memory.

Question 7.7: Given an array of integers. Check if the array can represent the preorder traversal of a BST. For example, if the given array is `{15, 40, 30}`, the answer is true, because the given array can represent the preorder traversal of the below tree:

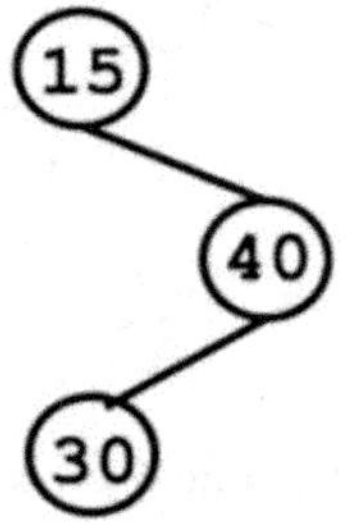

But if the array is `{15, 40, 10}` then it cannot represent the preorder traversal of a BST.

Solution: The brute-force solution of this problem is to do the following for each element of the array:

```
FOR i=0 TO n-1
  j = FIRST INDEX ON RIGHT OF i WHERE arr[j] >arr[i]
  IF ( ALL VALUES AFTER arr[j] ARE GREATER THAN arr[i]
        AND (arr[i] TO arr[j-1]) IS ALSO BST
        AND (arr[j+1] TO arr[n-1]) IS ALSO BST )
    RETURN true
  ELSE
    RETURN false
```

The complexity of above code is $O(n^2)$.

A better solution is to use stack to find the next greater element in the array like we did in Question 2 of Chapter-4 as shown in the below code:

```
bool isBST(int* arr, int n)
{
  stack<int> s;
  int root = INT_MIN;

  for(int i=0; i<n; i++)
  {
    if(arr[i] < root)
      return false;
    while(!s.empty() && s.top() < arr[i]){
      root = s.top();
      s.pop();
    }
```

```
        s.push(pre[i]);
    }
    return true;
}
```

Threaded Binary tree

If there are `N` nodes in a binary tree, then it can be verified that total number of `NULL` pointers are `N+1`. These pointers are just occupying the memory and are not used for any purpose. One way to use these `NULL` pointers is to set them in a way that make the traversals easier. Let us consider the below binary tree

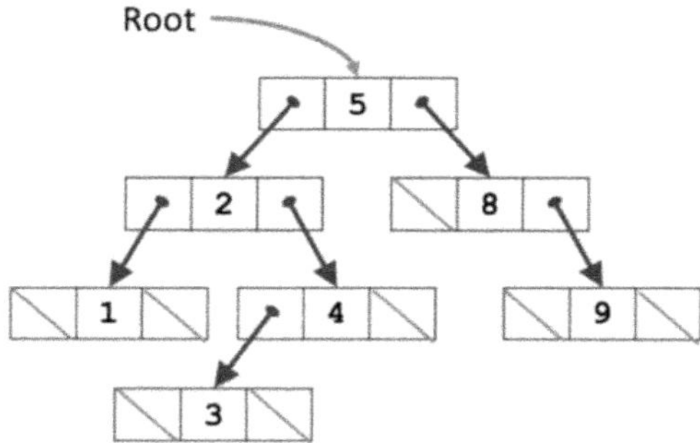

As we can see, there are 7 nodes and 8 `NULL` pointers in the above tree. Let us say, whenever `right` pointer in a node is `NULL`, we store the inorder successor of that node in the `right` field (instead of `NULL` pointer). Then the memory of the above tree looks like below:

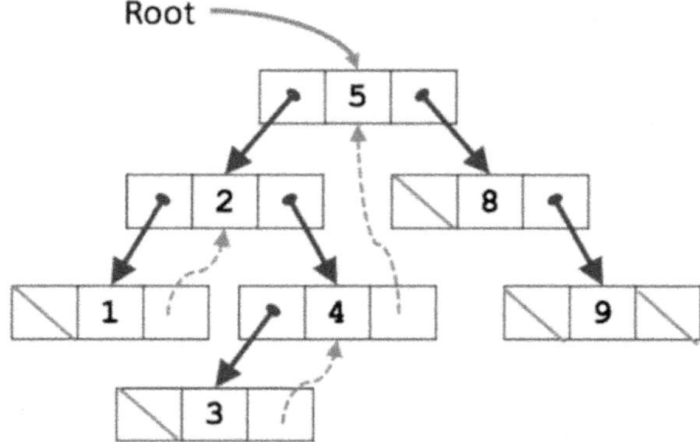

Now the `right` field of a node can store either a pointer pointing to the right child of that node, or it can store a thread (again a pointer) pointing to the inorder successor. There should be a way to distinguish whether the `right` field stores a pointer or a thread. This can be achieved by adding one more field to the structure of `Node`.

```
struct Node{
  int data;
  Node* left;
  Node* right;
  bool isRightThread;
};
```

the value of `isRightThread` is `true` if `right` field is storing the address of inorder successor of the Node. Value of field `isRightThread` is `false` if the `right` field holds pointer to the right child.

This is just one-way threading. Similarly, we can set the left pointer to point to the inorder predecessor when the `left` field is `NULL`.

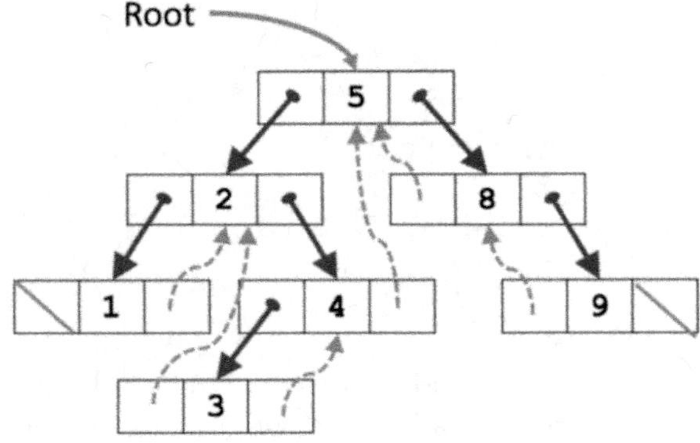

For this, we should add one more field to the structure of Node to tell us whether the `left` field of the node holds a pointer to the left child or to the preorder predecessor.

When we maintain only a single thread (either left or right) the tree is called "Single Threaded Binary Tree". If we have both the threads then the tree is called, "Double Threaded Binary Tree".

We can convert a normal binary tree to a threaded binary tree (assuming that the structure of Node is same as that of threaded binary tree) by doing an inorder traversal of the tree and keeping a track of the previous node visited. The right thread of previous node is set to the current node and the left thread of current node is set to the previous node and repeat it for each node of the tree. The right thread is set only if right pointer is `NULL`, similarly, the left thread is set only if the left pointer is `NULL`.

Question 7.8: Given a right-threaded binary tree. Write code to traverse the tree in inorder traversal.

Solution: If the right threads are set correctly, we can do the inorder traversal iteratively (rather than recursively).

Because `right` pointer is used for two purposes, the boolean field `isRightThread` is used to indicate whether the `right` pointer points to the right child or inorder successor.

The logic for inorder traversal is simple. If the right pointer holds thread, move to the right thread. Else get the leftmost node in the right subtree and move to that node.

Let us write a helper function that gives us the leftmost node in a subtree

```
Node* getLeftMost(Node *r)
{
  if(r == NULL){ return NULL; }

  while(r->left != NULL)
    r = r->left;

  return r;
}
```

Now, the inorder traversal of the tree is simple. The first node to start with is the left-most node in the binary tree.

```
void inOrder(Node *r)
{
  Node *r = leftmost(r);    // STARTING NODE
  while(r != NULL)
  {
    printf("%d ", r->data);

    if(r->rightThread)
      r = r->rightThread;
    else
      r = leftmost(r->right);
  }
}
```

This function takes `O(n)` time and constant extra memory.

Trie

A trie data structure is tree-based data structure used to implement dictionaries. It is very helpful in prefix-based search. A string `str1` is called prefix of string `str2` if string `str2` starts with `str1`. For example, `CAR` is prefix of both `CARTOON` and `CART`.

When we type in the search box of Google search engine, then as soon as we type something, Google suggest the strings that we may have in our mind.

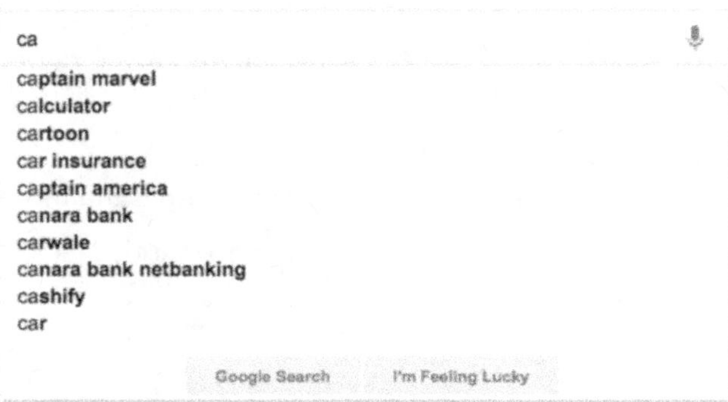

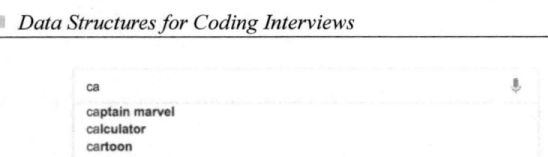

As we keep typing, the suggestions also keeps refining

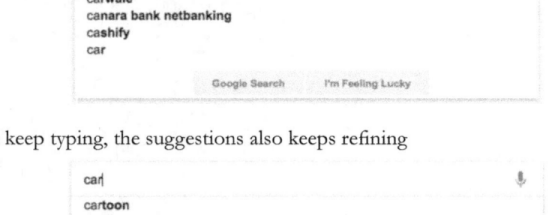

Essentially, Google takes what we have typed and suggest the strings which have this as the prefix.

How can google search so fast from so many strings in the pool? The answer (may be) lies in the trie data structure. A trie is used to store the collection of strings. If two strings have a common prefix, then the prefix is stored only once. Let us try to store following string in the trie

`CAR, CARB, CARRY, CARTOON, CAP, CRYPT, CRY`

The trie storing all the strings is shown on the right side.

The circled nodes indicate that this node is the end of a valid word. For example, `CRY` is a valid word but we can go further down in the tree and form another word `CRYPT`.

Once all the given strings are pre-processed and stored in a trie, searching for a valid string having a particular prefix takes `O(k)` time where `k` is the length of the string.

Each node of the trie typically stores one character and each node can possibly have 26 children. A node may either be the end of a string or not. Below structure represents a typical node of trie.

```
typedef struct trieNode
{
  struct trieNode* child[26];
  bool isEndOfWord;
}TrieNode;
```

We are using a `typedef` so that we can directly use `TrieNode` in later code instead of typing `struct trieNode` everytime.

`child` field in the `TrieNode` is used to hold the address of children of the Node. If `child[i]` is not `NULL`, it means that there is a valid string in that path and the character represented by `child[i]` is `('A' +i)`. We are considering only English alphabets here. If you want the structure to be more generic, you may use a Map as shown below:

```
typedef struct trieNode
{
  map<char, trieNode*> child;
  bool isEndOfWord;
}TrieNode;
```

We will stick with the array for simplicity. The previous Trie has words starting with letter C only. Actually, a trie looks something like below:

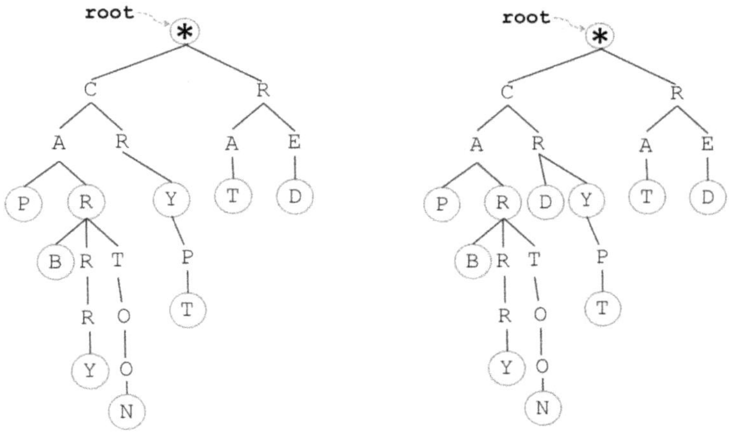

A typical Trie **After inserting string "CRD"**

Trie is typically used to implement dictionary. We may also use a hash table to implement dictionary, but it will not be possible to do a prefix-based search in a hashtable. Also, trie is space-optimized, because we are storing the common prefixes only once.

A trie can also be used to print the words in alphabetic order.

Inserting a new string in a Trie.

Let us say we want to add a new word, CRD in the given trie. Consider each character of the string one-by-one and move to that character in the Trie.

The first character in CRD is C, Node for C is already present, Move to the Node with value C in the first level.

Next character in CRD is R. R is already present in the children of C, there is no need to insert R and just move to R under C.

Next character in CRD is D. There is no child of R with value D. We insert a new node and because this is the last character in the word, mark it as end of word by setting isEndOfWord to true.

CART is a valid word, but according to our trie, this is not a valid word, because the word is not terminating at T. If we want to insert CART in the given trie, then there will be no physical insertion. Only the isEndOfWord field of T Node under R is set to true.

```
void insert(TrieNode* r, char* str)
{
  // FOR EACH CHARACTER IN THE STRING
  for(int i=0; str[i] != '\0'; i++)
  {
    int idx = str[i] - 'a';
    if (r->child[idx] = NULL)
      r->child[idx] =
                 (TrieNode*)malloc(sizeof(TrieNode));
    r = r->children[idx];
  }
  r->isEndOfWord = true; // END OF WORD MARKER
}
```

Searching for a word in the Trie

Searching for a word in the trie is even simpler. We just need to follow the path of characters in the string and if that ends at a node which represent an end of word, the word is valid else not.

```
bool search(TrieNode *r, string str)
{
  for (int i=0;r != NULL str[i] != '\0';i++)
    r = r->children[str[i] - 'a'];

  return (r!=NULL && r->isEndOfWord);
}
```

Deleting a node from a Trie can be done in a bottom up manner using recursion.

Balanced Binary Trees

A binary tree where all leaf node sare at a similar distance (may not be exactly same) from the root is called Balanced Binary Tree.

The definition of being at a similar distance is different for different balancing schemes. There are different realizations of a balanced binary tree concept.

AVL Tree, Red-Black Tree, Balanced BST, BTree are all examples of balanced binary trees. The advantage of having a balanced binary tree is that all major operations in a balanced tree takes `O(lg(n))` time.

One of the major one being the height-balanced binary tree that is used in the AVL tree.

According to that scheme, a binary tree rooted at `r` is considered height-balanced if it is either empty or satisfies the below conditions:

1) Left subtree of `r` is balanced.
2) Right subtree of `r` is balanced.
3) The difference between heights of left subtree and right subtree is not more than 1.

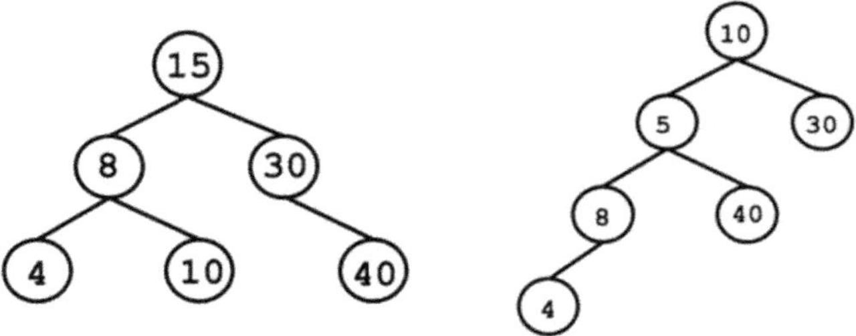

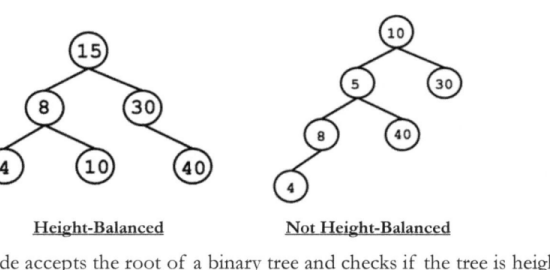

Height-Balanced **Not Height-Balanced**

The following code accepts the root of a binary tree and checks if the tree is height-balanced or not.

```
bool isHeightBalanced(Node* r)
{
  if(r == NULL){ return true; }

  // HEIGHT OF LEFT & RIGHT SUBTREE
  int leftH = height(r->left);
  int rightH = height(r->right);

  if( abs(leftH-rightH) <= 1 &&
      isHeightBalanced(r->left) &&
      isHeightBalanced(r->right) )
    return true;
```

```
    else
      return false;
}
```

It uses the `height` function that computes the height of a tree. The `abs` function returns the absolute value.

The function takes `O(n²)` time, because it is computing the height for every node. In a more optimized version, we can compute the height of the tree in the same recursion instead of calling the `height()` function separately. I leave it as an exercise for the reader.

AVL Tree

AVL tree is a self-balancing binary search tree. It is height-balanced and for any node in the tree, the height of left and right subtrees cannot differ by more than one.

The meaning of a self-balancing tree is that when a new node is inserted, it is inserted in a way that the tree remains height-balanced (along with maintaining the BST property).

Following two single rotation operations can be performed on a BST without violating the BST property of the tree:

1. **Single Left Rotation (LL Rotation)**

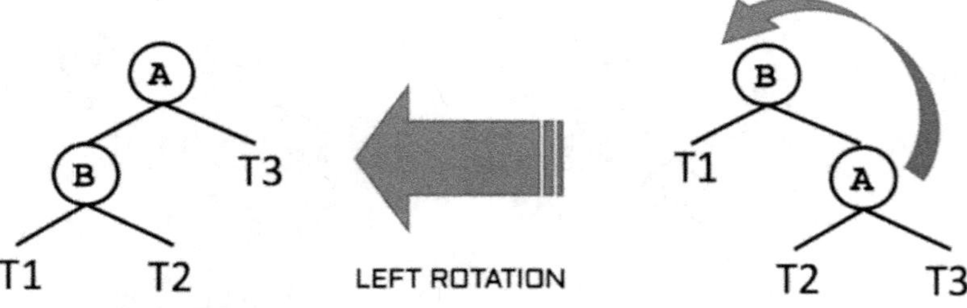

2. **Single Right Rotation (RR Rotation)**

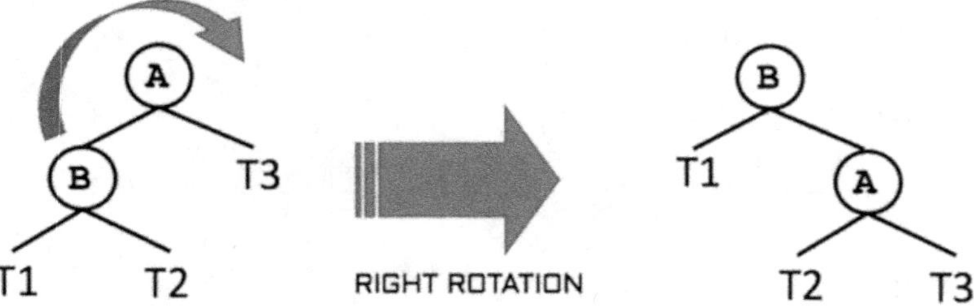

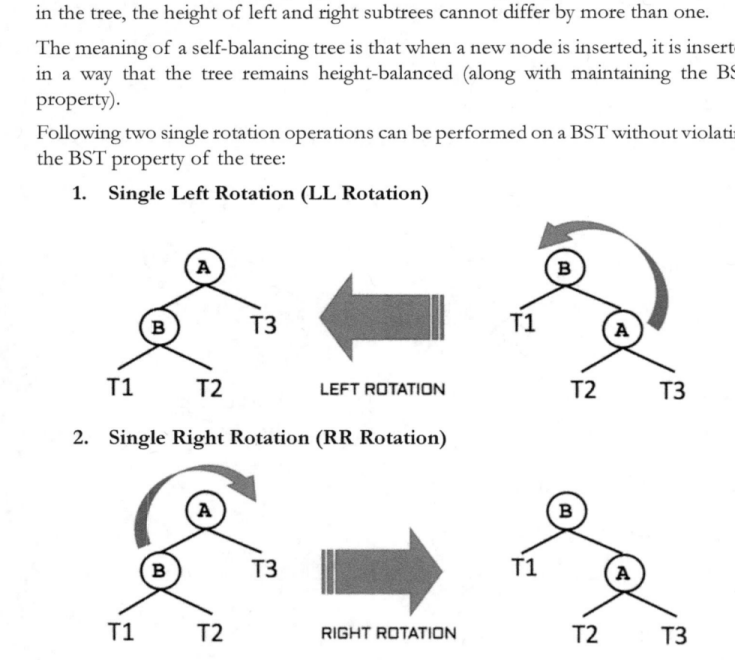

In the given tree in the diagram, `A` and `B` are the nodes, `A` is greater than `B` because it is the left child of `B`. `T1`, `T2` and `T3` are the subtrees of the given BST. Because of the given structure, we can say that all the nodes in subtree `T1` are less than B and all the nodes in subtree T3 are greater than A. All nodes in T2 are greater than B, but less than A. Note that the rotation of the tree is not violating the BST property of the tree in any way.

Let us see the examples of left and right rotations on simple imbalanced BST:

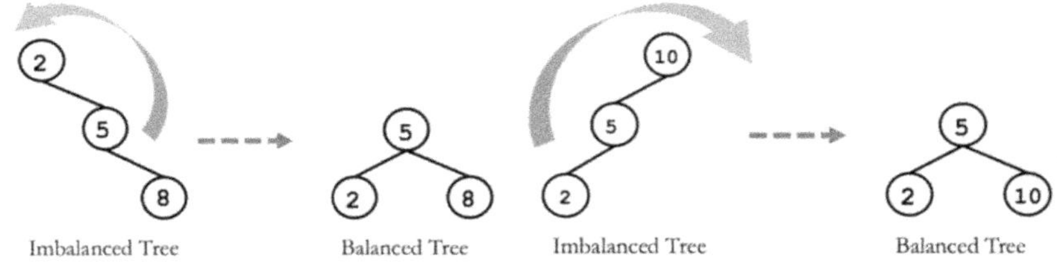

Similarly, following two double rotation operations can be performed on the BST without violating the BST property:

1. **Left-Right Rotation (LR Rotation)**

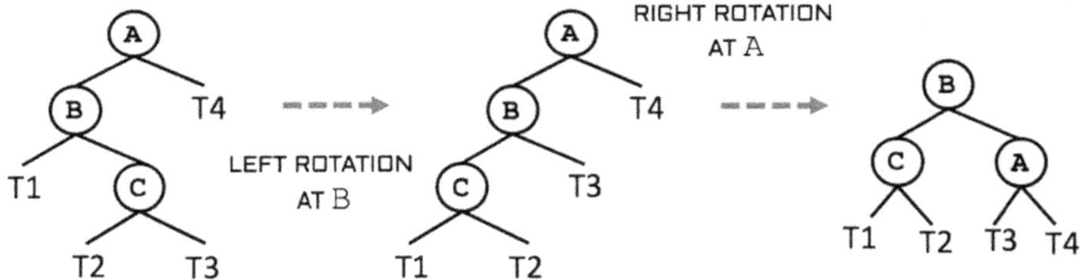

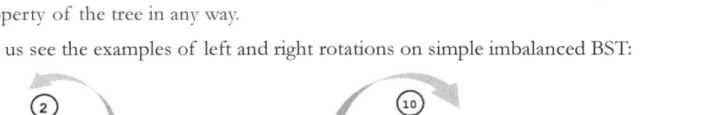

2. **Right-Left Rotation (RL Rotation)**

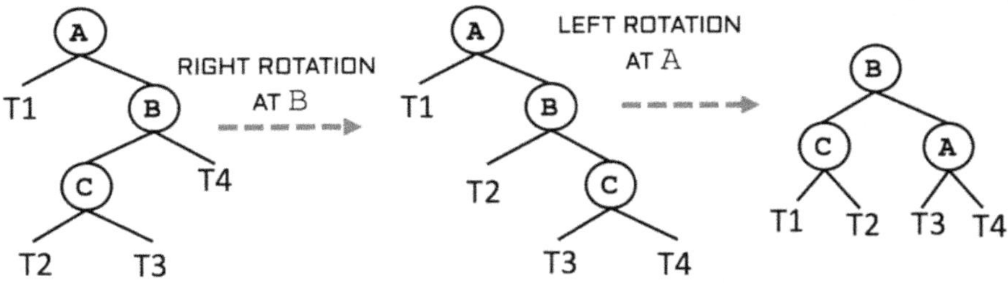

Below is the code for all above rotations:

```
Node* rr_rotation(Node* r){
  Node* temp= r->right;
  r->right = temp->left;
  temp->left = r;
  return temp;
}
```

```
Node* ll_rotation(Node* r){
  Node* temp = r->left;
  r->left = temp->right;
  temp->right = r;
  return temp;
}

Node* lr_rotation(Node* r){
  Node* temp= r->left;
  r->left = rr_rotation(temp);
  return ll_rotation(r);
}

Node* rl_rotation(Node* r){
  Node* temp= r->right;
  r->right = ll_rotation(temp);
  return rr_rotation(r);
}
```

The below helper function return the difference between left and right subtrees:

```
int diff(Node* r){
  return height(r->left)- height(r->right);
}
```

Below is the main function that insert a value in the AVL tree

```
Node* avlInsert(Node* r, int val){
  if(r == NULL){
    r = (Node*)malloc(sizeof(Node));
    r->data = val;
    r->left = r->right = NULL;
    return r;
  }else if(val < r->data){
    r->left = avlInsert(r->left, val);
    r = balance(r);
  }else if(val >= r->data){
    r->right = insert(r->right, val);
    r = balance(r);
  }
  return r;
}
```

The `balance` function will rebalance the tree after insertion. It first checks what kind of rotation is required and then calls the relevant function:

```
Node* balance(Node* r){
  int bal_factor = diff(r);
  if(bal_factor > 1){
```

```
    if (diff (r->left) > 0)
      r = ll_rotation(r);
    else
      r = lr_rotation(r);
  }else if (bal_factor < -1){
    if (diff (r->right) > 0)
      r = rl_rotation(r);
    else
      r = rr_rotation(r);
  }
  return r;
}
```

The new value is inserted in the BST using the normal logic of BST-insertion. After inserting the new node, if the tree becomes imbalance, it is re-balanced using one of the four rotations.

Assignment 7.2: Given an unbalanced BST, convert it into a balanced BST that has minimum possible height.

More practice questions

Question 7.9: Given a binary matrix (each value is either 0 or 1). Print all unique rows of that matrix. For example, if the input matrix is:

```
int arr[3][5] = {{0, 1, 0, 0, 1},
                 {1, 0, 1, 1, 0},
                 {0, 1, 0, 0, 1},
                 {1, 1, 1, 0, 0}};
```

Then the output should be

```
0 1 0 0 1
1 0 1 1 0
1 1 1 0 0
```

Solution: The brute-force solution is to check if there exist a duplicate row for each row in the matrix. If order of matrix is `N*M`, then the algorithm is

```
for(int i=0; i<N; i++)
{
  for(int j=i+1; j<N; j++)
  {
     // CHECK IF Row-i AND Row-j are SAME
  }
}
```

Another method is to consider each row as a sequence of bits and find the corresponding decimal number (formed from the binary representation of bits in a row).

```
   arr[0] = 01001  = 9
```

```
arr[1] = 10110  = 22
arr[2] = 01001  = 9
arr[3] = 11100  = 28
```

Now our challenge is reduced to printing unique values from the given decimal numbers. All of these numbers can be stored in a Hashtable. When a row is converted to decimal, see if that value is already present in the Hash and if it is present, then just ignore it and move forward, else insert that value in the Hashtable.

A variant of Trie data structure can also be used. Each `TrieNode` will have only two children, one for `0` and other for `1`. While inserting a row, see if it is already present in the Trie. If it is present, do not do anything, else print the row. This logic takes `O(N*M)` time.

Question 7.10: A dictionary (set of words) and a character array is given. Print all words present in the dictionary that are formed using characters from the array only (a character may be repeating in the word). For example, If the dictionary and character array are as given below:

```
string dictionary[]= {"go","bat","me","eat","boy"};
char arr[] = {'e','o','b', 'a','m','g', 'y'};
```

Then the output should be:

```
go, me, boy
```

Solution: After reading the word 'Dictionary' in the question, our mind should start giving us clues that we may be using Trie in the solution. Follow the following steps:

Create a Trie and insert the dictionary words into it. Look for only those paths in the Trie for which character is present in the character array.

Since we are doing a lot of look-up in the character array, it makes sense to store the characters in a HashTable. An array can be used as the hash in this case, because the number of characters are very limited.

```
int hash[ALPHABET_SIZE] = {0};
```

If `hash[i]` is zero, character `('a'+i)` is not present in array `arr`. So let us populate the hash:

```
for(int i=0 ; i < n; i++)
  hash[arr[i]-'a'] = 1;
```

Now check for each path in the trie that starts with a character present in the array.

```
void PrintAllWords(char Arr[], TrieNode *r, int n)
{
  int hash[SIZE] = {0};
  for (int i = 0 ; i < n; i++)
    hash[char_int(Arr[i])] = 1;

  TrieNode *pChild = r ;
  char str[100];
```

```
  for (int i = 0 ; i < SIZE ; i++)
  {
    if (hash[i] == 1 && pChild->child[i] )
    {
      str[0] = 'a'+i;
      checkWord(pChild->child[i], hash, str, 1);
    }
  }
}
```

Function `checkWord` is a recursive function that checks if the following path has characters from array `arr`. Below is the implementation of `checkWord`

```
void checkWord(TrieNode* r, int* hash, char* str)
{
  if (r->isEndOfWord == true)
  {
    str[pos] = '\0';
    cout << str << endl ;
  }

  for (int K =0; K < SIZE; K++)
  {
    if (hash[K] == true && r->child[K] != NULL )
    {
      char c = int_to_char(K);
      checkWord(r->child[K], hash, str+c);
    }
  }
}
```

Chapter 8

Heap

Complete Binary Tree (full binary tree)

You may find different authors using different terminologies for complete binary tree and almost complete binary tree. Let us define what we are referring to as a complete binary tree. If you choose to use a different name for the same, feel free to do that.

A complete binary tree is full up to the last level. All the below binary trees are complete binary trees (**some authors call it full-binary tree**).

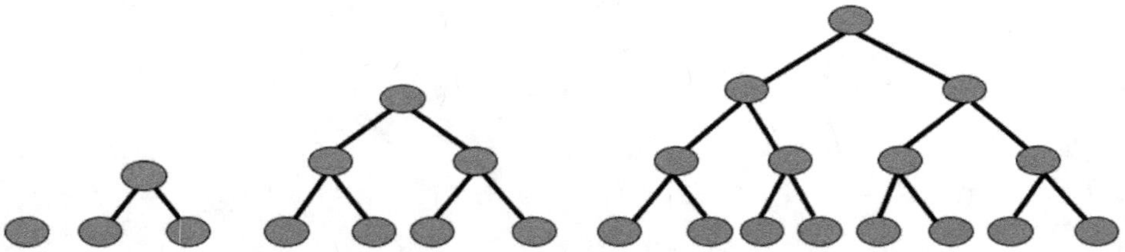

If the height of a complete binary tree is given, the number of nodes can be determined, and vice-versa. The total number of nodes in level-x of a complete binary tree are 2^x.

Question 8.1: Given a pointer to the root of a binary tree, check if that tree is complete binary tree or not?

Some interview candidates put the following check as a solution:

If all nodes in the tree either have 0 or 2 children, then the tree is complete binary tree.

This check is a necessary condition for a tree to be complete, but this is not the sufficient condition. Below tree qualifies this condition but it is not a complete binary tree.

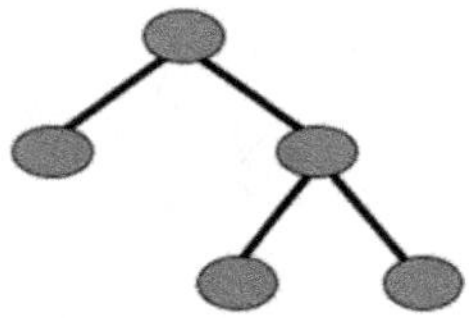

This check is for a Strictly Binary tree and not for Complete Binary tree. A tree is strictly binary when each node has zero or two children. A complete binary tree needs to satisfy the following two conditions:

1. All the nodes have either `0` or `2` children (i.e.it is a Strictly Binary tree).
2. All leaf nodes are at the same level.

The straightforward way to code this is to have two separate functions, one each for above two conditions. If both the functions return `true` only then the tree is complete binary tree, otherwise not.

To check if a tree is strictly binary tree, traverse the tree in any order and at each node see if number of children are either zero or two.

```
bool isStrictly(Node* r)
{
  // NULL OR LEAF NODE
  if(r==NULL || (r->left==NULL && r->right==NULL))
    return true;

  if(r->left == NULL || r->right == NULL)
    return false;
  else
    return isStrictly(r->left) &&
           isStrictly(r->right);
}
```

To check if all leaves are at the same level, take a reference variable to hold the level of leaf nodes. Traverse the tree in any order traversal, when the first leaf node is found, set this variable to the level of that node. When next leaf node is encountered, check if the level of that node is same as the level of first leaf node, if they are same then continue traversing the tree, else return `false`.

```
bool sameLevelLeaves(Node* r, int lvl, int* leafLvl)
{
  if(r == NULL){ return true; }

  if(r->left == NULL && r->right == NULL)
  {
    if(*leafLvl == -1)  // FIRST LEAF NODE
      *leafLvl = lvl;
    else if(lvl == *leafLvl)
      return true;
    else
      return false;
  }

  return sameLevelLeaves(r->left, lvl+1, leafLvl)
```

```
        && sameLevelLeaves (r->right, lvl+1, leafLvl);
}
bool leafAtSameLevel(Node* r)
{
   int leafLevel = -1;     // NO LEAF FOUND
   return sameLevelLeaves(r, 0, &leafLevel);
}
```

The main function should call `leafAtSameLevel` function which in turn calls recursive function `sameLevelLeaves` where all the processing happens. `leafAtSameLevel` declares `leafLevel` variable to hold level of first leaf node.

Both the checks takes `O(n)` time, because we are traversing the tree once.

Another way to check if a binary tree is complete or not is by traversing the tree in level order (apply BFS on given tree) and check for the following:

1. All the non-leaf nodes have both left and right child.
2. After the first leaf node, all the nodes are also leaf nodes.
3. All the leaf nodes are at the same level.

A check for the second condition is optional because it is already covered in the first and third conditions. The code for this logic can be written with a simple modification in the level-order traversal of binary tree described in code 6.2.

To check if all the leaf nodes are at the same level either insert the level number along with the node (while inserting them in Queue) or insert some sentinel node (may be with value `NULL`) at level boundaries.

Below are some interesting facts about a Complete binary tree.

1. If `T` is a complete binary tree with `N` levels (maximum level is `N-1`, root at level 0), then total number of nodes in the tree are 2^N-1. This further implies that total number of nodes in a complete binary tree are always odd. For any level `X`, total number of nodes from root up to, and including level `X` is $2^{X+1}-1$.
2. For each level `X` (root being at level 0), the total number of nodes in that level are 2^X.
3. If `T` is a complete binary tree with `N` nodes, then number of levels are $\lceil lg(N+1) \rceil$.
4. If `T` is a complete binary tree with `L` leaves, then there are $\lceil lg(L) \rceil + 1$ levels.
5. If there are `n` leaf nodes, then there are `n-1` non-leaf nodes. All leaf nodes are at the last level.

Memory efficient storage for a Complete Binary Tree

In practice, `data` field of a binary tree node is a complex data structure. `Node` of a binary tree that holds employee records is:

```
struct Node
{
  Employee data;
  Node*left
  Node* right;
};
```

If each employee record takes `1KB` memory and a pointer variables take `4` bytes, then for each node only `0.8%` of memory is being used to store left and right pointers, rest `99.2%` of node memory is used to store the actual employee data. This is a very good memory utilization by all standards.

But, if data of node is a `char` and `Node` structure is:

```
struct Node
{
  char data;
  Node*left
  Node* right;
};
```

Then, out of total `9` bytes of memory allocated to a node, only one byte is used for actual data, rest `8` bytes are used for storing the meta data (`left` and `right` pointers). It means only `11.1%` of memory stores actual data and `88.9%` stores the metadata. This is probably not a good memory utilization. Let us discuss a way to improve this utilization for complete binary trees.

Data of a complete binary tree can be stored inside an array without disturbing the hierarchical structure of the tree. If nodes of a binary tree are labelled in their level

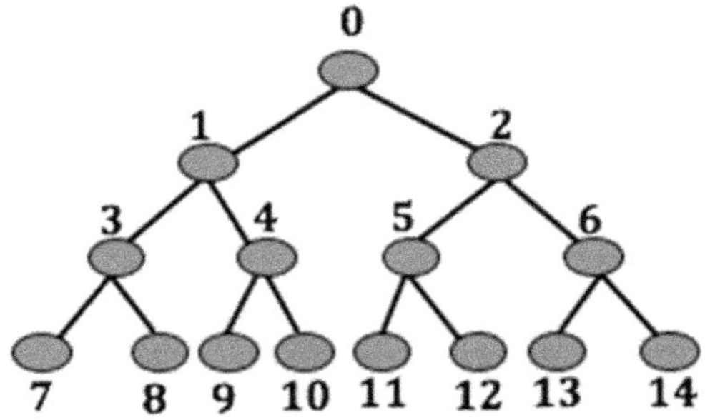

order traversal, it is called **canonical labelling of the tree**

Above tree has `15` nodes, labelled `0` to `14`. Let us store data of these nodes in an array such that data of the node labelled `k` goes to k^{th} index in the array. The label shown next to the node represents the index at which this node is stored. Figure 8.1 shows a complete binary tree and the corresponding array:

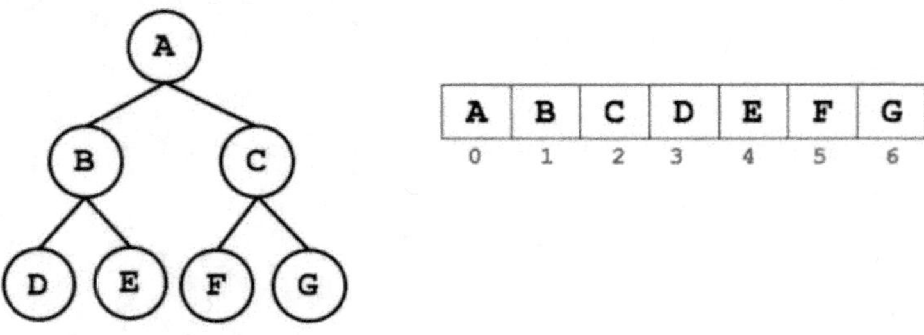

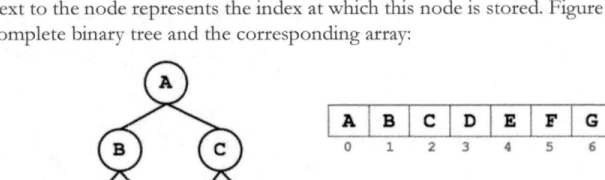

Figure 8.1: Complete binary tree and corresponding array

The idea is to visualize data as if it is stored in a binary tree, when actually it is stored in an array. Now we need to map the operations of a tree on array.

- Array has a name and each element in the array is accessed directly unlike a binary tree where we move to the node using the `left` and `right` pointers. Let `arr` be the name of the array storing data of a tree.
- Binary tree has a `root` pointer, pointing to root node of the tree. When tree is stored in an array, `root` pointer can be replaced by an integer holding index of root element. This index is `0` for the root of the tree, because root of tree is always stored at `arr[0]`.
- If pointer to a node is given, we can extract three values for that node:
  ```
  r->data
  r->left
  r->right
  ```
  If element pointed to by `r` is stored at index `i` in the array, then these three values are:
  ```
  r->data    arr[i]
  r->left    i*2+1
  r->right   i*2+2
  ```
  A pointer in case of node is an index in case of an array. Root node value is stored at index `0` in the array. `left` and `right` children of a particular index `i` are at index `2i+1` and `2i+2` respectively.

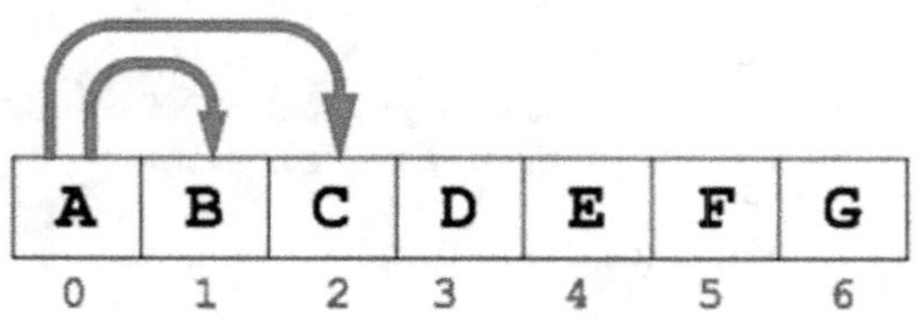

- Next is a check for NULL. A pointer in a node can be NULL, indicating absence of corresponding child. If 2i+1>=n, then the left child is NULL, and if 2i+2>=n, then the right child is NULL.

In Figure 8.1, the left part where tree is stored in the form of nodes, requires 63 bytes of memory (considering four bytes for a pointer). The right part where the same tree is stored in an array, takes just 7 bytes memory. A lot of memory is saved, without compromising on any functionality.

In fact, there is an added advantage of storing tree in an array. In a binary tree, it is not possible to go from a node to its parent directly, in constant time. But when the tree is stored in an array, parent of node arr[i] is at index (i-1)/2. Following constant time function getParent returns parent of a node in constant time

```
char getParent(int *arr, int n, int i)
{
  if(i == 0)
    return '\0';        // ROOT DOES NOT HAVE A PARENT
  else
    return arr[(i-1)/2];
};
```

Below table shows the index of a node, index of its parent node and index of left and right children.

Index of Node	0	1	2	3	4	5	6
Index of parent	-	0	0	1	1	2	2
Index of left child	1	3	5	-	-	-	-
Index of right child	2	4	6	-	-	-	-

We saw in chapter-6, the code to traverse a tree in inorder traversal when it is stored in the Node form.

```
void inOrder(Node* root)
{
  if(root == NULL){ return; } // TERMINATING COND.

  inOrder(root->left);          // TRAVERSE LEFT
  printf("%c ", root->data);    // PRINT DATA
  inOrder(root->right);         // TRAVERSE RIGHT
}
```

If a binary tree is stored in an array as shown in Figure 8.1, the code for inorder traversal is

```
void inOrder(int root, char* arr, int n)
{
  if(root >= n){ return; }      // TERMINATING COND.
```

```
    inOrder(2*root+1, arr, n);// TRAVERSE LEFT
    printf("%c ", arr[root]);// PRINT DATA
    inOrder(2*root+2, arr, n);// TRAVERSE RIGHT
}
```

`arr` is the array storing tree having `n` nodes. Logic of above two inorder traversals is exactly same, by using recursion print left subtree in inorder traversal, then print data stored at root and finally print right subtree in inorder traversal. The difference in two codes is because of the data storage.

When tree is stored inside an array, array need to be passed along with (sometimes optional) index of root element.

Any tree algorithm that can be implemented using `Node` and `Node*` for a complete binary tree, can be implemented for array with minor modifications.

Question 8.2: Given a Binary tree, write two functions. First function should write the binary tree in a file and second function should read that file and construct the original binary tree.

Syntax of writing to a file is not important, idea is to persist a Binary tree both in form and data. Think of it like storing the binary tree in a string and then constructing exactly same binary tree from that string.

For example, if we have an array in place of tree

```
int arr[] = {2, 3, 1, 7, 5, 6}
```

We can write this array to a file one element at a time linearly, then we can read the file one element at a time and construct the exact same array as it was originally. A complete binary tree can also be persisted by storing the level order traversal of that tree in the array form.

Doing the same thing for any general binary trees may be a little difficult. Consider the following binary tree:

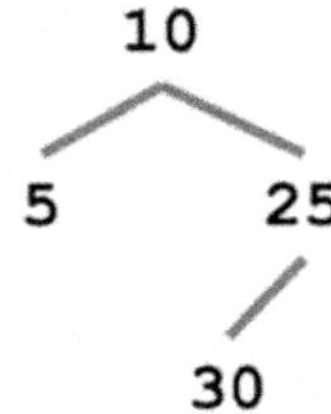

If we decide to write its level order traversal to a file, then that file will have

```
10 5 25 30
```

Even if know that this file stores level order traversal of a binary tree, it is not possible to know whether `30` was child of `5` or `25`, or whether `5` is the left or right child of `10` (what if `10` does not have a left subtree).

The logic works for complete binary tree because there is no gap in between and there is a one-to-one mapping between tree and array.

Solution: One of the solutions is to use the fact that a binary tree can be unambiguously constructed when both preorder and inorder traversals (or postorder and inorder traversal) of that tree are given.

Following is the logic of both the operations, writing binary tree to a file and creating the tree from that file.

Writing binary tree to a file

- Traverse given tree in preorder and write this traversal to the file.
- Insert a new line after the above traversal.
- Traverse given tree in inorder write this traversal to the file.

Creating tree from the File

- Read first line, it is preorder traversal of tree.
- Read next line, it is inorder traversal of tree.
- Construct the tree from preorder and inorder traversals.

Note that the tree cannot be constructed when only preorder and postorder traversals are given.

Using level-order traversal

From above discussions, we know that if a binary tree is either complete or almost complete binary tree (discussed next), we can just store its level order traversal in an array and construct the entire tree from that traversal. The level order traversal of the binary tree in Figure 8.2 is stored in the file as

```
11 10 7 9 5 6 4 8 2 3 1
```

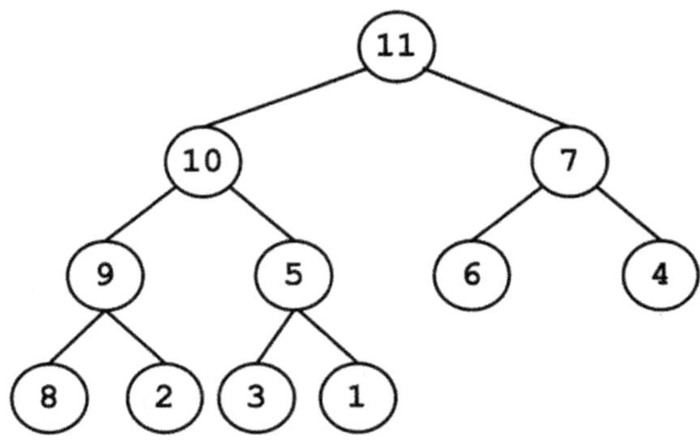

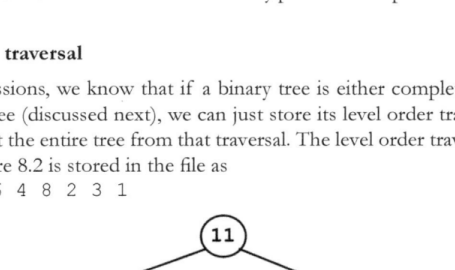

Figure 8.2: Almost Complete Binary Tree

When we read back from the file, we know the first element is root, then next 2 elements are from level-1 in left-to-right order, next 4 elements from level-2 in left-to-right order, and last 4 elements are the starting 4 elements (starting from left side) of level-3. Hence we can easily construct the tree.

But the tree given in the question is neither complete nor almost complete. There may be missing nodes in any (or all) level(s). Left child of 30 is missing in the binary tree of Figure 8.3.

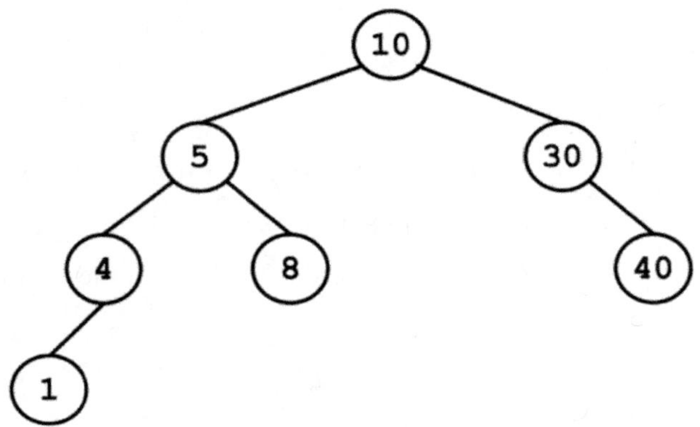

Figure 8.3: Binary tree (Not Almost Complete)

Following algorithm can be followed to persist any binary tree:

- Traverse the tree in level order and write each node.
- Wherever the node is missing, write some stub indicating empty position. If tree node has only positive integers, then `-1` can be the stub that indicates empty position.

For tree given in Figure 8.3, the data stored in the file is
`10 5 30 4 8 -1 40 1 -1 -1 -1 -1 -1 -1 -1`

We are actually storing the tree shown in Figure 8.4

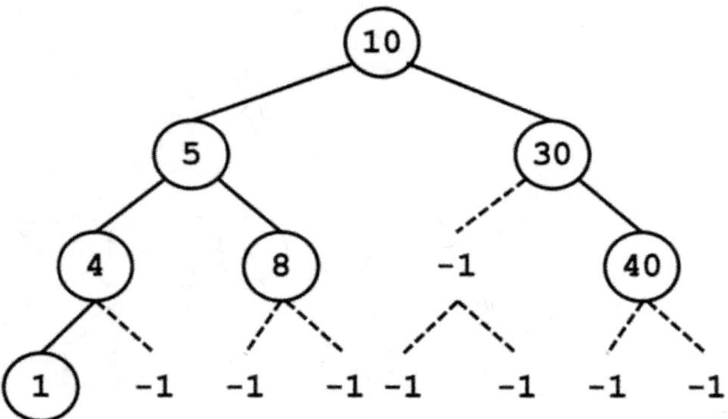

Figure 8.4: Storing Binary tree

Note that we are considering all places where a node could have been if the given tree is a complete binary tree. The values may either represent the actual data of node or a stub value indicating missing node. Nodes with value `-1` are dummy nodes inserted as place holders.

While reading from the file, create the `Node` of binary tree for actual values and ignore the `-1` values found in the file.

Following code writes the tree to an array and then, later, we create the same tree back from that array. We are avoiding the File I/O for simplicity.

```
void treeToArrayRec(Node* r, int* arr, int pos)
{
  if(r == NULL){ return; }
  arr[pos] = r->data;

  if(r->left != NULL)
    treeToArrayRec(r->left, arr,2*pos+1);

  if(r->right != NULL)
    treeToArrayRec(r->right, arr,2*pos+2);
}
void treeToArr(Node* root, int* arr, int maxNodes)
{
  // INITIALIZE ALL VALUES IN ARRAY TO -1
  for(int i=0; i<maxNodes; i++)
    arr[i] = -1;
  treeToArrayRec(root, arr, 0);
}
```

`treeToArray` initialize entire array to `-1`. `treeToArrayRec` is a recursive function much like preorder traversal. It traverses the given tree in preorder traversal and write nodes to their corresponding positions in array `arr`. If array is not pre-allocated, we can create the array on heap inside function `treeToArray`, size of array can be determined from height of the tree. Positions that does not have corresponding node, retain their initial `value -1`.

Next is the code that receives an array and construct the corresponding binary tree

```
void arrToTreeRec( Node* r,int* arr,int n, int pos)
{
  if(r==NULL || arr==NULL || n==0){ return; }

  // SETTING LEFT SUBTREE
  int newPos = 2*pos+1;
  if(newPos<n && arr[newPos] != -1)
  {
    r->left=new Node(arr[newPos]);//USE malloc IN C
    arrToTreeRec(r->left, arr, n, newPos);
  }
  // SETTING RIGHT SUBTREE
```

```
  newPos = 2*pos+2;
  if(newPos < n && arr[newPos] != -1)
  {
    r->right=new Node(arr[newPos]);//USE malloc IN C
    arrToTreeRec(r->right, arr, n, newPos);
  }
}
Node* arrayToTree(int* arr, int n)
{
  if(arr==NULL || arr[0]==-1)
    return NULL;

  Node* root = new Node(arr[0]); // USE malloc IN C
  arrToTreeRec(root, arr, n, 0);
  return root;
}
```

Almost complete Binary Tree

An almost complete binary tree of height `h` is complete up to height `h-1` and the last level has all its nodes on the extreme left side without any gap in between. All the trees shown below are almost complete binary trees:

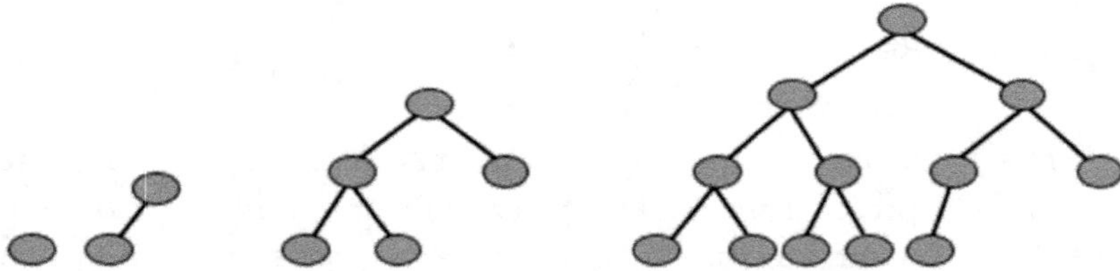

Below trees are not almost complete.

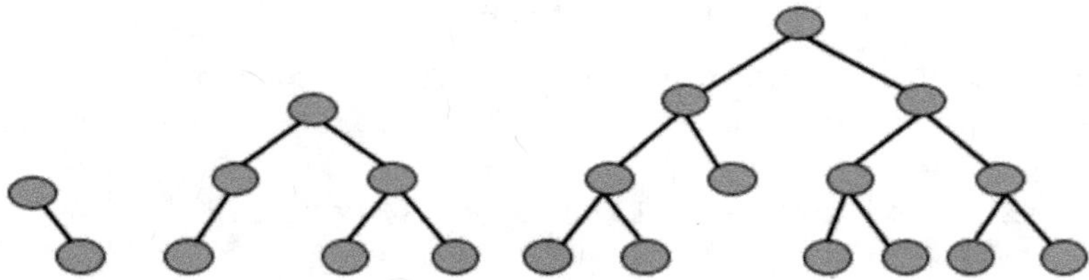

Note that a complete binary tree is also almost complete, but vice-versa is not correct.

Nodes of an almost complete binary tree can be canonically labeled in the level order traversal from left to right as shown below because there is no gap. Ability to canonically label nodes make it possible to store an almost complete binary tree in an array as shown in Figure 8.1.

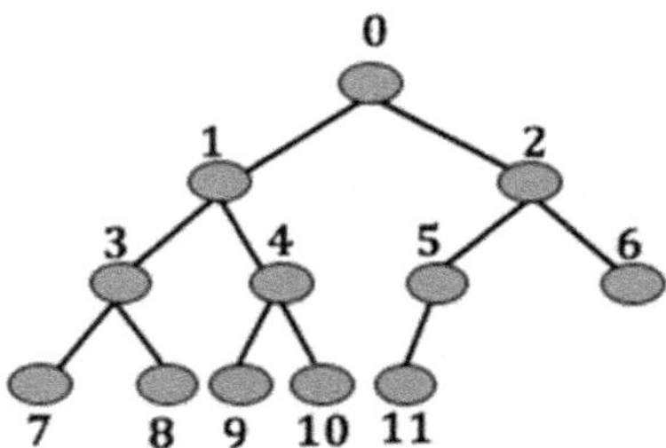

Another definition of an Almost complete binary tree can be, "*An almost complete binary tree is a tree made of first x nodes of a canonically labeled complete binary tree*".

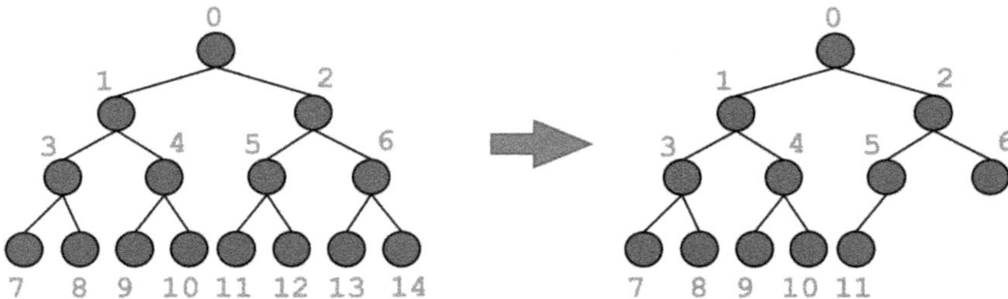

If last three nodes of the complete binary tree on the left side (in above diagram) are removed, we get the almost complete binary tree on the right side. For a tree to be almost complete, if the node at number x is removed, then all nodes at number greater-than x are also removed.

Assignment 8.1: Write a function that receives a pointer to the root of a binary tree and checks if that tree is an almost complete binary tree or not.

Priority Queue

Sometimes, we want to get the element with maximum (or minimum) value from the collection.

For example, consider a hospital where a patient needs to take a token before he is seen by the doctor. Tokens are given according to the emergency. If a patient's situation is critical, he is given a token with a large number. The Doctor always sees the patient with the maximum token number first. If patients are stored in an array, then they will always be removed from the array in decreasing order of token numbers.

This is a typical Priority Queue. The main operations in a priority queue are:

- Insert a new element in the queue.
- Remove an element with maximum (or in some cases minimum) priority.
- Check if the queue is empty or not.

- Get the highest priority element, without removing it from the queue.
- Find the size (total number of elements) of the queue.

There can be multiple approaches to implement a priority queue data structure. We can either use an array, linked list or Binary tree to store the data. Once we decide on the data structure, we can then decide whether we will maintain the order while inserting the element or while removing the element.

For example, if we are storing the data in an array, then one way is to always insert a new element at the end, and while removing, first find the element with max (or min) value and then remove it by shifting other elements toward the left. In this arrangement, insertion always takes constant time, but removing an element is a O(n) time operation.

Another option is to always maintain the array in sorted order, a new element is inserted at its appropriate position. This makes insertion an O(n) time operation, but elements can be removed in constant time.

Similar options can be used with data being in a linked list. If we maintain an ordered linked list, then insertion will be O(n) time operation, but removing the maximum (or min) element may just need to remove the head node (constant time). For an unordered list, insertion can always be at head, and removing an element will be O(n) time.

We may also choose BST to store the values, in which case, all the operations will take O(lg(n)) time. Obviously, we are talking about balanced binary search tree.

A heap data structure provides the implementation of priority queue and allows elements to be inserted and the smallest (or largest) element to be found and/or removed. There are many concrete implementations of a heap data structure, like Binary heap, Fibonacci heap, Binomial heap, Leonardo heap, etc. Binary heap is the most popular of them all, let us discuss it in detail:

Binary Heap Data structure

In a binary heap we can find the maximum (or minimum) element in constant time and can perform other operations in O(lg(n)) time. There are two types of binary heap:

1. Max-Heap
2. Min-Heap

A max-heap is an almost complete binary tree where, value at each node is greater than values of all the nodes below it.

Similarly, a min-heap is an almost complete binary tree where, value at each node is less than values of all the nodes below it.

Because we are dealing with almost complete binary trees, the implementation of both max-heap and min-heap may use array to store the actual values, those arrays represent an almost complete binary tree. Figure 8.5 shows both max-heap and min-heap in both tree and array representations.

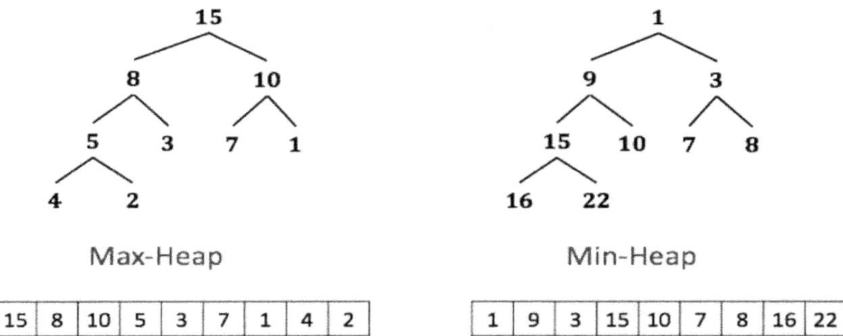

Figure 8.5: Max-heap and min-heap in tree and array representations

Note that an array being a max-heap or min-heap does not mean that it is sorted. However, an array sorted in ascending order is a min-heap and an array sorted in descending order is a max-heap. The array being sorted is a sufficient condition, and not a necessary condition.

Heap definition is applicable on the hierarchical relation (the way it is in binary tree) of values, rather than their linear relation (the way it is in array).

Largest element in a max-heap is always at the root. Similarly, the smallest element of min-heap is always at the root (first element in the array), but we cannot say that second element in the array is the second smallest.

Out of the below two arrays, `arr1` is neither a min-heap nor a max-heap. But, `arr2` is a min-heap.

```
int arr1[] = {15, 19, 10, 7, 17, 16}
int arr2[] = {5, 14, 10, 27, 17, 11}
```

All root-to-leaf paths in a max-heap gives elements in non-increasing order and all root-to-leaf paths in a min-heap gives elements in non-decreasing order. Values in the left and right subtrees are unrelated and can be in any order.

If there are n elements in a heap, $\lceil n/2 \rceil$ of them are leaf nodes.

Question 8.3: Given an array holding an almost complete binary tree. Check if it is a max-heap or not.

Solution: The simplest solution is to recursively check, for each node, if that node is greater than its direct left and right children.

```
bool isMaxHeap(int *arr, int n, int r)
{
  // LEAF NODE NEVER VIOLATE HEAP PROPERTY
  if(r > n/2){ return true; }

  // LEFT SUBTREE VOILATE HEAP
```

```
  if(arr[r]<arr[2*r+1] || !isHeap(arr, n, 2*r+1))
    return false;

  // RIGHT SUBTREE EXIST AND VOILATE HEAP
  if( 2*r+2 < n && (arr[r] < arr[2*r+2] ||
      !isHeap(arr, n, 2*r+2)))
    return false;

  return true;
}
```

This solution takes `O(n)` time because we are visiting each node only once.

Heapify operation

Given an array representing a max-heap. Let there be one (and only one) element in the array that violates the heap property. We want to fix that element. For example, if the given array is

`{16, 4, 10, 14, 7, 9, 3, 2, 8, 1}`

Then 4 is the only element because of which this array is not a heap. If 4 can be fixed, it will become a heap.

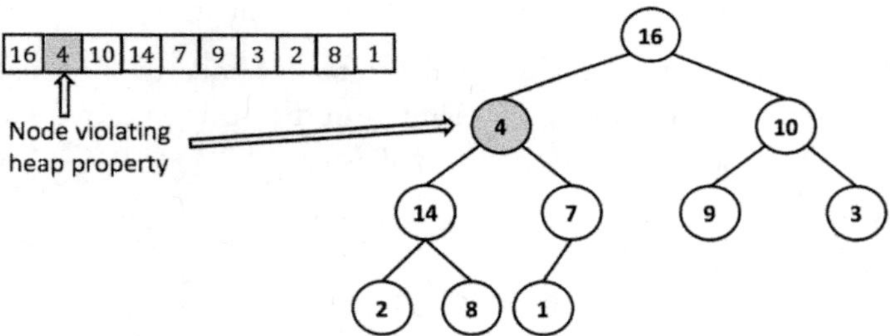

For fixing this, we just need to analyse subtree rooted at 4. The problem, now reduces to fixing the heap property of a (sub)tree where the property is violated only at the root node. This may be a simpler and more general problem to solve. **Heapify** operation fix this specific irregularity.

Compare root of subtree that violates heap property with values of its left and right child. Swap the maximum of these three values with the root.

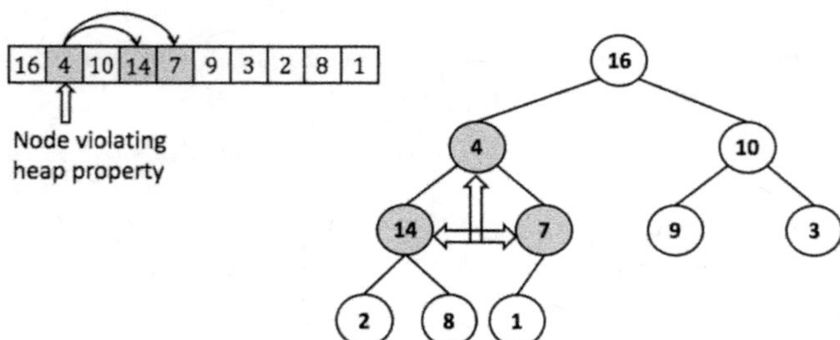

Data is stored in array and actual operations are also performed on array, but in our mind, we should visualize it happening in a tree.

In given array, compare values at indices, `r`, `2r+1` and `2r+2` and swap the maximum of these three values with value at index `r`. If no swapping happens (i.e. value at `r` is maximum), just stop there, else continue the same with node whose value is swapped with `r` (in this case `14`) as shown below. After this swapping, the array becomes `{16, 14, 10, 8, 7, 9, 3, 2, 4, 1}`

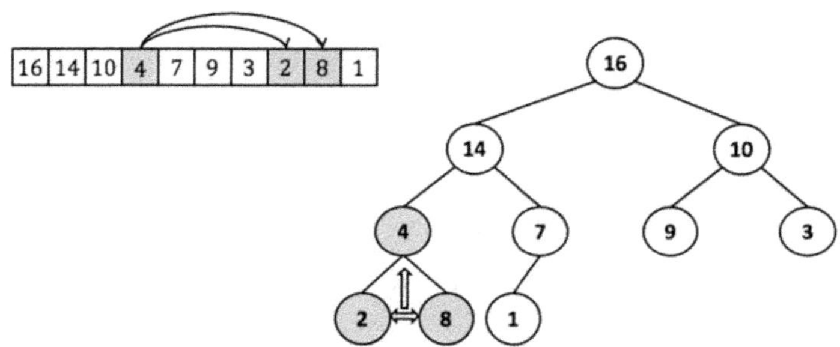

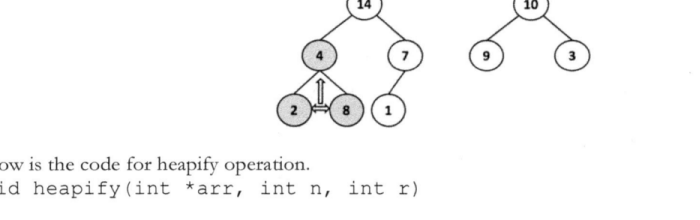

Below is the code for heapify operation.

```
void heapify(int *arr, int n, int r)
{
  if(root <= n/2) // LEAF NODES ALREADY HEAPIFIED
  {
    int max=r,left=2*r+1,right=2*r+2;

    if(arr[left] > arr[max])
      max = left; // LEFT CHILD IS MAXIMUM

    if(right < n && arr[right] > arr[max])
      max = right; // RIGHT CHILD IS MAXIMUM

    if(max != r)
    {
      swap(&arr[r], &arr[max]);
      heapify(arr, n, max);
    }
  }
}
```

Code 8.1

`swap` function is given in Code 2.3. The algorithm takes constant time at each level. In worst case, we may have to go thru all the levels and take `O(lg(n))` time. It also takes `O(lg(n))` extra memory because of recursion. Space complexity can be easily brought down to constant by removing recursion.

Heapify operation on a min-heap compares and find the minimum element out of the three and swap it with the root. Everything else remains the same.

Converting an array to heap

Given an array of random numbers, shuffle the elements, such that this array becomes a max-heap. If the given array is:
`{11, 4, 10, 14, 7, 9, 3, 2, 8, 1}`

One of the multiple max-heaps that can be created from these numbers is
`{14, 11, 10, 8, 7, 9, 3, 2, 4, 1}`

Remember, this is just one of the many max-heaps that can be formed from given array. We know that sorted array is also a heap, the most straight-forward way of converting an array into a max-heap is to sort it in decreasing order. Sorting takes `O(n.lg(n))` time. Let us see an algorithm that takes less than `n.lg(n)` time.

There is a linear time algorithm to convert an array into a heap. Traverse all non-leaf nodes in backward order and call `heapify` (Code 8.1) for all of them. Heap is an almost complete binary tree and in an almost complete binary tree, last $\lceil n/2 \rceil$ elements are leaves. Ignore last $\lceil n/2 \rceil$ elements and call `heapify` function for all others as shown below.

```
void buildHeap(int *arr, int n)
{
  for(int i=n/2; i>=0; i--)
    heapify(arr, n, i);
}
```

It can be proved that running time of above algorithm is `O(n)`. We are able to build a heap out of a random array in linear time.

Inserting an element in the heap

To insert a new element in the heap, append the new element at the end of heap. In array implementation, store the new element at `arr[n]` (assuming array has enough space) and in node implementation, add a new node in the almost complete binary tree.

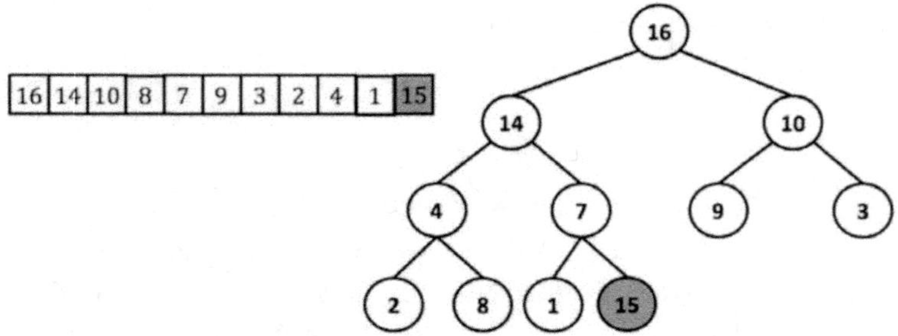

This newly added element may destroy the heap property. Fix the heap property by comparing the newly inserted element with its parent, and moving it upward if needed as shown below:

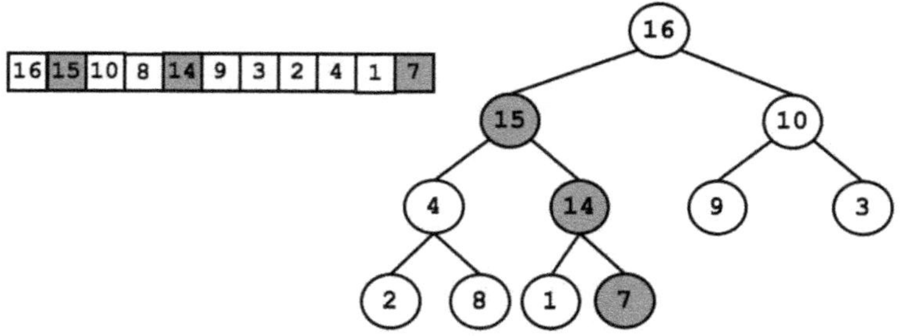

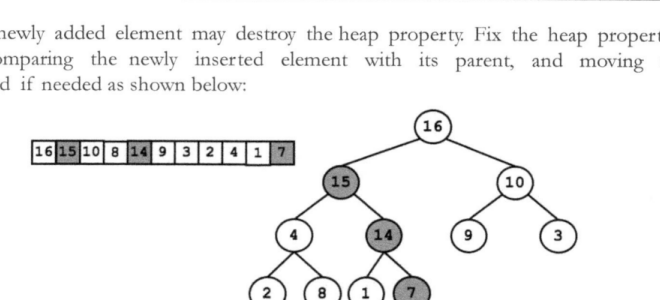

15 is compared with its parent and moved up till parent's value is less than 15. 7 and 14 are moved down and 15 is moved up. The heap property is restored. Finally, the heap looks like:

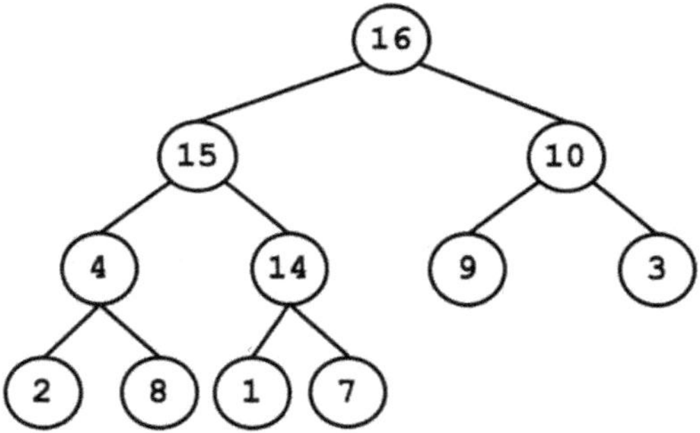

Following code inserts an element in the heap. It assumes that there is sufficient space in the array. There are n elements in the array, but actual size of array is large enough to hold the newly inserted element.

```
void insertInHeap(int *arr, int n, int value)
{
  arr[n] = value; // APPEND ELEMENT
  while(n > 0)
  {
    int parentIndx = (n-1)/2; // PARENT NODE INDEX
    if(arr[parentIndx] < arr[n])
      swap(&arr[parentIndx], &arr[n]);
    else
      break; // STOP WHEN ROOT IS LARGER
    n = parentIndx;
  }
}
```

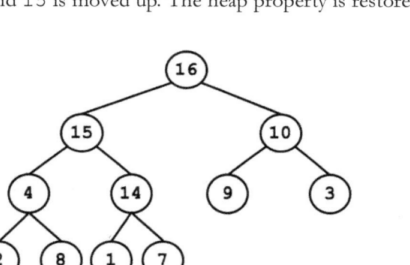

In worst case, the newly inserted element can move up to the root, the total time taken by this operation in the worst case is `O(lg(n))`.

Deleting element from a heap

Deleting an element from a max-heap means deleting the maximum element and deleting element from a min-heap means deleting the minimum element.

In both the cases, the root element is deleted. The algorithm to delete the root element from a heap is,

1. Swap root element (`arr[0]`) with last element (`arr[n-1]`).
2. Decrease size of heap by one (`arr[n-1]` is in the array, but it is not part of the heap. Size of array and heap are different).
3. At this point, heap property can only be violated at the root. Call `heapify` at root to restore the heap property.

Following pictures show step-by-step process of deleting an element from a heap.

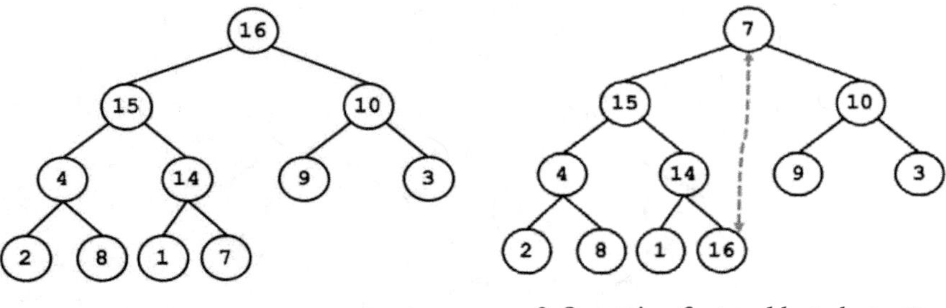

1. Original Heap **2. Swapping first and last elements**

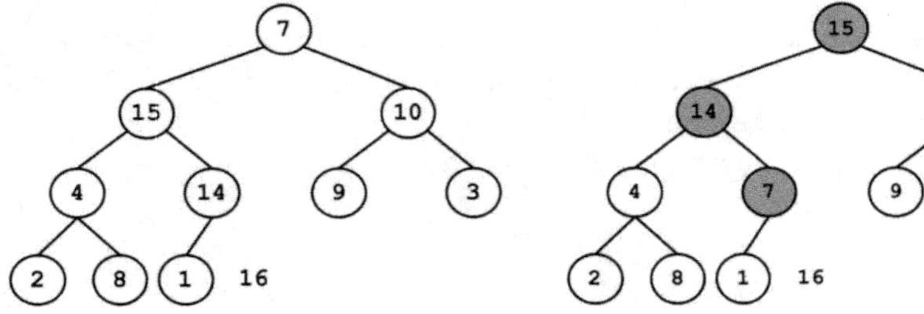

3. Remove last element from heap **4. Heapify the root node**

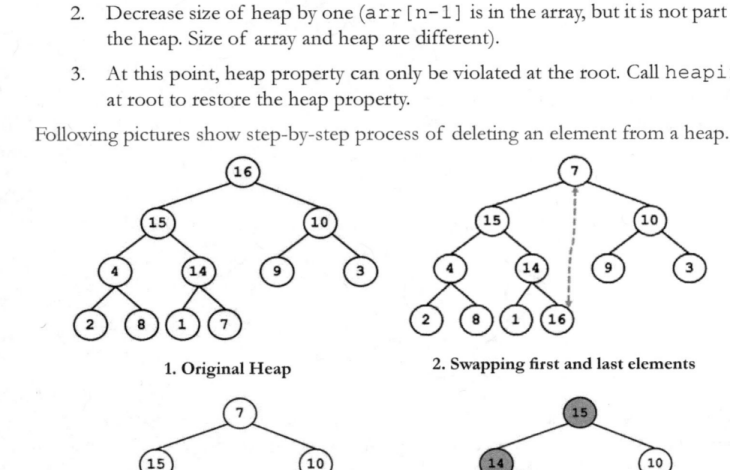

The code is straight forward as shown below.

```
// SIZE OF HEAP WILL BE DECREASED BY 1.
// REMOVED ELEMENT WILL BE AT arr[n-1].
void removeFromHeap(int *arr, int n)
{
  if(n<=0){ return; }

  // SWAP FIRST AND LAST VALUE
```

```
  swap(&arr[0], &arr[n-1]);
  heapify(arr, n-1, 0);
}
```

Other heap data structure

The heap implementation we saw in the previous section is called, Binary Heap. This heap is primarily used in heap sort algorithm. In general, a heap is an implementation of a Priority queue, an element with the highest priority resides at the root (or first position) and gets deleted first.

An array or a linked list sorted in decreasing order can also be used as a priority queue. In this case, operations like inserting and removing elements is more time consuming (linear time). A Heap is different from an array in its implementation of priority queue because it is just partially sorted and operations take less time.

There are other ways also to implement a heap data structure, the popular ones being Binomial heap and Fibonacci heap, both, in a way are extensions of Binary heap.

Use of Heap in Heap sort

Heap sort uses binary heap to sort an array. Essentially, it uses the property that maximum element in a max-heap is at root. Below is the algorithm of heap sort:

1. Build max-heap from array (in-place).
2. FOR i = N-1 DOWN TO 0
    a. Swap maximum element with element at position i. i.e. swap first and last elements of heap.
    b. Decrease size of heap. i.e. remove last element from heap.
    c. Call heapify on root, `arr[0]` and restore heap property of the array.

Let us take an example and see heap sort happening step-by-step.

Input Array: `{9, 7, 10, 20, 16, 14}`

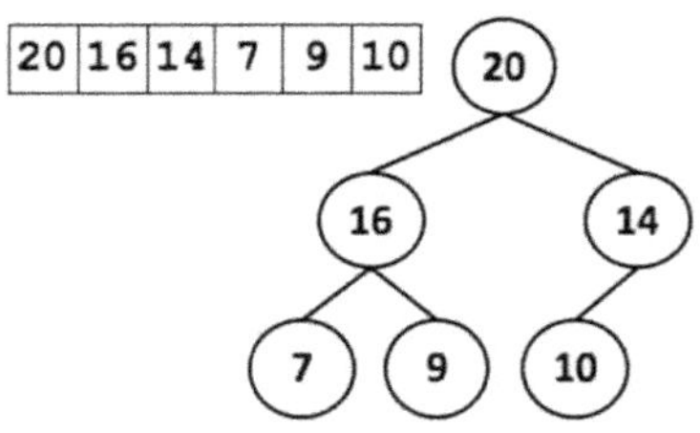

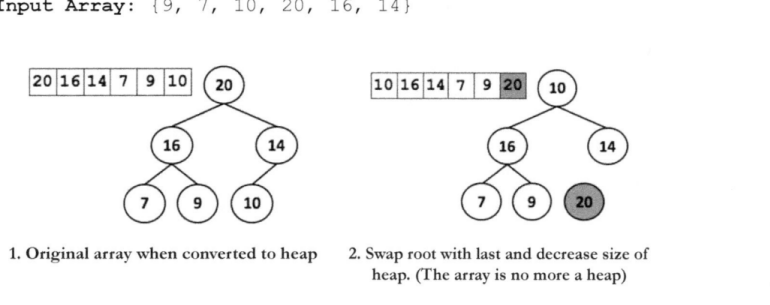

1. Original array when converted to heap

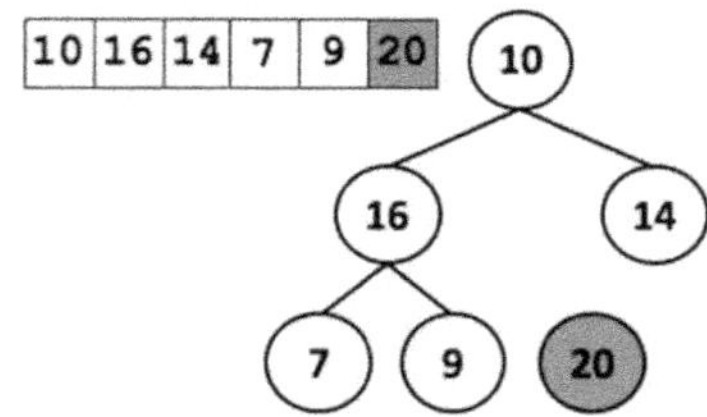

2. Swap root with last and decrease size of heap. (The array is no more a heap)

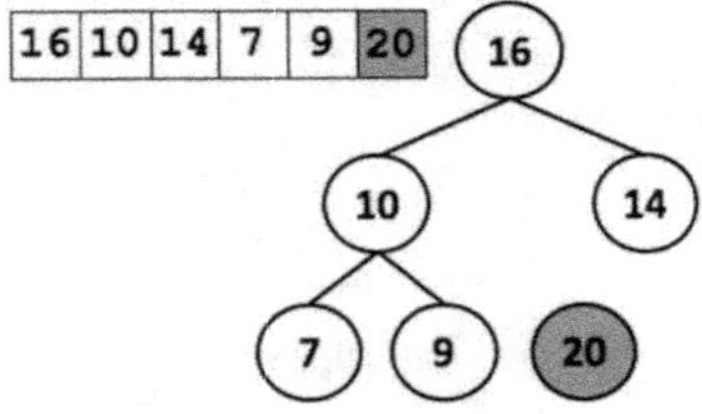

3. Heapify at root to make it heap

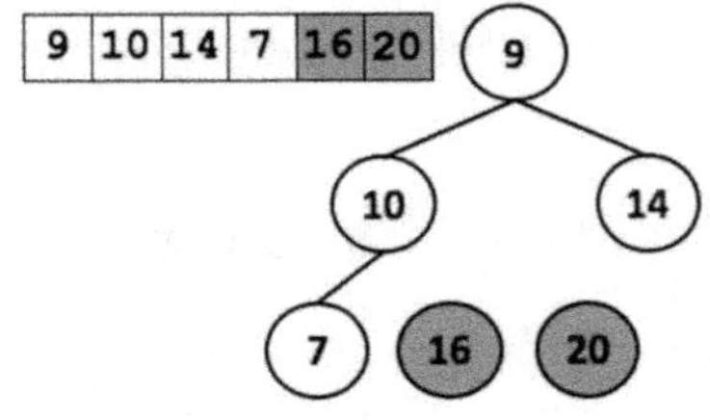

4. Swap first & last and decrease heap size.

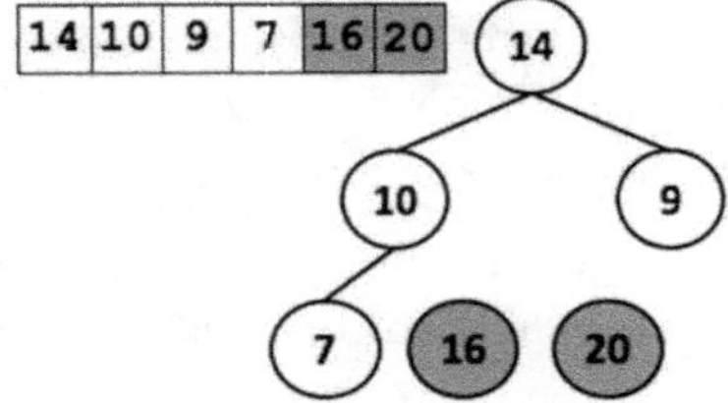

5. Heapify at root

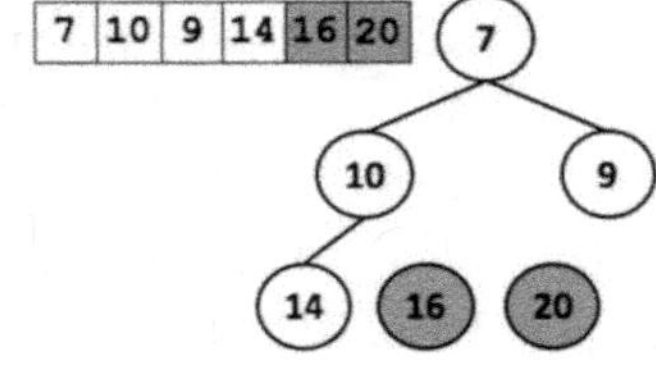

6. Swap first & last element

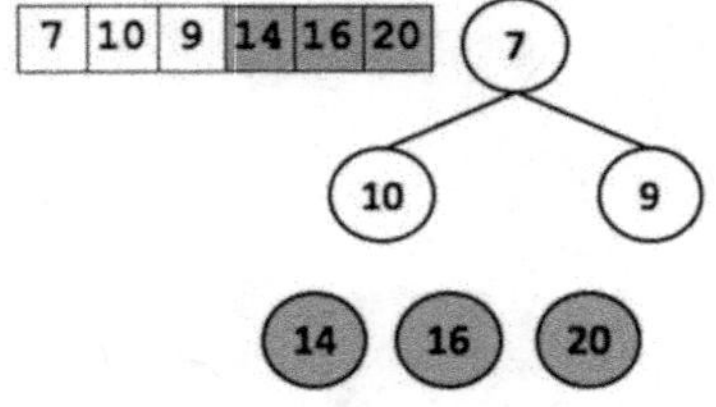

7. Remove last element from heap by decreasing heap size

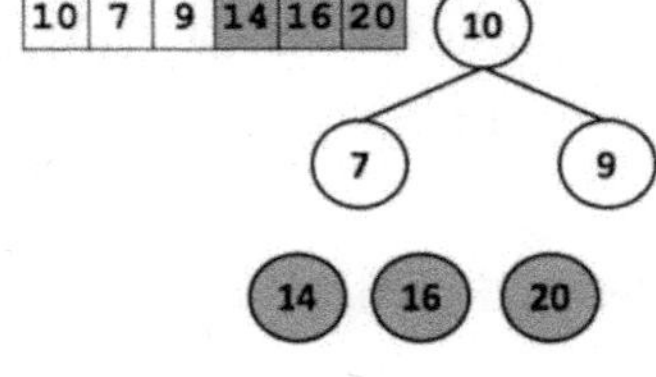

8. Heapify at root.

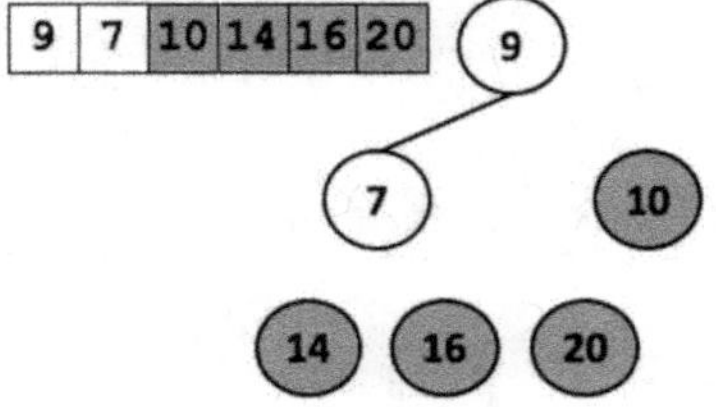

9. Swap first & last and decrease heap size.

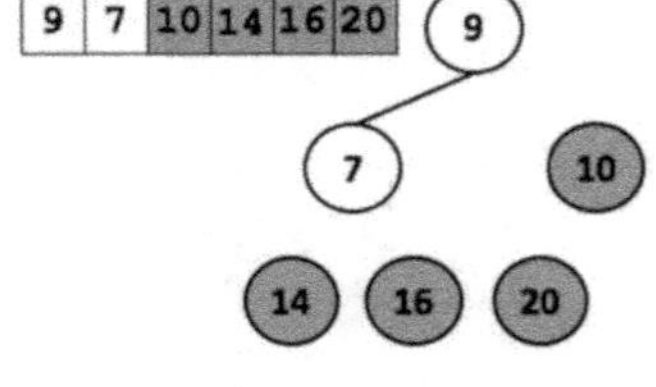

10. Heapify at root (does not change)

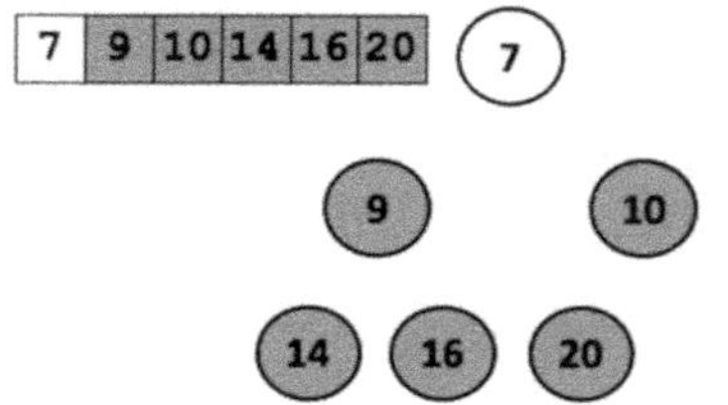

Stop when array size reduces to one. The array is now sorted in increasing order.

Below is the actual code of Heapsort.

```
void heapSort(int *arr, int n)
{
  buildHeap(arr, n);   // CONVERT THE ARRAY TO HEAP.

  for(int i = n-1; i >= 1; i-- )
  {
    swap(&arr[0], &arr[i]); //SWAP 1st & LAST ELEMENT
    heapify(arr, i, 0);
  }
}
```

`buildHeap` is a linear-time operation and `heapify` takes `O(lg(n))` time. We are doing `heapify` `n-1` times, intuitively heap sort takes `O(n.lg(n))` time.

Heap as a problem-solving tool

Heap sort is rarely used because quick sort and merge sort gives better performance than heap sort. But concept of heap is a powerful problem-solving tool used in many questions not related to sorting. In this section, we discuss some interview questions that use heap data structure.

Question 8.4: Given `m` sorted arrays, each of size `n`. Merge all of them and print their sorted output. For example, if `m=3, n=4` and input arrays are:

```
{1, 6, 12, 20}
{3, 5,  7,  9}
{2, 8, 25, 49}
```

Then output should be: `1 2 3 5 6 7 8 9 12 20 25 49`

Solution: Let us take input in a two-dimensional array of order `m*n`. Each row of this 2-dim array represents one array and is sorted in ascending order.

```
int arr[3][4] =  { {1, 6, 12, 20},
                   {3, 5,  7,  9},
                   {2, 8, 25, 49}};
```

The simplest solution is to create a one-dimensional output array of size `m*n`, copy all arrays into this array appended one after another and then sort the output array. But, it will take `O(mn.lg(mn))` time and `O(mn)` extra memory.

Using min-heap, we can print sorted output in `O(mn.lg(m))` time using `O(m)` extra space. The algorithm is as below:

1. Create a min heap of size `m` and insert the first element of each array into this heap. Also, insert the array number to which each element in the heap belongs (along with the element).
2. Repeat the below steps `m*n` times.
   a. Print the minimum element of heap (root element) to the output.
   b. Replace current root element with the next element of the same array to which the root element belongs. If that array does not have any more elements, then put infinity at the root.
   c. Heapify the root.
3. Stop when element at root is Infinity.

Each element of the heap is a structure, having three fields, data, array-number that this data belongs to and index of next element in that array

```
struct MinHeapNode
{
  int data;
  int x;     // ARRAY THAT THIS ELEMENT BELONGS TO
  int y;     // INDEX OF NEXT ELEMENET TO BE PICKED
};
```

For below code to work, we need to modify the `buildHeap` and `heapify` functions defined in the last section.

```
void swap(MinHeapNode *a, MinHeapNode *b)
{
  MinHeapNode temp = *a;
  *a = *b;
  *b = temp;
}
void heapify(MinHeapNode *arr, int n, int root)
{
  if(root <= n/2) // LEAF NODES ALREADY HEAPIFIED
  {
    int min=root, left=2*root+1, right=2*root+2;

    if(arr[left].data < arr[min].data)
      min = left; // LEFT CHILD IS MAXIMUM

    if(right<n && arr[right].data<arr[min].data)
      min = right; // RIGHT CHILD IS MAXIMUM
```

```
    if(min != root)
    {
      swap(&arr[root], &arr[min]);
      heapify(arr, n, min);
    }
  }
}

void buildHeap(MinHeapNode *arr, int n)
{
  for(int i=n/2; i>=0; i--)
    heapify(arr, n, i);
}

void printMerged(int arr[ ][N], int m)
{
  // CREATING HEAP
  MinHeapNode *heap = ( MinHeapNode*)malloc(m *
                        sizeof(MinHeapNode));

  // STORE FIRST ELEMENT OF EACH ARRAY IN HEAP
  for(int i=0; i<m; i++)
  {
    heap[i].data = arr[i][0];
    heap[i].x = i; heap[i].y = 1;
  }
  buildHeap(heap, m);

  for (int cnt = 0; cnt < N*m; cnt++)
  {
    MinHeapNode root = heap[0];
    printf("%d", root.data);
    if(heap[0].y < N)
    {
      heap[0].data = arr[root.x][root.y];
      heap[0].y++;
    }
    else
    {
      heap[0].data = INT_MAX;
    }
    heapify(heap, m, 0);
  }
  free(heap);
}
```

Assignment 8.2: A constant stream of numbers is coming. At any point of time, you have to return k[th] largest element seen till now. The question can be asked to print all the k largest elements seen till now.

Hint: Use heap of size k.

Assignment 8.3: Given a two-dimensional matrix, with rows and columns individually sorted in non-decreasing order. Write code to print all the elements of the matrix in sorted order.

Question 8.5: Each element in an array of size `n` is at most `k` positions away from its final position in the sorted array. How will you sort this array?

Solution: We know that, Insertion sort will take `O(nk)` time and constant extra space in the worst case to sort such an array.

This time complexity can be improved if we use a heap data structure. Below is logic to sort such an array:

1. Create a min-heap of size `k+1` and add first `k+1` elements to that heap. These `k+1` positions in the input array are now empty and can be used for storing results.
2. Repeat the below steps `n` times:
    a. Remove (minimum) an element from the heap and put it back at first empty position in the array.
    b. Add a new element to the heap from remaining elements. Position of the newly added element is now empty. If there is no element in the array, then add infinity to the heap.
    c. Heapify to restore the the heap again.

Total time taken by above algorithm is `O(n.lg(k))`. You may choose to sort last `k` elements using insertion sort.

In most cases, there is space-time trade-off while choosing between data structures and algorithms. In this case too, we use `O(k)` extra memory and reduce time taken from `O(nk)` to `O(n.lg(k))`.

Question 8.6: Given a string in which characters may be repeating. Write code to rearrange characters such that adjacent characters are never the same. If such an arrangement is not possible, print "`IMPOSSIBLE`". Below are the sample inputs and corresponding output:

INPUT	**OUTPUT**
aaabc	abaca
aaaxx	axaxa

aaa	IMPOSSIBLE
aab	aba

Solution: A Greedy approach of putting character with the highest frequency first works in this case.

- Traverse the string and create a count table of frequency of each character in the string.
- Print the character with highest frequency and character with next highest frequency alternately.
- Also, keep updating the frequency of an element when it is printed.

If we store frequencies in an array, searching for element with max frequency takes `O(n)` time. Instead, we can keep them in a max-heap. Below is the detailed algorithm:

1. Create a max-heap. Each node stores the character and its frequency. Maintain heap property on frequency, i.e. character with max frequency appears at the root.
2. While max-heap is not empty
    a. Delete element from max-heap and print the character in it.
    b. Decrement the frequency of deleted element.
    c. Insert the element stored in `temp` variable (if any) into the heap.
    d. Assign deleted element to the `temp` variable.
3. If `temp` has an element with frequency `>1`, print `IMPOSSIBLE`.

Question 8.7: Given an array of `n` random numbers, and two integers `x` and `y`. Find sum of all elements that are greater than x^{th} smallest element and less than y^{th} smallest element (`x<y<n`). For example, if given input is

```
int arr[] =  {18, 7, 56, 9, 6, 4, 8, 13, 12, 5},
             x = 3,  y = 7
```

If we sort the given array then it becomes:

```
{4, 5, 6, 7, 8, 9, 12, 13, 18, 56}
```

3rd smallest element is `6` and 7th smallest element is `12`. Sum of all element between `6` and `12` is `7+8+9 = 24`.

Solution: Brute force way of solving given problem is to first sort the array and then add elements from index `x` to `y-2` (both included). This takes `O(n.lg(n))` time.

Alternately, we can create a min-heap of all elements in the array in `O(n)` time. Remove first `x` element from heap and then remove and add next `(y-x-1)` elements. This approach takes `O(n+y.lg(n))` time.

Question 8.8: Given an array of `n` single digit integers. Form two numbers by appending these digits, such that sum of these two numbers is minimum. For example, if the array is
`{3, 6, 1, 2, 0, 8}`

The two numbers are `026` and `138` with their sum as `164`. Any other numbers formed using these digits is greater than or equal to `164`.

If you look at it closely, what we are doing is, get minimum element from array and append it to two numbers alternately. Minimum element in the array is 0, it is appended to the first number. Next minimum, 1 is appended to the second number, next minimum 2 is again added to the first number, it becomes 02, and so on. The logic is:

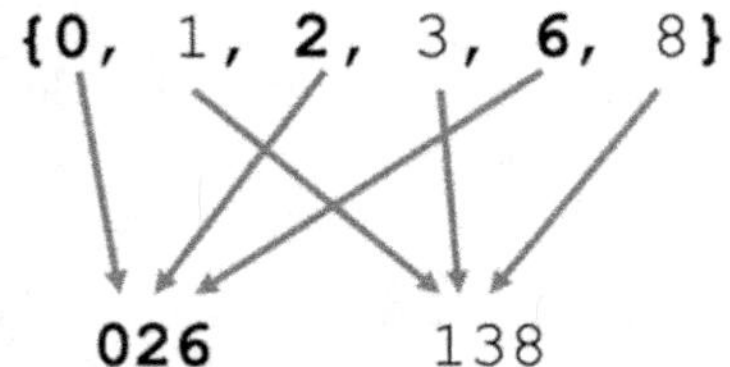

1. Create min-heap from all elements of the array.
2. While heap is not empty
   a. Remove two elements from heap and store them in variables `x` and `y`.
   b. Append `x` to first number and `y` to second number.

You must have got the hint, heap as a data structure has many more applications than just heap sort. You can find umpteen number of interview questions that has nothing to do with sorting but still use heap data structure.

INTERVIEW TIP

Whenever our logic requires us to find maximum or minimum from array multiple times, think o using a heap. If the logic demands finding k largest or k smallest elements in a large pool of data, thin

If the logic demands to find k largest or k smallest elements in a large pool of data, think of using a heap.

Leonardo Heap

Leonardo numbers, defined as below are very similar to Fibonacci numbers

$$L(n) = 1 \qquad\qquad if\ \ n = 0,1$$

$$= L(n-1) + L(n-2) + 1 \qquad\qquad if\ \ n > 1$$

`1, 1, 3, 5, 9, 15, 25, 41, 67, 109,`... are the Leonardo numbers. An important thing about Leonardo numbers is that

Any number, `n` *can be written as sum of* `lg(n)` *distinct Leonardo numbers.*

Leonardo tree is a binary tree defined recursively as below:

- L_0 = Single node
- L_1 = Single node
- L_n = A node whose left subtree is L_{n-1} and right subtree is L_{n-2}

Figure below shows some Leonardo trees. Every Leonardo tree represents a max-heap.

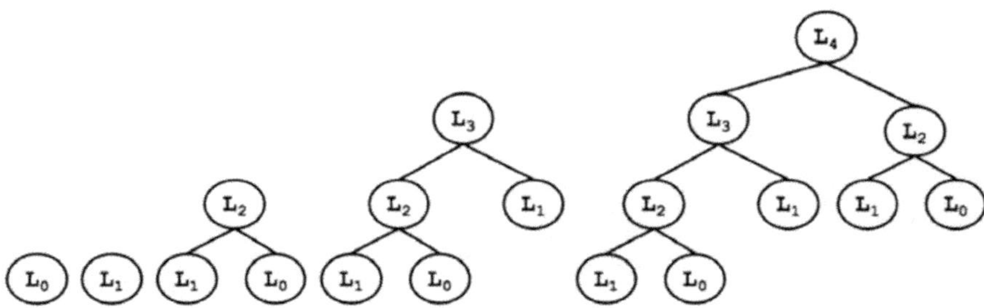

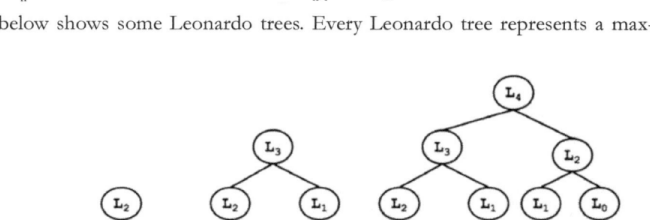

Leonardo heap is an ordered collection of Leonardo trees (max-heaps). Size of each tree is unique and is strictly decreasing. Values in root nodes of the trees are in ascending order from left to right. Figure given below represent a Leonardo heap. Notice that the maximum element is always at the root of the smallest heap (rightmost).

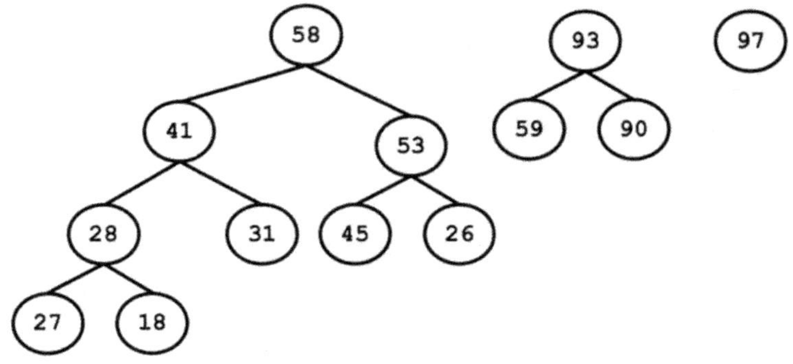

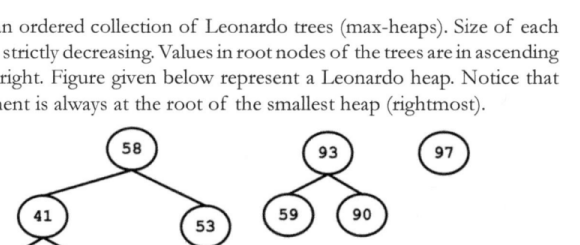

While inserting a new element in Leonardo heap we need to ensure that its properties are not disturbed. Time taken to insert a new element in Leonardo heap is between `O(1)` and `O(lg(n))`. Similarly time taken to delete an element from Leonardo heap also varies between `O(1)` and `O(lg(n))`.

Below is the algorithm to insert a new element in Leonardo heap:

- If last two trees in the heap are of adjacent order, merge them into a new tree.

- If order of last tree is more than 1, add a new tree or order 1.
- Else add a new tree of order 0.

Following diagram shows the sequence of operations performed to create Leonardo heap from the below array
`{5, 2, 6, 3, 12, 10}`

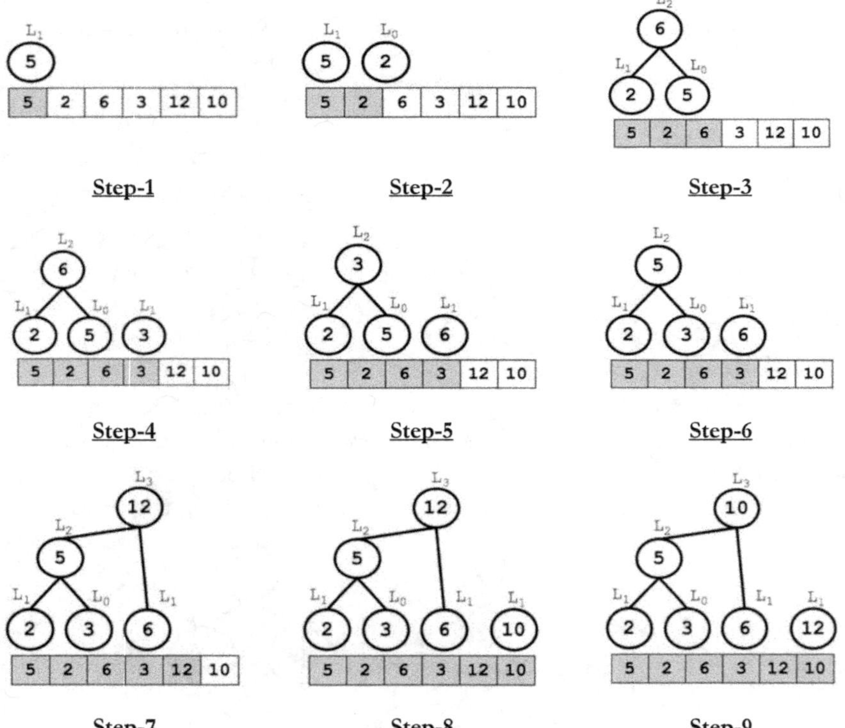

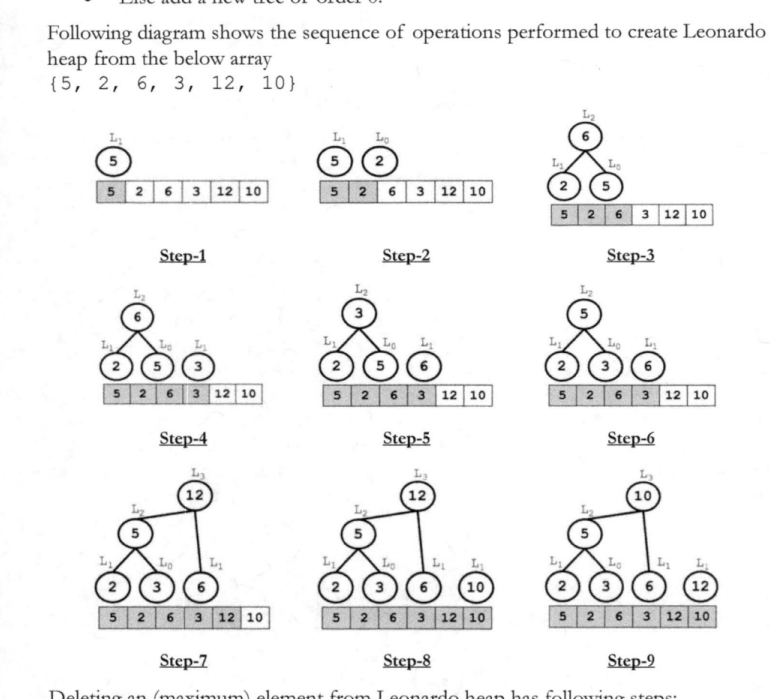

Deleting an (maximum) element from Leonardo heap has following steps:

- Remove the topmost node from the rightmost tree.
- If it has no children, do not do anything and return.
- Else, fix the left and right of two trees.

Chapter 9

Hashing

If you noticed, throughout the book, we are essentially trying to optimize following three operations:

1. **Insertion** (storing information in a data structure).
2. **Searching** (retrieving information stored in the data structure).
3. **Deleting** (removing the stored information from data structure).

And we are looking for a data structure where these three operations take minimal time. In an array, searching takes `O(n)` time or `O(lg(n))` time depending on whether it is linear or binary search. Deleting an element takes linear time because other elements need to be shifted. Insertion takes more time when the array is full because size of the array cannot be increased after allocation (irrespective of the memory area where the array resides). When array is on the stack frame of a function, it is not possible to increase its size at all, but when the array is on heap memory, its size can be increased by reallocating a new bigger memory and copying the contents of current array to the new location and deallocating the current array. In any case, its time consuming Inserting in a sorted array also takes linear time in worst case.

Linked list has an advantage when it comes to changing size dynamically, but it cannot be searched using binary search. We may use variations like jump search, but worst case of these are no batter than linear search. Insertion in a linked list can be constant-time operation if we always insert at the head (or tail), but removing an element requires searching for that element and hence take `O(n)` time.

A balanced binary search tree is a really good data structure where all the above three operations take `O(lg(n))` time.

Hashing is one step further where the aim is to make all the three operations take constant time. Hashing system stores data in a **Hashtable**. Location of an element in the Hashtable is computed using a constant time function called **Hash function**. The Hash function accepts a key and returns an address in the Hashtable table on which the value for that key can be stored. It may look easy, but a hash system can be very tricky to implement.

Hashing is a method to store and retrieve information and is used to speed up searching. However, hashing as a technique is used in many interview questions.

In a special arrangement, an array can be used as a Hashtable, where the Hash function can be as basic as address of `i` is `i` itseslf (the value of key `i` is stored at `arr[i]`.

Question 9.1: Given two strings `str1` and `str2`, print all the characters in `str2` that are not present in `str1`.

```
INPUT:
    char str1[] = "ritambhara";
    char str2[] = "technologies";
OUTPUT:
    c e g l n o s
```

For simplicity, assume strings to be of only small alphabets.

Solution: The brute-force solution takes `O(n²)` time each character of `str2` is searched in `str1`. Effectively, we are performing `O(n)` searches, this is where hashing comes to rescue.

Use two hash tables. Traverse string `str1` linearly and for each character, see if it is already present in the first hash, if not, add that character to the first hash table.

After adding all unique characters of `str1`, traverse second string and check, for each character of `str2`, if it is present in the first hash. If not, add it to the second hash. Later print contents of the second hash. If we do not use the second hash table, then those characters of `str2` which are not present in `str1` and repeat multiple times in `str2` get printed multiple times.

An integer array of size `26` (number of alphabets) can be used as a hashtable for this problem. We can use the same array (hashtable) for both the strings..

```
int hash[26] = {0};
```

Initially, all elements of `hash` are zero. `hash[i]` is set to `1` if `i`th character is present in `str1`.

```
char *s1 = str1;
while(*s1 != '\0'){
  hash[*s1-'a'] = 1;
  s1++;
}
```

Now traverse the second string and check if the current character is present in the hash, if not, set it to `-1`.

```
char *s2 = str2;
while(*s2 != '\0'){
  if(hash[*s2-'a'] == 0)
    hash[*s2-'a'] = -1; // ONLY IN str2
  else if(hash[*s2-'a'] == 1)
    hash[*s2-'a'] = 2; // IN BOTH str1 & STR2
  s2++;
}
```

same `hash` is used for both. If `i`th character is only present in `str2`, then `str1` does not change `hash[i]`. At the end, value of `hash[i]` indicates the following for the `i`th character:

- **-1 :** This character is only present in `str2` and not in `str1`.
- **0 :** This character is neither present in `str1` nor `str2`.
- **1 :** This character is only present in `str1` and not in `str2`.
- **2 :** This character is only present in both `str1` and `str2`.

At this point scan the hash and print characters that has `-1` value in hash.

```
for(int i=0; i<26; i++){
  if(hash[i] == -1)
    printf("%c", 'a'+i);
}
```

We earlier said that insert, delete and search are constant time operations in a hash table. It can be verified from this example.

The hash function in this question is as below:

```
int hash(char ch){
  return ch-'a';
}
```

But the hash function may not always be this simple.

Assignment 9.1: Given two string, print union and intersection of characters present in these strings.

INTERVIEW TIP

The idea of hashing is to have a direct access table, for most interview questions, where you need to use a hash, try to take an array and map index of array with the key. In practice, however, implementation of hash table (or hash map) is not this simple.

Question 9.2: Given an array of single digit non-negative integers. Write a function to print all the duplicates in the array. For example, if the array is

```
int arr[] = {1, 2, 7, 1, 4, 2, 8, 0, 1, 6, 2, 7};
```

Then the output should be `1 2 7`, because they are repeating in the array.

Solution: The brute-force solution is to traverse the array `10` times for `0` to `9`, in the ith traversal check if `i` is repeating in the array.

Another solution is to use a hashtable to store the frequency of each number. Since there can be `10` different numbers (from `0` to `9`), we need a hashtable of size `10` only. Below code takes an array and use it as a hash. The hash function that takes a number and return an address in the hashtable for that number is as simple as

```
int hashFun(int x){
  return x;
}
```

we may not use this hash function and can directly use the number itself. Let us write the function to print the duplicates:

```
void printDuplicates(int *arr, int n)
```

```
{
  int hash[10] = {0}; // HASH TABLE

  // STORE FREQUENCY OF NUMBER x IN hash[x]
  for(int i=0; i<n; i++)
    hash[arr[i]]++;

  // TRAVERSE THE HASH AND PRINT NUMBERS
  // WHOSE FREQUENCY IS > 1
  for(int i=0; i<10; i++)
    if(hash[i] > 1)
      printf("%d ", i);
}
```

In question 9.2 we already knew the range of numbers. What if the input array can have any integer. In that case also the solution remains same, but the implementation of hashtable should be changed because we do not know the range of numbers.

Coming up with a 1-on-1 mapping between the number and array index may not be possible. Let us say, we use the hash array of size `10` and value corresponding to a number x is stored at index `x%10`. The hash function in this question is as below:

```
int hash(int x){
  return x%10;
}
```

It will work fine until we get two numbers `x` and `y` for which `x%10 = y%10`.

For example, the hash value for `11` and `91` will come out to be same, i.e 1. It means the frequency of both of these numbers should be stored at `hash[1]` position.

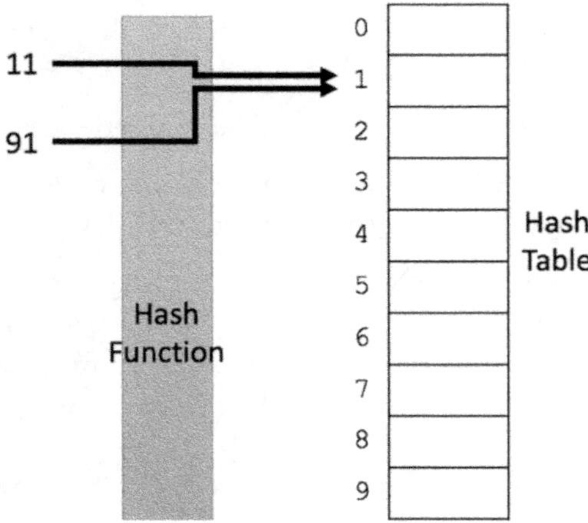

Such a situation is called, **Collision**. A good hash function is a one which results in minimum collisions, but, we cannot say that collision will never happen because it is practically impossible to write a hash function that returns unique address for all possible values.

While designing a hashing system, we must have an explicit plan for **Collision handling**. There are two popular ways of handling collisions:

1. Separate Chaining
2. Open addressing

Separate Chaining

In this method, each entry in the hashtable points to the head of a linked list that store all the values colliding at that address. If the hash function returns the same address for two values, then both of these values are stored at that address in a linked list:

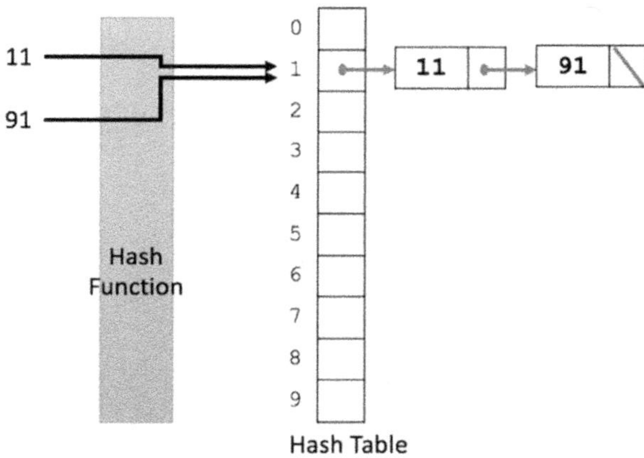

Separate chaining is easy to implement and is also easily scalable. We can add as many values in the linked list as we want. It is also less sensitive to the load factors. If we do not know anything about the values and number of entries to be stored in the hashtable, it is a good option to go with.

If there is a collision between all the values that we want to store in the hash, then the hashtable will reduce to a linked list and the search time will be O(n). But, practically this never happens, unless your hash function is really-really bad.

Open Addressing

In Open Addressing, all the elements are stored in the hash table itself. There is no separate memory allocated outside the hashtable (like linked list in separate chaining). When there is a collision, search (probe) for an empty space and put the new value there.

A corollary to this implementation is that at any point, size of the hash table must be greater than or equal to total number of values that are stored inside the hashtable.

While searching for an element, follow the same searching logic (probing) that was used while inserting the elements, and keep searching till we either find the key or an empty space is reached.

Delete operation is interesting. If we delete a key, then it may leave an empty space in between and our search may fail. One solution is to mark the slots of deleted keys as "deleted". Insertion operation can insert an item in a slot marked as "Deleted", but the search does not stop at a deleted slot.

Open addressing is not necessarily faster than separate chaining. It is used when memory is expensive. Open addressing also provides better cache performance as everything is stored in the same table.

There are many ways of searching the next empty location in a hash table. One of the simplest ways is to search linearly.

Consider, inserting the following values in the hashtable: 15, 46, 27

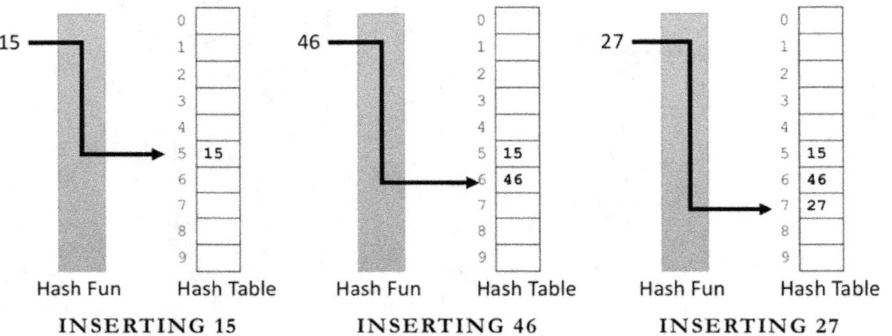

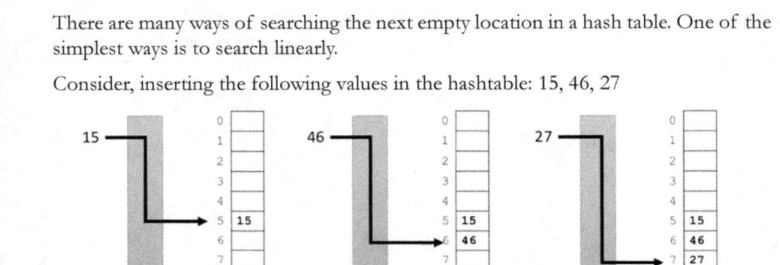

Now if we want to insert 75 in this hash, the hash function produces index value 5 which is looked into the table. Position 5 is already occupied. The next empty position is searched in the table. The next two positions, 6 and 7 are also occupied, and they cannot be used. Searching the table linearly, the next empty and available position is 8. This searching is called **Linear Probing**.

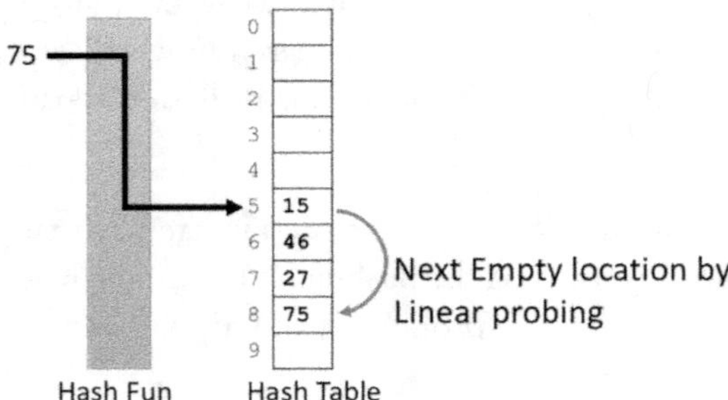

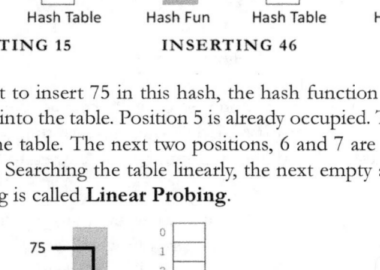

The problem with linear probing is that it can cause clustering that can slow down searching in the hashtable. To avoid clustering, we can use other probing techniques like **quadratic probing and double hashing**.

An Important thing in a hashing system is to define a good hash function, and handle collisions efficiently. Hash table design is outside the scope of this book; next we discuss a case study based on `HashMap` class in Java.

Case Study: Hash Map performance improvement in Java

Java language's `utils` package has `HashMap` class that implements `Map` interface using a hash table. It implements two (almost) constant-time functions `put` and `get` that insert and remove values in `HashMap`. `HashMap` stores information in Key-Value pair as demonstrated in the Code below.

```
import java.util.*;
public class RTDemo
{
  public static void main(String args[ ])
  {
    HashMap<Integer,String> hash = new
    HashMap<Integer,String>();
    hash.put(1000, "Ritambhara");
    hash.put(2000, "Moksha");
    hash.put(3000, "Radha");
    System.out.println("1000:" + hash.get(1000));
    System.out.println("2000:" + hash.get(2000));
    System.out.println("3000:" + hash.get(3000));
  }
}
```

The output is

```
1000: Ritambhara
2000: Moksha
3000: Radha
```

`HashMap<Integer,String>` declares a `HashMap` with `String` values stored on `Integer` keys. Obviously, the implementation cannot keep an array of size `3000` to store these values. The `key` is passed to internal hash function that generates a smaller value indicating index in the actual hash table, something like.

```
Index = hashFunction(1000);
```

let us assume that size of hash table is two, and `hashFunction` is as below

```
int hashFunction(int n){
  return n%2;
}
```

Index for `1000`, `2000` and `3000` are `1`, 2 and `1` respectively. Both `1000` and `3000` generate same index. This is called Collision in hashing. There are multiple ways to handle collision, one of them is thru chaining.

Hash table holds a pointer to the head of linked list of records having same hash value.

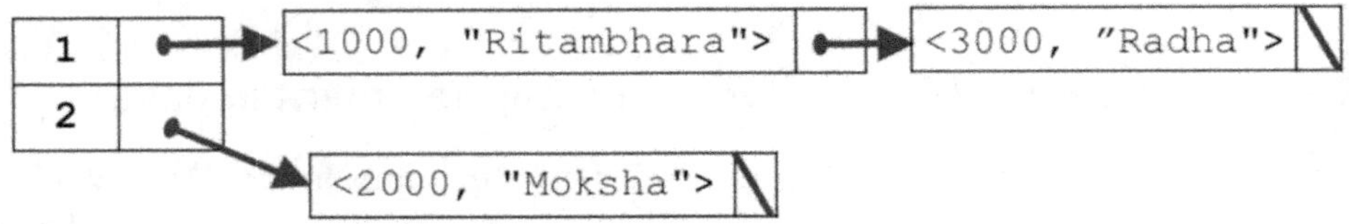

Calling `get` function with key `3000`, retrieves the value in two steps:

1. Find the index of hash table that stores key `3000` using the same hash function. This value comes out to be 1. It means that value for this key will be on this address only and if it is not on this address, then this key is not present in the hash at all.
2. Traverse the list of index 1 in the hash table to find the record corresponding to key `3000`.

In the worst case, all the keys may result in the same hash value and correspond to the same index. Searching in such a hash table takes `O(n)` time. But this is an almost impossible situation for a good hash table implementation. However, there are possibilities that individual linked list of a hash index has multiple nodes.

As you can see each collision has a direct impact on performance. Since searching in a linked list is linear, it takes `O(k)` time in the worst case, where k is number of nodes in the linked list. In Java-8, when a linked list has more nodes, they replace it with a balanced binary tree dynamically, improving the next search time from `O(k)` to `O(lg(k))`.

This simple improvement makes `HashMap.get()` function in Java-8 almost, `20%` faster than the same function in Java-7.

Questions based on hashing

Question 9.3: Given an array of size `n` and two numbers `low` and `high`. Find all numbers in range `[low..high]` which are not present in the array.
Input: `Array={4, 1, 6, 12, 57, 7, 10}, low=4, high=12`
Output: `5, 8, 9, 11`

Element `4, 6, 7, 10` and `12` are present in the array.

Solution: There can be multiple ways of solving this problem. The brute-force way is to search the entire array using linear search for each value in the range. If that value is not present, print it, otherwise move forward and search for the next value. This takes $O(n^2)$ time.

An improvement is to sort the array and use binary search. Sorting takes `O(n.lg(n))` time, search for `low` in array takes `O(lg(n))` time. Then just traverse the array linearly for the range till `high`. Total time taken is `O(n.lg(n))`.

```
void printMissing(int *arr, int n, int low, int high)
{
  quickSort(arr, n); // SORT THE ARRAY
  int i = binarySearch(arr, n, low);
  int x = low;
  while (i < n && x<=high)
  {
    if(arr[i] != x) // x DOES NOT PRESENT
      printf("%d ", x);
    else // x IS PRESENT
      i++;
    // MOVE TO NEXT ELEMENT IN RANGE [low..high]
    x++;
  }

  // PRINT LEFT OVER ELEMENTS
  while(x <= high)
  {
    printf("%d ", x);
    x++;
  }
}
```

If we use hash table, extra memory of size `high-low+1` is needed but the problem can be solved in linear time as shown below.

```
void printMissing(int *arr, int n, int low, int high)
{
  int HASH_SIZE = high-low+1;
  int hash[HASH_SIZE];

  // INITIALIZING HASH
  for(int i=0; i<HASH_SIZE; i++)
    hash[i] = 0;

  // POPULATING HASH
  for(int i=0; i<n; i++)
    if(arr[i]>=low && arr[i]<=high)
      hash[arr[i]-low]++;

  // TRAVERSING HASH TO SEE MISSING ELEMENT
  for(int i=0; i<HASH_SIZE; i++)
    if(hash[i] == 0)
      printf("%d", i+low);
}
```

`hash[i]` is used to store the number of times `low+i` appears in the array. Usually, there is a trade-off between execution time and memory. This method takes `O(n)` extra memory, but has brought down the execution time to `O(n)` without distorting the original array.

```
Whenever you need to search for some element in a
collection, know that if data is stored in the hash then
you can search in constant time.
```

Assignment 9.2: Given array of `n` integers, how will you check if an arrangement of numbers in the array forms an arithmetic progression. Return `true` if they form an AP series and `false` otherwise.

```
Input : Array = {3, 11, 5, 9, 7}
Output: true
```

Question 9.4: Given two integer arrays, write code to check if one is permutation (arranging the same numbers differently) of elements present in the other. For example, the below two arrays are permutations of each other:

```
int a[5] = {1, 2, 3, 4, 5};
int b[5] = {2, 1, 4, 5, 3};
```

Solution-1: Use multiplication and addition

- Multiply all elements of first array, call it **mul1**.
- Multiple all elements of second array, call it **mul2**.
- Add all elements of first array, call it **sum1**.
- Add all elements of second array, call it **sum2**.

If (`mul1== mul2` AND `sum1 == sum2`), the two arrays are permutations of each other else, they are not.

```
bool checkPermutation1(int* a, int m, int* b, int n)
{
  // IF TWO ARE NOT OF SAME SIZE.
  if(m != n){ return false; }
  if(a==NULL && b==NULL){ return true; }

  int mul1 = 1, mul2 = 1;
  int sum1 = 0, sum2 = 0;

  for(int i=0; i<m; i++){
    mul1 *= a[i];
    mul2 *= b[i];

    sum1 += a[i];
    sum2 += b[i];
  }
```

```
  return (mul1 == mul2 && sum1 == sum2);
}
```

Multiplication and addition may result in overflow (or underflow) of integers. Moreover, the multiplication will become zero if there is even one zero in the array, it may not give us the right answer in that case.

Solution-2: Sort both the arrays, if they become exactly same, then they are permutations of each other, else not:

```
bool checkPermutation2(int* a, int m, int* b, int n)
{
  if(m != n){ return false; }
  if(a==NULL && b==NULL){ return true; }

  quickSort(a, m);
  quickSort(a, n);

  for(int i=0; i<m; i++){
    if(a[i] != b[j]){
      return false;
    }
  }

  return true;
}
```

Solution-3: First create the hash using the first array, then iterate the second array, and for each element in the second array do the following:

Search the element in the hash, if it is not present, return `false`. Else, remove the element from the hash and continue.

If the hash becomes empty after the second array traversal, it means that both are permutations of each other.
This method takes linear time, but it also uses `O(n)` extra memory for the hash. There is usually a trade-off between time and memory.

Question 9.5: If letters of a word can be arranged to form another word, the two words are called Anagrams. e.g. `MARY` and `ARMY` are anagrams because both of them are formed from the same characters.

Below are some interesting anagram sentences (Not considering space as part of the string):

```
graduation = out in a drag
a decimal point = I'm a dot in place
A diet = I'd eat
Admirer = Married
Debit card = Bad credit
```

```
dormitory = dirty room
mother-in-law = woman Hitler
```

The list will go on and on… Write a function which will accept two strings and return `true` if they are anagrams.

Solution: Let the two string be string-1 and string-2. The Bruite-force solution is to traverse in string-1 and search each character of string-1 in string-2. If the character is not present in string-2, the two strings are not anagrams and return `false`. Else, mark that character as visited (or replace it with some special character like `'-'`) and continue.

At the end check if string-2 is empty (all characters are visited). Let us see a sample run of the above algorithm

```
MARY, ARMY
    Char 'M', Found and removed from string-2.
MARY, AR-Y
    Char 'A', Found and removed from string-2.
MARY, -R-Y
    Char 'R', Found and removed from string-2.
MARY, ---Y
    Char 'Y', Found and removed from string-2.
MARY, ----
    Second string is Empty (all '-'), Return True.
```

Below is the code for the same:

```
int isAnagram(char *str1, char* str2)
{
  if(str1 == NULL && str2 == NULL){ return true; }

  if(!str1 ||!str2 ||strlen(str1)!=strlen(str2))
    return false;

  for(int i=0; str1[i] != '\0'; i++)
  {
    int found = false;
    for(int j=0; str2[j] != '\0'; j++)
    {
      if(str2[j] == str1[i])
      {
        str2[j] = '-';
        found = true;
        break;
      }
    }
    if(!found)
      return false;
```

```
  }

  return true;
}
```

This code takes `O(n²)` time and constant extra memory. In the process, it is also distorting the string. Let us consider, some better methods.

Method-2: Using Hashing (count characters)

Create a `count` array which will keep the count of number of times a character appears in the string. Here is the algorithm:

```
int count[256] = {0}.

for(i=0; i<strlen(str1); i++)
  count[str1[i]]++

for(i=0; i<strlen(str2); i++)
  count[str2[i]]--

for(i=0; i<256; i++)
  if(count[i] != 0)
    return FALSE;

return TRUE
```

Below is the code for above algorithm

```
int isanagram2(char *str1, char* str2)
{
  if(str1 == NULL && str2 == NULL)
    return true;

  if(str1==NULL || str2==NULL ||
                  strlen(str1)!=strlen(str2))
    return false;

  int count[256] = {0};

  for(int i=0; i<strlen(str1); i++)
    count[str1[i]]++;

  for(int i=0; i<strlen(str2); i++)
    count[str2[i]]--;

  for(int i=0; i<256; i++)
    if(count[i] != 0)
```

```
      return false;

  return true;
}
```

It takes `O(n)` time and constant extra space.

Method-3: Use Sorting

It is not better than the previous method. But one way to solve the problem is to sort the two arrays and then compare the corresponding elements.

```
int isanagram2(char *str1, char* str2)
{
  if(str1 == NULL && str2 == NULL)
    return true;

  if(str1 == NULL || str2 == NULL ||
                          strlen(str1) != strlen(str2))
    return false;

  quicksort(str1);
  quicksort(str2);

  for(int i=0; str1[i] != '\0' ; i++)
    if(str1[i] != str2[i])
      return false;

  return true;
}
```

Assignment 9.3: Given an integer array, print the number appearing the maximum number of times in the array.

Assignment 9.4: Given a string, print the first repeating character in the string.

Assignment 9.5: Given a string, find the number of substrings of the string that starts and end with the same character. For example, if the input string is:

```
"ABA"
```

Then following are the 3 substrings that start and end with the same character:

```
"A", "B", "ABA"
```

The output should be 3.

Disadvantages of HashTable over BST

Values stored in a BST can be traversed in sorted order by doing an Inorder Traversal. This is not a natural operation in a Hash Tables.

Finding the kth largest/smallest element in a BST is easy. Similarly, finding the range of elements is easier in a BST. HashTable does not provide any advantage for such order statistics algorithms.

Hashtable is relatively complex to implement as compared to a BST. In a Self-Balancing BST, all operations are guaranteed to take `O(lg(n))` time. But with Hashing, nothing is guaranteed. Operations in a hashtable take constant time on an average but in some situations, they may be costly.

Chapter 10

Graphs

A Binary tree is a specialization of Graph data structure. In theory, probably it makes more sense to study Graph first and then start learning the specialized Graph, i.e. Binary tree data structure. But in practice, it helps to learn the implementation of a binary tree before learning about the Graphs.

A Graph is a collection of Nodes connected with Edges. A Node in a Graph is called a **Vertex** and connection between two vertices is called an **Edge**.

Unlike binary trees, where an edge can only be from a parent to its children. In a Graph, any vertex (Node) can connect to zero or more vertices (no fixed hierarchy) and there is no limit on the number of vertices that a vertex can connect to. The connection can be uni-directional (like binary trees) or bi-directional. When connections are uni-directional, edge is called **directed edge** and graph is called **directed graph** (or di-graph). When connections are bi-directional, edge is called **un-directed edg**e and the graph is called **undirected graph**. There is no hierarchy among vertices of a Graph.

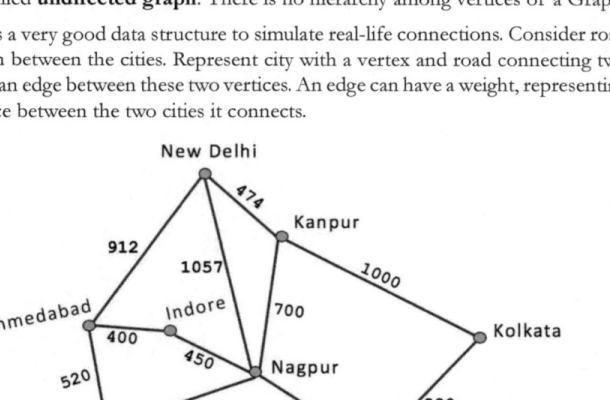

A Graph is a very good data structure to simulate real-life connections. Consider road connection between the cities. Represent city with a vertex and road connecting two cities with an edge between these two vertices. An edge can have a weight, representing the distance between the two cities it connects.

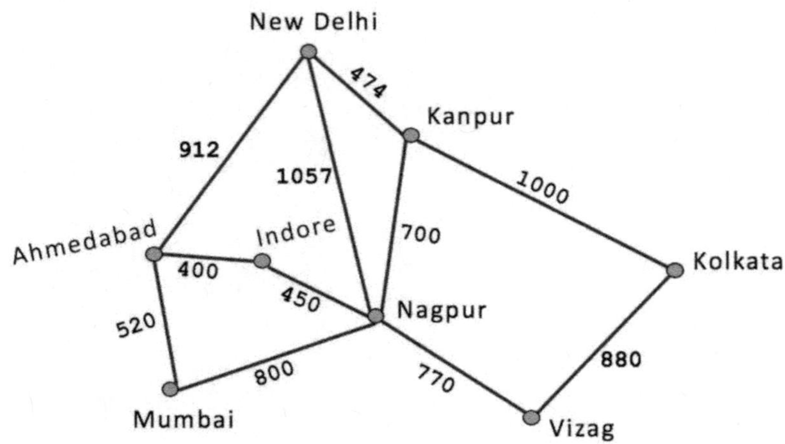

Two vertices are adjacent if they are end-points of the same edge. An edge is a **self-loop** if both the end-points are at the same vertex. **In-degree** of a vertex is the number of edges terminating at that vertex. **Out-degree** of a vertex is the number of outgoing edges from that vertex. The **Degree** of a vertex is sum of its In-degree and Out-degree.

Path in a graph is a sequence of vertices where each pair of consecutive vertices is connected by an edge. A path that starts and ends on the same vertex is called a cycle. Figure 10.1 shows different types of Graphs.

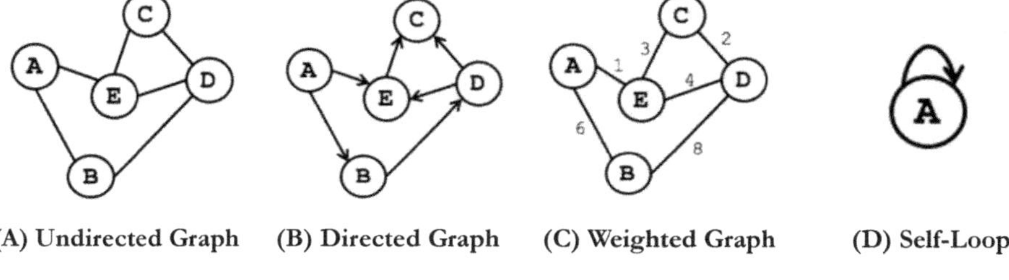

(A) Undirected Graph **(B) Directed Graph** **(C) Weighted Graph** **(D) Self-Loop**

Figure 10.1

If all nodes are connected, then the graph is **connected graph**. If there exist two nodes in the graph, with no path connecting them, then it is called **un-connected graph**.

If there is at least one cycle in the graph, then the graph is called **cyclic** graph. All the graphs in Figure 10.1 are cyclic graphs. If there is no cycle in the graph, then the graph is **acyclic**. Both the below graphs are acyclic

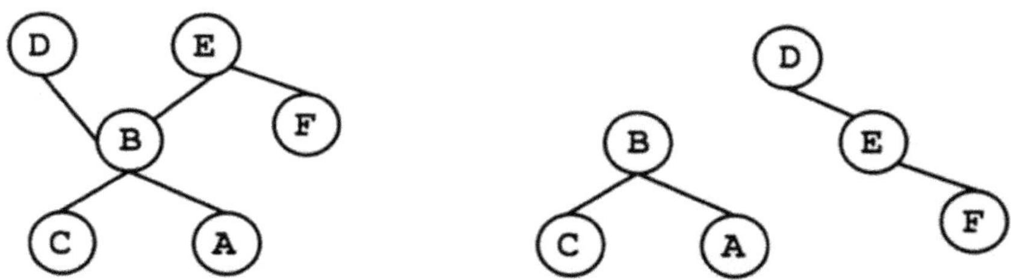

The graph on the right side is also unconnected. A binary tree is a connected acyclic and directed graph that is rooted and hierarchical.

Graph Representations

It is not practical to represent Graph like binary trees using Node structure similar to what we saw in chapter 6. There are two major representations of graph in computers.

1. Adjacency-matrix representation

In adjacency matrix implementation, a two-dimensional array is taken of order `N*N`, where `N` is the number of vertices. Cell `(i,j)` is `true` if there is an edge from Vertex-`i` to Vertex-`j` in the graph (Vertex-`i` and Vertex-`j` are adjacent). For undirected graph shown in Figure 10.1(A), adjacency matrix looks like:

	A	B	C	D	E
A	0	1	0	0	1
B	1	0	0	1	0
C	0	0	0	1	1
D	0	1	1	0	1
E	1	0	1	1	0

`false` is represented by `0` and `true` by `1`. Since each cell stores a binary value, using a bit-matrix can save space.

For weighted graph (each edge has a weight), a cell can store weight of the corresponding edge, `0` representing absence of an edge. Below is adjacency matrix for graph in Figure 10.1(C).

	A	B	C	D	E
A	0	6	0	0	1
B	6	0	0	8	0
C	0	0	0	2	3
D	0	8	2	0	4
E	1	0	3	4	0

Both of these matrices are symmetric and diagonal elements have no meaning. It is a fit case to use sparse matrix to save memory. For directed graphs, we have to keep the complete matrix.

There are two most used operations in any graph:

1. Given two vertices, check if they are connected
2. Given a vertex, print all its adjacent vertices

In our example, vertices represent cities. A city can have a lot of information (name, population, socio-economic indices, etc.). To simulate a real-life situation, let us keep vertex array separate from adjacency matrix. It gives the flexibility to change city information without changing the connections. To start with, assume that a city has only name. `vertex` array stores information about cities and `edges` matrix store the connections between them

```
char vertex[N] =     {'A', 'B', 'C', 'D', 'E'};
int edges[N][N] =    { {0, 1, 0, 0, 1},
                     {1, 0, 0, 1, 0},
                     {0, 0, 0, 1, 1},
                     {0, 1, 1, 0, 1},
                     {1, 0, 1, 1, 0} };
```

We have kept city names to be single character for simplicity. But `vertex` array can very well be of a complex structure type.

`edges[i][j]` shows connection between `vertex[i]` and`vertex[j]`. With this structure, we have decoupled vertex information from their connections. This decoupling is important for, changing city name does not affect connection, because vertex-id does not change.

Let index of a city in `vertex` array represents `ID` of that city. Below function receives a city by name and returns its `ID`

```
int getVertexId(char v)
{
  for(int i=0; i<N; i++)
    if(vertex[i] == v)
      return i;
  return -1;
}
```

Function `isAdjacent` below returns `true` if two vertices are adjacent.

```
bool isAdjacent(char v1, char v2)
{
  int i = getVertexId(v1), j = getVertexId(v2);

  // INVALID VERTEX
  if(i == -1 || j == -1) { return false; }
  return edges[i][j];
}
```

If `getVertexId` function is implemented as a constant-time function (using hash) then `isAdjacent` function takes constant time. This is the biggest advantage with adjacency matrix implementation.

Let us now print all adjacent nodes of a given vertex

```
void printAllAdjescent(char v1)
{
  int i = getVertexId(v1);
  if(i == -1){ return; }    // INVALID VERTEX
  for(int j=0; j<N; j++)
    if(edges[i][j] != 0)
      printf("%c", vertex[j]);
}
```

This function traverses the entire row and takes `O(N)` time. In a directed graph, an edge can be from a vertex to itself also (self-loop).

2. Adjacency list representation

A linked list with one node for each adjacent node of the given vertex is called adjacency list of that vertex.

In this representation, we keep an array of head pointers of size `N`. Each element of the array points to adjacency list of the corresponding vertex as shown below.

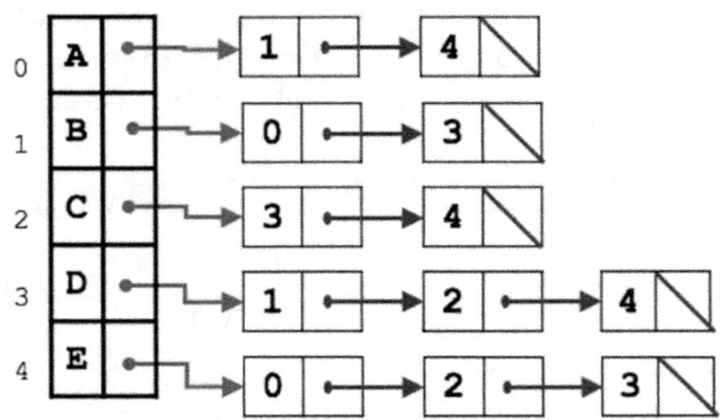

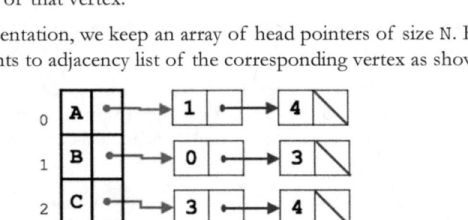

Figure 10.2 Adjacency List representation

Adjacency list holds vertex-ids and next pointers, vertex-id is the index of a vertex in the array. Let `LinkNode` define the node of adjacency list

```
struct LinkNode
{
  int data; // INDEX OF ADJACENT VERTEX
  struct LinkNode* next;
};
```

Each element in the array on the left side of Figure 10.2 has two fields, data (City information) and a pointer to adjacency list of that vertex. Array `vertex` is defined as below:

```
struct VertexInfo
{
  char data;
  struct LinkNode *head;
};
struct VertexInfo vertex[N];
```

Assume that `vertex` array is initialized and all adjacency lists are created as shown in Figure 10.2. The functions to check for adjacency of two vertices and printing the adjacency list of a vertex are defined below:

```
bool isAdjacent(char v1, char v2)
{
```

```
    int i = getVertexId(v1);
    int j = getVertexId(v2);

    // INVALID VERTEX
    if(i == -1 || j == -1) { return false; }

    // TRAVERSE ADJ. LIST OF Vertex-i
    LinkNode* head = vertex[i].head;
    while(head != NULL)
    {
      if(head->data == j)
        return true;
      head = head->next;
    }
    return false;
}

void printAllAdjescent(char v1)
{
    int i = getVertexId(v1);
    if(i == -1){ return; } // INVALID VERTEX

    // TRAVERSE ADJ. LIST OF Vertex-i
    LinkNode* head = vertex[i].head;
    while(head != NULL)
    {
      printf("%c", vertex[head->data].data);
      head = head->next;
    }
}
```

Traversals

Unlike an array or a linked list, Graph is a non-linear data structure, and unlike a Binary tree, the graph is also non-hierarchical. There is no root node in the graph. But, to traverse a given graph, we have to start from some point, so let us pick one vertex as the source node.

Traversals of a data structure are important because, for many questions, we just traverse all vertices and perform a certain action on the vertex. For example, if we want to search if there is a vertex with a particular value in the graph, then we can visit each vertex of the graph and check if that is the vertex we are searching for.

There are two popular ways of traversing a graph, Breadth First Search and Depth First Search.

Breadth First Search

While traversing a graph in BFS, start from a selected node and traverse the graph layer-wise visiting nodes which are directly connected to the starting node first and then move to the next layer. The graph is traversed breadth-wise, horizontally.

Consider the following graph:

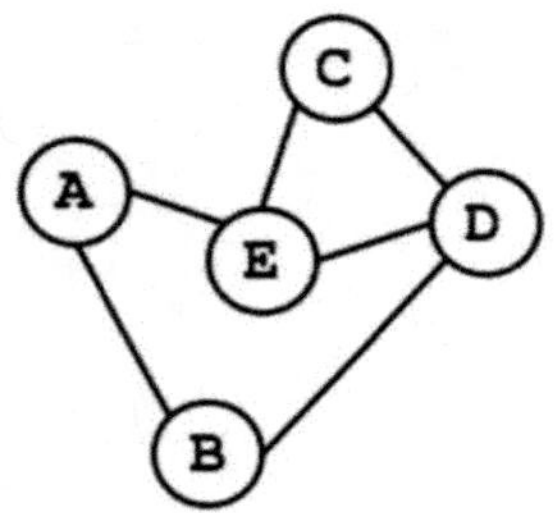

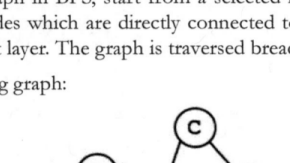

If we start from vertex A, then the graph is traversed as shown in the diagram:

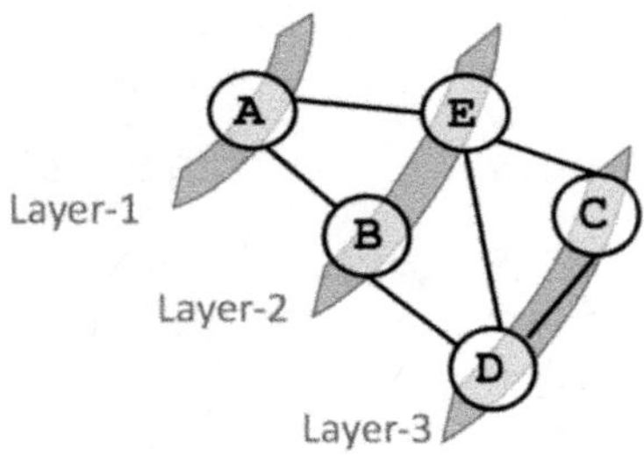

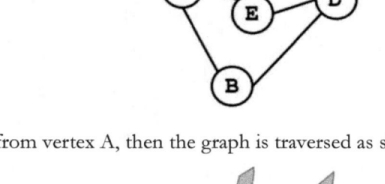

This is exactly same as the level-order traversal of a binary tree (actually, it should be the other-way around, the level-order traversal is same as BFS).

Remember the level-order traversal of a Binary tree?

1. Insert starting vertex in the queue

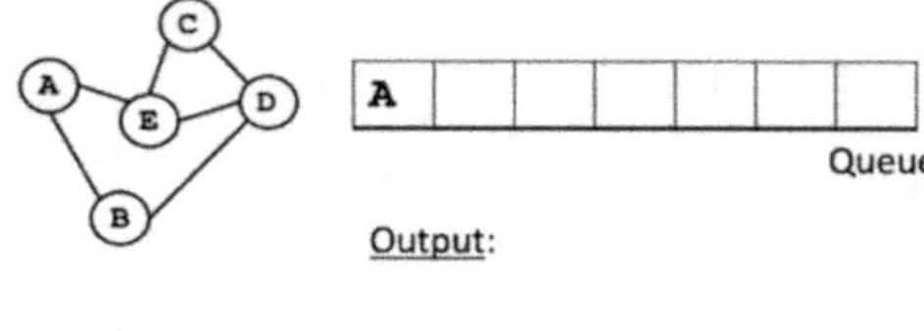

2. Pop an element from the queue, print its data to the output and insert its adjacent nodes in the queue.

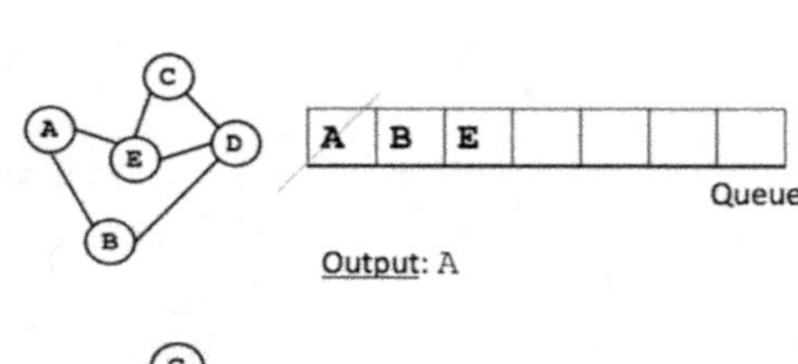

3. Pop element from the queue, print its data to the output and insert its adjacent nodes in the queue.

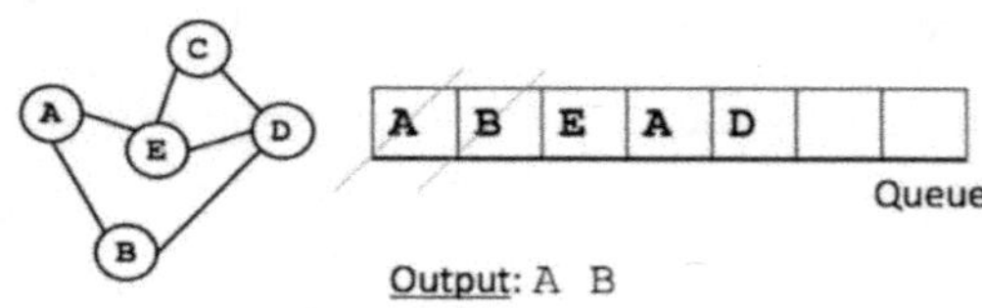

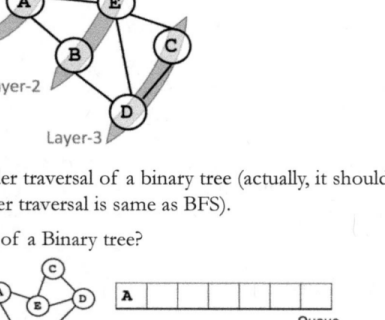

Oops! We have inserted Vertex-A again in the queue. If we follow this algorithm, we will end up in an infinite cycle. In a binary tree, connections are uni-directional from parent to its child node. In a Graph, there may be a cycle, which means that one node can be reached from multiple paths. Also, if B is adjacent to A, then A can also be adjacent to B.

We need a way to mark the already visited nodes so that a node is not visited multiple times. Let us keep track of visited nodes (either change the Node structure or keep a hash of visited nodes).

Insert only those adjacent nodes in the queue, which are not yet visited. And once a node is inserted in the queue, mark it as visited.

Below is the algorithm for BFS:

```
BFS (G, s) // G-GRAPH, s-STARTING VERTEX
  Queue q = new Queue
  q.insert(s)
  MARK s AS VISITED
  while( !q.isEmpty())
  {
    v = q.remove()
    for( ALL ADJESCENT VERTICES w OF v)
    {
      if( w IS NOT VISITED)
      {
        q.insert(w)
        MARK w AS VISITED
      }
    }
  }
```

The time taken by the BFS algorithm is `O(n)` where `n` is the number of vertices in the Graph, because each vertex is inserted in the queue only once.

Let us write the code for BFS in a graph. For this code, we have modified the structure of `VertexInfo` by adding one more field, `visited`. The value of `visited` field for all the vertices is set to `false` before calling the BFS code. When a node is inserted in the queue, its `visited` field is set to `true`, so that it is not visited again.

Below code creates the sample graph shown in Figure 10.2.

```
#define N 5
struct LinkNode{
  int data; // INDEX OF ADJACENT VERTEX
  LinkNode* next;

  LinkNode(int val){// CONSTRUCTOR
    data = val;
```

```
    next = NULL;
  }
};

struct VertexInfo{
  char vName;
  LinkNode *head;
  bool visited;
};
VertexInfo vertex[N]; // KEEPING GLOBAL

// GET THE ID OF VERTEX WHOSE NAME IS vName
int getVertexId(char verName){
  for(int i=0; i<N; i++)
    if(vertex[i].vName == verName)
      return i;
  return -1;
}

// RETURN THE verted FOR ID = verId
// VERTEX WITH ID = verId IS THE ELEMENT AT INDEX
VertexInfo* getVertexInfo(int verId){
  return &vertex[verId];
}

// CREATE SAMPLE GRAPH
void createSampleGraph(){
  vertex[0].vName = 'A';
  vertex[1].vName = 'B';
  vertex[2].vName = 'C';
  vertex[3].vName = 'D';
  vertex[4].vName = 'E';

  vertex[0].head = new LinkNode(1);
  vertex[0].head->next = new LinkNode(4);

  vertex[1].head = new LinkNode(0);
  vertex[1].head->next = new LinkNode(3);

  vertex[2].head = new LinkNode(3);
  vertex[2].head->next = new LinkNode(4);

  vertex[3].head = new LinkNode(1);
  vertex[3].head->next = new LinkNode(2);
  vertex[3].head->next->next = new LinkNode(4);
```

```
  vertex[4].head = new LinkNode(0);
  vertex[4].head->next = new LinkNode(2);
  vertex[4].head->next->next = new LinkNode(3);
}
```

This code is written in C++. We have added a constructor in `LinkNode` so that we can initialize the `LinkNode` while allocating the memory using `new` operator. The `vertex` array is kept global to keep things simple. I completely understand the nuisance value of keeping things global.

There is a function,`getVertexId,` to get the id of a vertex when the name of the vertex is given. Similarly, there is a function that returns the `VertexInfo` of a vertex whose ID is given.

The function `createSampleGraph` can be used for creating the sample graph that can be used to unit test our code.

Next is the BFS function that traverses the graph in breadth first search order. It accepts the address of source node to start the traversal.

```
void bfs(VertexInfo* ver){

  queue<VertexInfo*> q;  // QUEUE FOR BFS

  ver->visited = true;// INSERTING SOURCE NODE
  q.push(ver);


  while(!q.empty())
  {
    // REMOVE FRONT ELEMENT FROM QUEUE
    VertexInfo* temp = q.front();
    q.pop();

    // PRINT THE NAME OF VERTEX
    cout<<temp->vName<<" ";

    int i=getVertexId(temp->vName); //GET VERTEX-ID
    if(i == -1){ return; } // INVALID VERTEX


    // CONSIDER ADJACENCY LIST OF Vertex-i
    for(LinkNode* h=temp->head; h!=NULL; h=h->next)
    {
      VertexInfo* ver = getVertexInfo(h->data);

      if(ver->visited == false){
```

```
            ver->visited = true;
            q.push(ver);
        }
    }
  }
}
```

Since this is the first major graph code, let us write the `main` function also; so, that you can see the above code running:

```
int main(){

  createSampleGraph();

  // SETTING visited PROPERTY OF ALL NODES
  for(int i=0; i<N; i++)
    vertex[i].visited = false;

  // TAKING 'A' AS SOURCE VERTEX
  bfs(&vertex[0]);
}
```

After we combine the above functions and execute the program, the output will be:

A B E D C

The nodes are traversed as per breadth first search. We have not considered till now the way to traverse the entire graph when there are multiple unconnected components.

This algorithm will just traverse all the nodes which are connected to the source node passed to the function. This can be fixed, by putting a check in the `main` function, if there are nodes which are left unvisited (traverse the `vertex` array and see `visited` field for each node), if there is an unvisited node, then that node is not connected to the subgraph of which the source node is a part. Call the `bfs` function again for un-visited nodes.

Also, note that `bfs` traverse the nodes in the order of their shortest distance from the source node. So, if we want to find the shortest distance of a node from the source node, then we can add one more field in the `VertexInfo` structure for the distance of that node from the source node. For source node the distance is 0. Inside the `while` loop when we are inserting the adjacent node in the queue, update the distance of that node as one-plus-shortest-distance of the current node.

Depth First Search

Let us re-look at the preorder traversals of a Binary tree. We first print the root node and then traverse the entire left subtree. Once all the nodes that can be reached from the left child are traversed, we come back to the root again and then traverse the entire right subtree, i.e we traverse all the nodes that can be reached via the right child of the root.

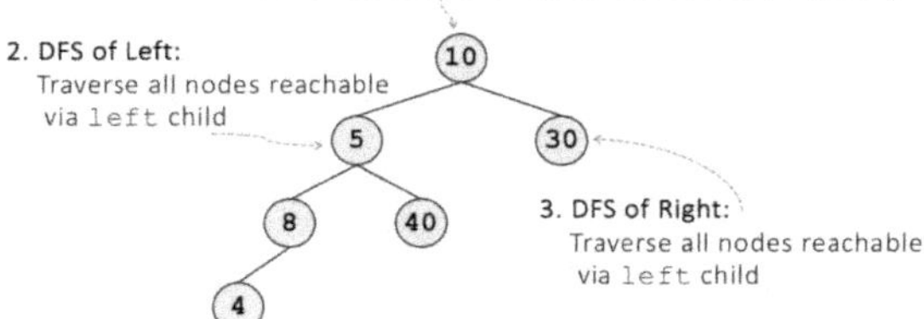

This is essentially the DFS algorithm. Start with a source node (root node in a Binary Tree), pick the first adjacent node (first child node in case of a Binary tree) and then traverse the graph assuming that child as the start node (call preorder for the left child). Once all the nodes connected to the first adjacent node are traversed (entire left subtree is traversed), look for nodes connected to the second adjacent nodes in the graph (traverse the right subtree).

We need to put some extra checks, because, for Graph, the right child (second adjacent node) may be reachable from the left child (first adjacent node). Consider below graph

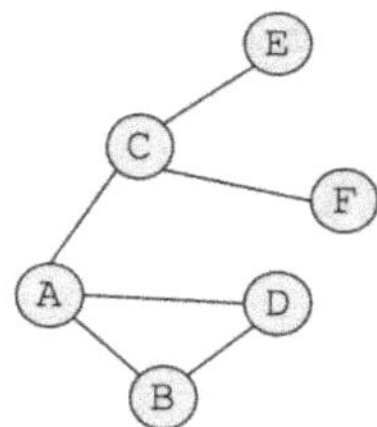

If we pick A as the starting point (source node), then, we essentially follow the following approach:

1. PRING VALUE OF NODE **A**
2. CALL DFS FOR EACH ADJACENT NODE. i.e.
    a. CALL DFS FOR **B**
    b. CALL DFS FOR **D**
    c. CALL DFS FOR **C**

As you might have noticed, this will result in infinite recursion because B and D are connected. In a graph, two child nodes (actually adjacent nodes) may be connected forming a cycle. This is not the case in a Binary tree.

We use the same approach as we used in BFS. When a node value is printed, mark the node as `VISITED`. And before call DFS for a child only when it is not already visited. Below is the code for Depth First Search.

```
void dfs(VertexInfo* ver)
```

```
{
  if(ver == NULL){ return; }

  cout<<ver->vName << " "; // PRINT CURRENT NODE
  ver->visited = true; //MARK CURRENT NODE AS VISITED

  // TRAVERSE ADJ. LIST OF Vertex-i
  for(LinkNode* h = ver->head; h!=NULL; h=h->next)
  {
    VertexInfo* ver = getVertexInfo(h->data);

    if(ver->visited == false)
      dfs(ver);
  }
}
```

Preorder, inorder, postorder and many other binary tree algorithms actually use DFS approach. We follow one child, once that is done, we back-track to the root and follow the second child. That's DFS.

Asymptotically, both DFS and BFS take same O(n) time and O(n) extra memory. BFS use the memory in the form of a Queue and DFS use recursion and hence use the memory Stack.

The applications of BFS and DFS are not only limited to graph data structure. Their use is rampant across other data structures also, as we see in the next section.

Questions based on traversals

Question 10.1: Given a matrix of order `M*N`, where each cell can have zero or one orange. The orange in the cell (if present) can be either fresh or rotten.

A rotten orange will rot the fresh oranges in all its nearby cells in a unit time. If a rotten orange is in cell `(i,j)`, then it will rot the fresh oranges in the four nearby cells

`(i+1, j) (i-1, j) (i, j+1) (i, j-1)`

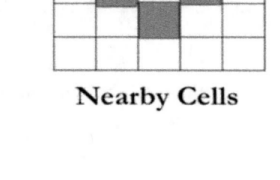

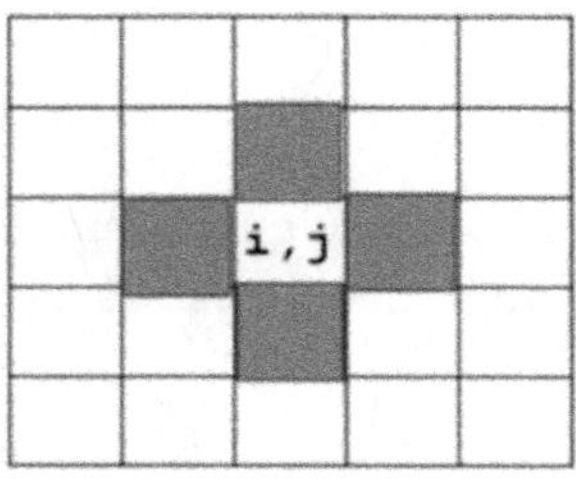

Nearby Cells

A fresh orange surrounded by all empty cells cannot ever be rotten at all. Consider below matrix

```
{{2, 1, 0, 2, 1}
 {1, 0, 1, 2, 1}
 {0, 0, 0, 2, 1} }
```

Value of a cell represents the following state:

0- Empty Cell | **1** - Fresh Orange | **2** - Rotten Orange

In the given matrix, all oranges will rot in unit time, because all fresh oranges are just one step away from their nearby rotten orange. For below matrix

```
{{2, 1, 0, 1, 1}
 {1, 0, 1, 1, 1}
 {1, 1, 1, 1, 1} }
```

It will take 8-unit time to rot all oranges in the matrix. The state of matrix will be the following after each unit time:

After 1 Unit Time

```
{{2, 2, 0, 1, 1}
 {2, 0, 1, 1, 1}
 {1, 1, 1, 1, 1}}
```

After 2 Unit Time

```
{{2, 2, 0, 1, 1}
 {2, 0, 1, 1, 1}
 {2, 1, 1, 1, 1}}
```

After 3 Unit Time

```
{{2, 2, 0, 1, 1}
 {2, 0, 1, 1, 1}
 {2, 2, 1, 1, 1}}
```

After 4 Unit Time

```
{{2, 2, 0, 1, 1}
 {2, 0, 1, 1, 1}
 {2, 2, 2, 1, 1}}
```

After 5 Unit Time

```
{{2, 2, 0, 1, 1}
 {2, 0, 2, 1, 1}
 {2, 2, 2, 2, 1}}
```

After 6 Unit Time

```
{{2, 2, 0, 1, 1}
 {2, 0, 2, 2, 1}
 {2, 2, 2, 2, 2}}
```

After 7 Unit Time

```
{{2, 2, 0, 2, 1}
 {2, 0, 2, 2, 2}
 {2, 2, 2, 2, 2}}
```

After 8 Unit Time

```
{{2, 2, 0, 2, 2}
 {2, 0, 2, 2, 2}
 {2, 2, 2, 2, 2}}
```

The following matrix can never be completely rotten because orange in the cell (2, 0) will remain fresh.

```
{{2, 1, 0, 2, 1}
 {0, 0, 1, 2, 1}
 {1, 0, 0, 2, 1} }
```

Write a function that accepts a two-dimensional matrix having values 0, 1 and 2 representing the state of oranges in the cell and returns the time in which entire matrix will rot. If all the oranges in the matrix cannot be rotten, then the function should return -1 to indicate the same.

Solution: The solution of this problem uses BFS algorithm. Use a Queue that can hold the following structure (Data in queue)

```
struct QElement{
     int i;
     int j;
     int time;
};
```

`(i,j)` represents the cell index and `time` in this structure is the time at which orange in this cell will rot.

Traverse the matrix and insert all the cell with rotten oranges in the Queue. The time field of all these cells is set to 0 (because they are already rotten). Now follow the following algorithm that is very much like BFS

```
MaxTime = 0;
WHILE (Queue is not empty)
  DataElement temp = Queue.remove()
  MaxTime = temp.time
  FOR EACH cell ADJACENT TO (temp.i, temp.j)
     IF cell HAS FRESH ORANGE
        Queue.Insert(cell.i, cell.j, temp.time+1)
IF MATRIX STILL HAS FRESH ORANGE
  RETURN -1
ELSE
  RETURN MaxTime
```

Time of the last cell removed from the queue represents the time to rot all oranges in the queue. If there are still any fresh oranges in the matrix (after the loop), then those oranges are unreachable from all rotten cells and remain fresh.

Question 10.2: Each element in a two-dimensional array is connected to eight neighboring elements. The cell with value 0 in below matrix is connected to all the other cells

```
1 1 1
1 0 1
1 1 1
```

Given a two-dimensional array of only `0`'s and `1`'s. Find the number of Islands. An island is a group of connected 1's. For example: below matrix has 4 islands.

1	1	0	0	0
0	1	0	0	1
1	0	0	1	1
0	0	0	0	0
1	1	1	0	1

Matrix

1	1	0	0	0
0	1	0	0	1
1	0	0	1	1
0	0	0	0	0
1	1	1	0	1

Islands

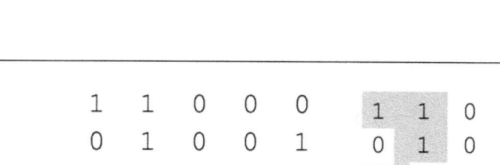

Write code that accepts a matrix and return the number of islands in it.

Solution: This problem is similar to the problem of connected components in a graph.

The solution is simple. Look for `1`'s in the matrix which are not connected to any other 1. When you find such a `1` search for matrix all the entries in this island by recursively searching the eight connected cells, and mark them connected.

Let us set the cell to `-1` to indicate that the node is already connected to an island.

Let's first write the recursive function to mark the connected elements of a given cell. It takes a position `(i,j)` and set it to `-1` and then look for nearby cells marking all set nodes as -1 recursively.

```
void markIsland(int arr[][5], int i, int j,
                              int m, int n)
{
  arr[i][j] = -1;

  if(i-1 >= 0){
    if(j-1 >= 0 && arr[i-1][j-1] == 1) // (i-1, j-1)
      markIsland(arr, i-1, j-1, m, n);

    if(arr[i-1][j] == 1)               // (i-1, j)
      markIsland(arr, i-1, j, m, n);

    if(j+1 < n && arr[i-1][j+1] == 1)  // (i-1, j+1)
      markIsland(arr, i-1, j+1, m, n);
  }

  if(i+1 < m){
    if(j-1 >= 0 && arr[i+1][j-1] == 1) // (i+1, j-1)
      markIsland(arr, i+1, j-1, m, n);

    if(arr[i+1][j] == 1)               // (i+1, j)
```

```
      markIsland(arr, i+1, j, m, n);

    if(j+1 < n && arr[i+1][j+1] == 1)   // (i+1, j+1)
      markIsland(arr, i+1, j+1, m, n);
  }

  if(j-1 >= 0 && arr[i][j-1] == 1) // (i, j-1)
    markIsland(arr, i, j-1, m, n);

  if(j+1 < n && arr[i][j+1] == 1)  // (i, j+1)
    markIsland(arr, i, j+1, m, n);
}
```

The main function that counts the islands will call this function to mark all the elements in the islands whenever a 1 is encountered.

```
int countIslands(int arr[][5], int m, int n)
{
  int count = 0;

  for(int i=0; i<m; i++)
    for(int j=0; j<n; j++)
      if(arr[i][j] == 1){
        count++;
        markIsland(arr, i, j, m, n);
      }

  return count;
}
```

The only problem with above code is that it changes all the `1`'s to `-1`. But that can easily be corrected by resetting all the `-1` values back to `1`. The logic takes `O(M*N)` time.

Other Graph Algorithms

In coding interviews, you are not expected to remember specific algorithms. It is usually a test of your problem-solving abilities and not memory. I have not seen interviewers asking questions like, implement Dijkstra's algorithm or Prims algorithm in the interview. You may end up talking about the concept of shortest path algorithm, but, most likely, in context of solving some other problem.

Having said that, I encourage you to learn these algorithms. Balanced binary tree as a concept is more useful in the interviews and specific function to insert in an AVL Tree may never (and should never) be asked. But, knowing the algorithms of AVL tree can help you in better understanding (and hence explaining) the concept of balanced binary trees.

I encourage you to learn the following algorithms on graphs:

- **Traversal algorithms**
  - A-Star Algorithm
- **Algorithm to arrange vertices**
  - Topological Sorting
- **Minimum Spanning Tree algorithms:**
  - Kruskal's Algorithm
  - Prim's Algorithm
- **Shortest Path algorithms**
  - Dijkstra's Algorithms
  - Bellman-Ford Algorithm

Question 10.3: There are n cities, which are not connected by road with each other. The government wants to connect all these cities to each other by road. But there is a cost associated with building roads. The cost of creating a road between cities is given. However, it is not possible to connect all cities, the cost is given for cities that can possibly be connected.

Let us say there are 5 cities, namely A, B, C, D and E. The below diagram shows the cost of each road joining these cities:

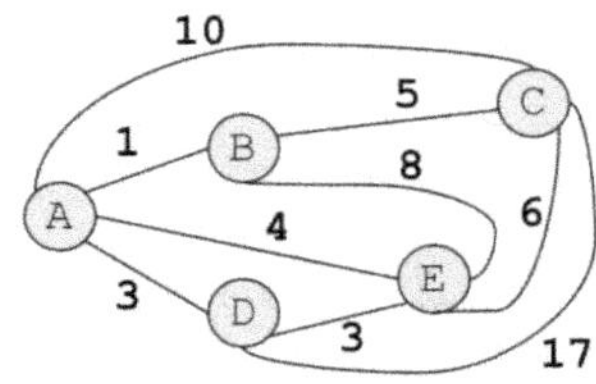

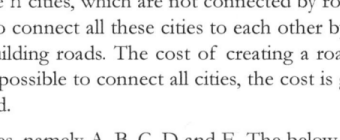

The government is looking for the minimal cost of connecting each city with the other. You are given the task to compute the minimum cost for the given graph of cities roads. For example, for the above graph, the minimum cost will be when only the following rods are constructed.

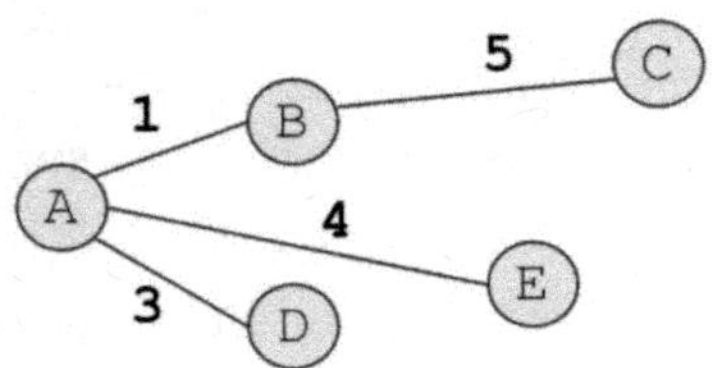

This will connect all the cities (if someone wants to go from C to E, he can take the road from C to B to A to E). The output of your function should be 5+4+3+1 = 13, the minimum cost.

Solution: The solution is to find the minimum spanning tree, and the minimum cost is the sum of the costs of edge values in the Minimum Spanning Tree (MST).

Let us follow the Prims algorithm to find the MST. Prim's algorithm builds a MST by adding one vertex at a time to the set of MST. We keep all the vertices in two sets. Let us call these two sets `S1` and `S2`. Let `S2` represent the final MST set. Initially, `S2` is empty and all the vertices are in set `S1`. Pick any node as the source node to start with and add it to set `S2`. Let us pick `A` as the source node.

Remove all the self-loops and parallel edges. If we have two edges, both directly connecting same two vertices, then they are called parallel edges. In given graph, we do not have any parallel edges or self-loops.

Check out all outgoing edges from vertices in set `S2` to vertices in set `S1` and choose the minimum value edge from them. Add this edge to the MST and bring the corresponding vertex in set `S2`.

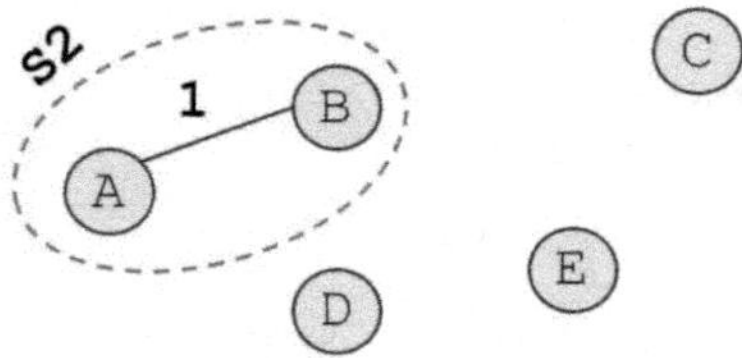

Now again we consider all the edges that move out from either vertex `A` or `B` and go to any of the vertices outside `S2`. Vertex-D is now added to set S2 and the corresponding edge is added to the MST.

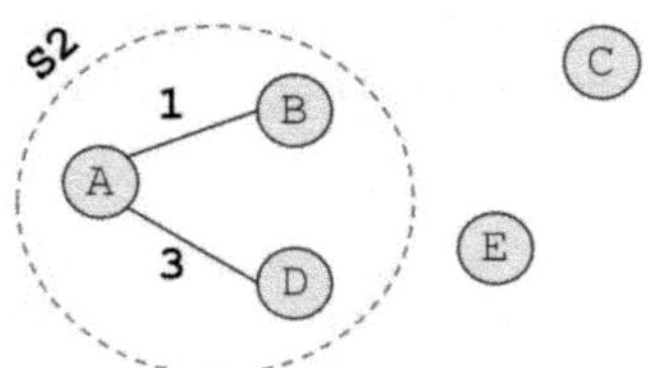

Following the same logic, we get the MST of the entire graph. We stop when all the vertices are added to set `S2`.

I think if you have read this entire book religiously and followed all the questions and assignments, you are good to face any interview of any company on data structure. Good luck !

Steve Uhlig

From the Traffic Properties to Traffic Engineering in the Internet

Steve Uhlig

From the Traffic Properties to Traffic Engineering in the Internet

Implications of the Traffic Properties on Traffic Engineering with BGP: The View from Stub ASes.

VDM Verlag Dr. Müller

Imprint

Bibliographic information by the German National Library: The German National Library lists this publication at the German National Bibliography; detailed bibliographic information is available on the Internet at http://dnb.d-nb.de.

Cover image: www.purestockx.com

Publisher:
VDM Verlag Dr. Müller Aktiengesellschaft & Co. KG , Dudweiler Landstr. 125 a, 66123 Saarbrücken, Germany,
Phone +49 681 9100-698, Fax +49 681 9100-988,
Email: info@vdm-verlag.de

Zugl.: Louvain-la-neuve, Universite catholique de Louvain, 2004

Produced in USA and UK by:
Lightning Source Inc., La Vergne, Tennessee, USA
Lightning Source UK Ltd., Milton Keynes, UK
BookSurge LLC, 5341 Dorchester Road, Suite 16, North Charleston, SC 29418, USA

ISBN: 978-3-8364-9330-7

Contents

List of Tables

List of Figures

Preface

The Internet routing system today is divided into two views: intradomain and interdomain. The interdomain Internet is made of autonomous systems (AS). Each autonomous system uses an exterior gateway protocol (EGP) to exchange reachability information between ASes. Autonomous systems are composed of routers interconnected through physical links. This internal topology of ASes is called the intradomain topology. Routers in a given AS exchange intradomain routing information through an interior gateway protocol (IGP) that distributes the whole map of the intradomain network to all routers of the AS.

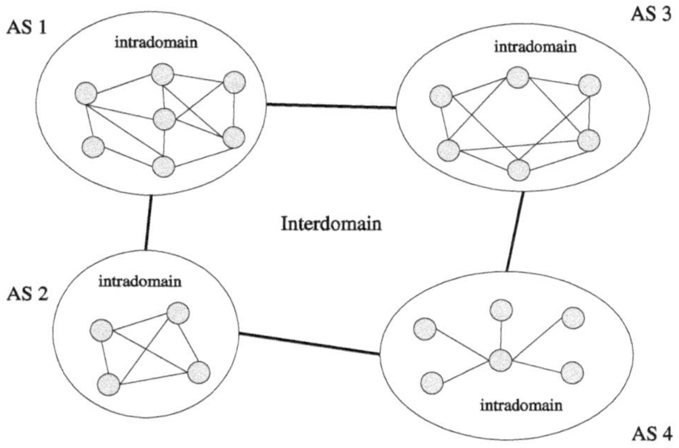

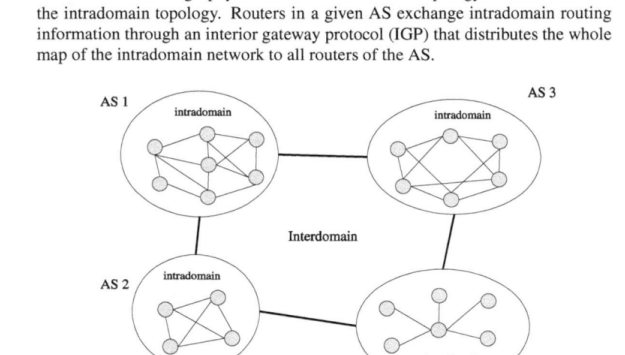

Figure 1: Intradomain and interdomain views of the Internet

Figure 1 illustrates those two topological levels in the Internet by showing four ASes, each having a particular intradomain topology with routers running at least an IGP protocol (e.g. OSPF or IS-IS) that distributes the whole map of the intradomain topology to all other routers of the AS. Each AS has a different

intradomain topology that is not seen by the interdomain view. The interdomain view only sees the four ASes and their interconnections.

Each AS knows its intradomain topology through the IGP protocol, and has full control over the routing information. Engineering the flow of the traffic within an AS by tweaking the intradomain routing protocols is thus possible [88, 87]. At the interdomain level on the other hand, the approach used today in the Internet is the one of a distributed control of the routing information. ASes do not have a complete map of the interdomain topology through the current interdomain routing protocol, BGP [175].

The objective of BGP is to distribute information about the reachability of IP prefixes (i.e. reachable networks) to other ASes. With BGP, an AS advertises to each neighboring AS all the networks it can reach. Among the IP prefixes that an AS advertises, some are internal prefixes that are reachable within this AS (internal to the AS) and others are prefixes that have been learned through its BGP neighbors (external to the AS). A key feature of BGP is that it allows each network operator to define its routing policies. Those policies are implemented by using filters [100]. A BGP filter is a rule applied upon receiving a BGP route from a BGP neighbor or before sending a BGP route to a BGP neighbor. BGP filters can prevent some routes from being accepted from or announced to neighboring ASes, and can also modify the attributes of the BGP routes on a per-AS basis so that some routes be preferred over others.

Internet hierarchy

The graph of the interdomain Internet can be thought as loosely hierarchical [176]. The core of the Internet is made of about twenty large ASes today, called tier-1's. These ASes provide global connectivity in the Internet and are densely interconnected. The main purpose of tier-1's is to provide transit service. Tier-1's carry IP packets that enter their network and whose destination is outside their network. Tier-1's constitute the top-level of the hierarchy. Then, we have smaller or less connected transit service providers that make the second level of the hierarchy. These smaller transit ASes are sometimes classified into different tier levels (e.g. tier-2, tier-3, and tier-4) to indicate their limited geographic coverage. Finally, we have stub ASes at the edge of the interdomain topology that do not provide transit service but that only generate and receive IP packets without allowing IP packets to otherwise transit through their network. It can host content or provide access to dialup or broadband users. This tier-structure of the Internet is quite loose: stub ASes can be connected to any type of transit provider, and transit providers also interconnect between one another without strict consideration of the "type" of the AS, rather based on mutual interests. The interconnections between ASes arise

for economical as well as geographical reasons. This partitioning of the ASes into the different types as presented in [176] cannot be considered as a definitive classification of ASes. More details can be found in [176]. We provide on Figure 2 an illustration of the hierarchical structure of the Internet. We show on this figure the classification of the ASes made by [176] based on BGP routing tables from January 9 2003. The whole topology is made of 14, 695 ASes and 30, 815 unique undirected inter-AS edges. The size of each oval on Figure 2 emphasizes the relative importance of each type of AS. We used dashed lines between tier-2, tier-3 and tier-4 on Figure 2 to insist on the loose classification of these three AS types.

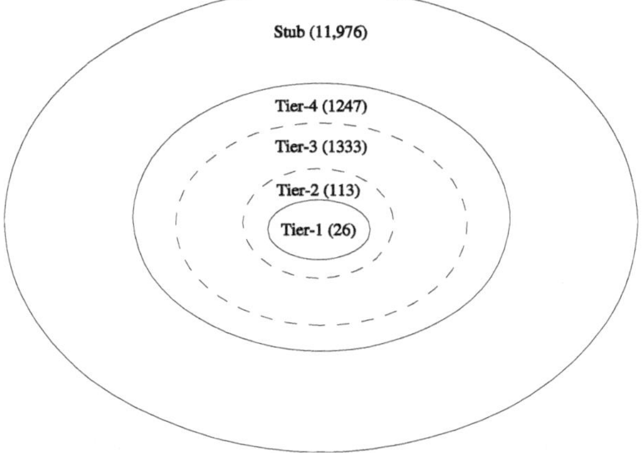

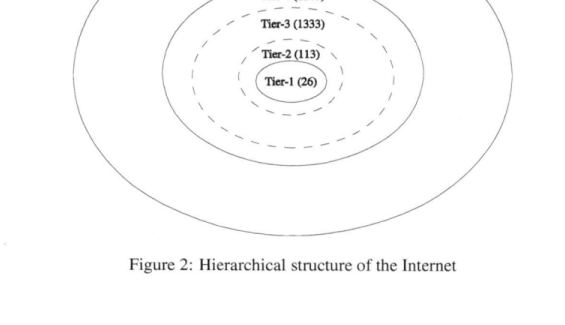

Figure 2: Hierarchical structure of the Internet

Traffic Engineering

The main purpose of this book is to evaluate interdomain traffic engineering with BGP, based on the behavior of real traffic. The term "traffic engineering" was defined in RFC 3272 [19] as

Internet traffic engineering is defined as that aspect of Internet network engineering dealing with the issue of performance evaluation and performance optimization of operational IP networks. Traffic Engineering encompasses the application of technology and scientific principles to the measurement, characterization, modeling, and control of Internet traffic.

Traffic engineering thus consists of all the available techniques whose purpose is to directly or indirectly allow to modify the behavior of the traffic to achieve certain objectives. Traffic engineering has received a lot of attention during the last few years [19, 207]. Initially, traffic engineering was considered as a solution to allow large transit providers to optimize the utilization of their network. In these large networks, there are typically several possible paths to reach a given destination or border router. Ideally, to achieve a good network utilization, the traffic should be spread evenly among all the available links. Unfortunately, this does not correspond to the way traditional IP routing protocols behave.

In most cases, the IP routing protocol is not aware of the load on the various parts of the network and selects for each destination the shortest path based on static metrics such as the hop count or the delay. This destination-based routing creates an uneven distribution of the traffic that may lead to periods of congestion inside the Internet service provider (ISP) backbone. Several techniques have been proposed to better spread the load throughout the entire network [19]. A first solution is to select appropriate link metrics based on a known traffic matrix [88, 206]. This solution can provide some interesting results if the traffic matrix is known and stable. A second solution is to rely on a connection-oriented layer-2 technology [21] such as ATM, MPLS or one of the emerging optical technologies. In this case, layer-2 connections can be established statically [66] or dynamically between distant routers and the layout of these connections can be optimized to achieve an even distribution of the traffic inside the network [19]. It is also possible to dynamically create new layer-2 connections in order to quickly respond to link failures or changes in the traffic pattern [19].

At the opposite of large transit providers, small providers and multi-homed corporate networks have different traffic engineering requirements. Their networks have usually a simple topology and are frequently over-provisioned. The traffic engineering solutions mentioned above are not really useful in such networks. For these networks, the costly resource that needs to be optimized with traffic engineering is their interdomain connectivity, i.e. the links that connect them to the rest of the Internet, as they typically have to pay to access the commercial Internet.

Most of the traffic engineering literature has focused on intradomain traffic engineering and large transit providers. This book on the other hand aims to improve our understanding of the issues related to interdomain traffic engineering for stub ASes. Throughout this book, our approach is to rely on real network traffic traces to evaluate various aspects related closely or remotely to interdomain traffic engineering. The only part of the book where we do not rely on traffic traces is Part 1 which is a prerequisite for all other parts.

Approach

Now let us explain the logic behind the structure of this book. Traffic engineering is about engineering the traffic. What is being done is thus *engineering* while *on what* it is being done is the *traffic*. To understand interdomain traffic engineering, the *traffic* part is as important as the *engineering* one. We thus partitioned the content into several parts that deal with different aspects of the interdomain traffic engineering problem.

In Part 1, we provide the necessary information related to BGP and its working for controlling the interdomain traffic. Reading thoroughly Part 1 is necessary to understand the complexity and issues of interdomain traffic engineering. This first part is largely protocol-centered, focusing on the *engineering* part of interdomain traffic engineering. In Part II, we focus on the *traffic* part of traffic engineering by closely looking at the properties of the interdomain traffic. Part III tackles the optimization of traffic objectives by modifying the choice of the BGP routes. Viewing interdomain traffic engineering as an optimization problem will show us the limitations of pushing BGP to its limits. Finally, Part IV gives a global evaluation of interdomain traffic engineering and discusses its future.

Throughout this book, our viewpoint will be largely biased towards the *traffic* part of traffic engineering. Our feeling when this work started (end of year 2000) was that while traffic engineering techniques were being designed, not much was known about the behavior of the interdomain traffic except its scaling properties (see Chapter 2). It is [193] where we evaluated the feasibility of using MPLS for carrying interdomain traffic that lead us to raising the issue of the feasibility of interdomain traffic engineering for stub ASes.

Due to concerns about resilience and performance [12, 13, 73], more and more non-transit ASes rely on multi-homing [10]. Multi-homing consists in connecting a network to the Internet at different access points. Multi-homing requires to control the traffic at those multiple interdomain access points. Interdomain traffic engineering tools and techniques are thus becoming more important for such networks [155, 25, 15]. One popular traffic engineering tool is referred as "route optimization" [15]. Route optimization techniques work on relatively short timescales and target multi-homed ASes. Their principle is to find, based on active and passive measurements, the "best" route to attain a destination or/and to be attained by external hosts. These techniques work on timescales in the order of a round-trip time or more and adapt the traffic flow to the current conditions of the network, i.e. they try to find the best upstream provider to send or receive traffic. Their objective is often to optimize the "instantaneous" quality of service (QoS) experienced by the traffic, by measuring the "quality" of the routes available from the upstream providers and trying to choose the best upstream in real time. As "route optimization" techniques are commercial solutions, no work in the scien-

tific literature has evaluated their internals or their actual impact on the traffic. One of the goals of this book is to give some insight into the issues of such techniques, by studying the traffic properties as well as the traffic engineering mechanisms.

The current state-of-the-art of interdomain traffic engineering is of manually changing the configuration of the routers to influence the traffic on a trial and error basis [77]. Interdomain traffic engineering is still in its infancy. Its evolution is hard to foresee. Understanding the dynamics of the interdomain traffic and the relationships between the topology of the traffic and BGP peering relationships is absolutely necessary to do it properly [17]. Our contribution in this book is to reduce the gap between the *traffic* part and the *engineering* part of traffic engineering. In addition, we give strong indications in Part III that designing interdomain traffic engineering is possible with a proper understanding of the traffic characteristics and interdomain routing.

Although we largely delay the discussion on the rationale of interdomain traffic engineering to Part IV, we nevertheless mention here very shortly a few aspects of the design of the Internet before getting into the content of this book. The design objectives during the early years of the Internet (called the ARPANET at the time) had in mind interoperability and robustness to failures, as it was a research internetwork. Roughly speaking, the issue of the ARPANET was to allow end hosts to communicate across the network even in the presence of router failures, and to ensure interoperability between heterogeneous devices. Today the Internet covers a large fraction of the planet, with millions of hosts connected to the network. While the ARPANET was a communication system, the Internet today is rather a virtual world over which agents interact. The ARPANET was a communication infrastructure that had to be kept working. The Internet nowadays is much more than just another medium to exchange data. Many businesses today depend on the Internet. Agents that pursue different potentially conflicting objectives share this new virtual world, and have expectations about its behavior and performance. The Internet has thus become a far more critical infrastructure than what it was 30 years ago.

Road map

As the reader may be interested in only specific aspects of the interdomain traffic engineering problem, we provide possible shortcuts through the content of thie book. Figure 3 illustrates the structure of the book and the interdependencies among its four parts. Part 1 is a prerequisite for all other three parts. Even readers that might think they understand the working of BGP should read Part 1, because understanding the BGP protocol is not sufficient to fully grasp the complexity of interdomain traffic engineering with BGP. Parts II, III and IV are relatively independent and any of them can be skipped without impeding the understanding of others parts.

Part I presents the working of the BGP protocol and the interactions between BGP and the control of interdomain traffic.

Part II presents the characteristics of the interdomain traffic, as seen from stub ASes. The first chapter of Part II presents the time dynamics of the interdomain traffic as seen at the access links of several stubs ASes (Chapter 2). The next chapter studies how this traffic seen at the access links of stub ASes is spread over the interdomain topology of the Internet. In the last chapter of this first part, Chapter 4, we combine the time dynamics and the topological spread on the interdomain topology.

In Part III, we apply our knowledge from Part I in order to control the traffic of stub ASes by tweaking the BGP routing protocol. We start in Chapter 5 by doing BGP-based traffic engineering to optimize the daily traffic of stub ASes. Then, we ask ourselves the question of how to minimize the number of BGP changes to optimize the daily traffic of stub ASes in Chapter 6. In the next chapter (Chapter 7), we push traffic engineering a bit more, by reducing the timescale at which the traffic is controlled, from days to minutes. The last chapter of Part III, Chapter 8, looks at the problem of traffic control when multiple objectives have to be optimized simultaneously.

In the last part of this book, Part IV, we evaluate the current state of inter-

domain traffic engineering, envision its future and provide a few guidelines for further work.

Sections that we consider more difficult or technical are marked with an asterisk (*), these might be skipped on a first reading without impacting too much on the understanding of the rest of the text. Note however that some important technical terms might be defined in these sections so that checking them whenever an undefined term is encountered is recommended.

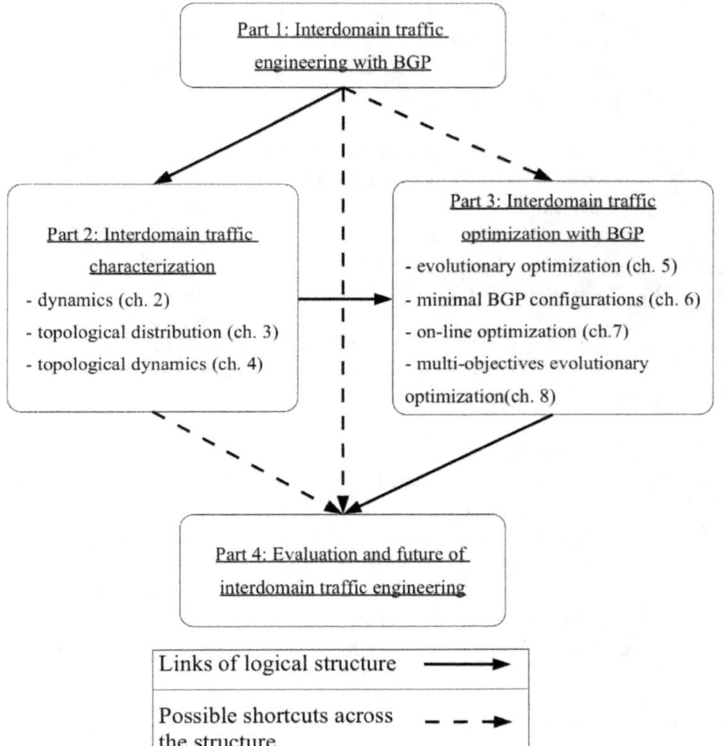

Figure 3: Interdependencies among the different parts of the monograph

Acknowledgments

Acknowledging every person that has had a direct or indirect impact on the realization of this book is not possible. I have benefited from too many persons to ever being able to mention all of them in such a limited space as this acknowledgement section. I shall thus try to keep short and praise only the first few whom I recall by the time of this writing.

The key person that made this monograph possible is my friend and coincidentally thesis advisor, Olivier Bonaventure. The freedom he gave me during the early years of my career as a researcher had a strong impact on how I think today. Our respective viewpoints have sometimes been conflicting, with the fortunate consequence of strengthening my personality as a researcher. Olivier never forced me to adopt his opinion, my stubborness would have made such a willingness futile anyway. Rather he pushed me to justify my viewpoint even if it had to go against his. I thus believe today that Science largely consists in disproving our beliefs, which in turn leads closer to the truth. I would like to thank him again for being part of this.

The next most important collaborator is Bruno Quoitin. First, Bruno co-authored several important papers related to this book [154, 155, 197, 153, 32, 200]. Second, Bruno provided simulations used in Part IV of the book. Finally, our many discussions contributed to reinforce my opinion on several subjects discussed in Part IV.

Sébastien Tandel co-authored [31, 201]. Additionally, he pushed me to provide the lengthy discussion on the tie-breaking rules of the BGP decision process in Chapter 1.

Chris Rapier has been a wonderful paper co-author to work and discuss with. Unfortunately, our contacts have almost exclusively occured through e-mail, except a short meeting at ACM SIGCOMM 2002 in Pittsburgh. Chris generously provided several traffic traces captured at the Pittsburgh Supercomputing Center

that helped make possible several papers [198, 155, 190, 199] as well as Chapters 2 and 4 of this thesis.

I cannot close this aknowledgement part without thanking my parents, Małgorzata Migalska and Pierre Uhlig, my sister Sophie Uhlig, and my beloved partner Almerima for their support, love and patience while my body and mind were busy doing research.

Before concluding this acknowledgement section, let the readers pardon me for a moment of humility. Recognizing our limitations is a skill I consider most valuable for any researcher in the search of truth. The following passage comes from a famous poem, the Bodhicaryavatara, written around the 9th century A.D. by the Buddhist monk Santideva. While vainly trying to pursue a long Buddhist tradition of humility, may this citation incense your mind during the journey throughout this book, as it incenses mine during my scientific journey...

"Nothing new will be said here, nor have I any skill in composition. Therefore I do not imagine that I can benefit others. I have done this only to perfume my own mind".

Santideva, The Bodhicaryavatara.

Part I

Background

Chapter 1

Interdomain traffic engineering with BGP

1.1 Introduction

In this first part and chapter we present the working of the interdomain routing protocol, BGP, and discuss the various ways in which interdomain traffic engineering is possible by tweaking BGP. A large fraction of the content of this chapter appeared in [155, 31].

This chapter is structured as follows. In section 1.2 we explain the working of BGP routing in the Internet. In section 1.3 we introduce the main feature of BGP, route filtering. In section 1.4 we explain the working of the BGP decision process. Finally, in section 1.5 we explain interdomain traffic engineering.

1.2 BGP routing in the Internet

Internet routing is handled by two distinct protocols with different objectives. Inside a single domain[1], link-state intradomain routing protocols (e.g. OSPF or IS-IS) distribute the entire network topology to all routers and compute the shortest path towards each router according to a metric chosen by the network administrator. Across interdomain boundaries, the interdomain routing protocol is used to distribute reachability information and to select the best route to each destination according to the policies specified by each domain administrator. For confidentiality and scalability reasons, the interdomain routing protocol is only aware of the interconnections between distinct domains, it does not know any information about the internals of external domains. A result of this two-level system is that a

[1]In BGP terminology, a domain is equivalent to an Autonomous System (AS).

router typically has a detailed knowledge of the topology within its domain while only very limited information about other domains.

The Border Gateway Protocol (BGP) [156, 175] is the current de facto standard interdomain routing protocol. BGP is a *path-vector protocol* that works by sending *route advertisements.* A route advertisement indicates the reachability of a network, namely a network address and a netmask representing a block of contiguous IP addresses. For instance, `192.168.0.0/24` represents a block of 256 addresses between `192.168.0.0` and `192.168.0.255`. Besides the reachable network and the IP address of the router that must be used to reach this network (known as the `BGP next-hop`), a route advertisement also contains the `AS-path` attribute which contains the list of all the transit ASes that must be used to reach the announced network. A route advertisement may also contain several attributes such as the `local-pref`, Multi-Exit Discriminator (`MED`) or `communities` attributes [156, 175]. An important point about BGP is that if a BGP router of `ASx` sends a route announcement for network N to a neighbor BGP router of `ASy`, this implies that `ASx` accepts to forward IP packets to destination N on behalf of `ASy`. The flow of BGP routing information and the flow of the traffic go in opposite directions.

Route advertisements are sent over *BGP sessions* between routers. There are two flavors of BGP sessions [156, 175]. The eBGP variant is used to announce the reachable prefixes on a link between routers that are part of distinct ASes. The iBGP variant on the other hand is used to distribute, inside an AS, the best routes learned from neighboring ASes. To distribute the best BGP routes inside an AS, each BGP router of that AS will establish an iBGP session with all the other BGP routers of the AS. A full-mesh of iBGP sessions is typically created inside small ASes, thus requiring $\frac{n \times (n-1)}{2}$ sessions. Large ASes on the other hand rely on route-reflectors or confederations [26, 186] to reduce the number of iBGP sessions.

The top part of Figure 1.1 presents a simple Internet topology composed of six ASes. Each AS contains routers named Rxy where x identifies the AS number to which the router belongs and y identifies the router in ASx. The bottom part of Figure 1.1 shows the eBGP and iBGP sessions established between the BGP routers. All routers of the simple Internet topology (top part of Figure 1.1) dot not need to run BGP. R_{12} for instance is used by the edge routers of AS1 as a transit router for the traffic between the edge routers. The BGP routers of AS1 are not aware of the existence of router R_{12} with BGP since no BGP session is established with R_{12}. They only know router R_{12} through IGP routing.

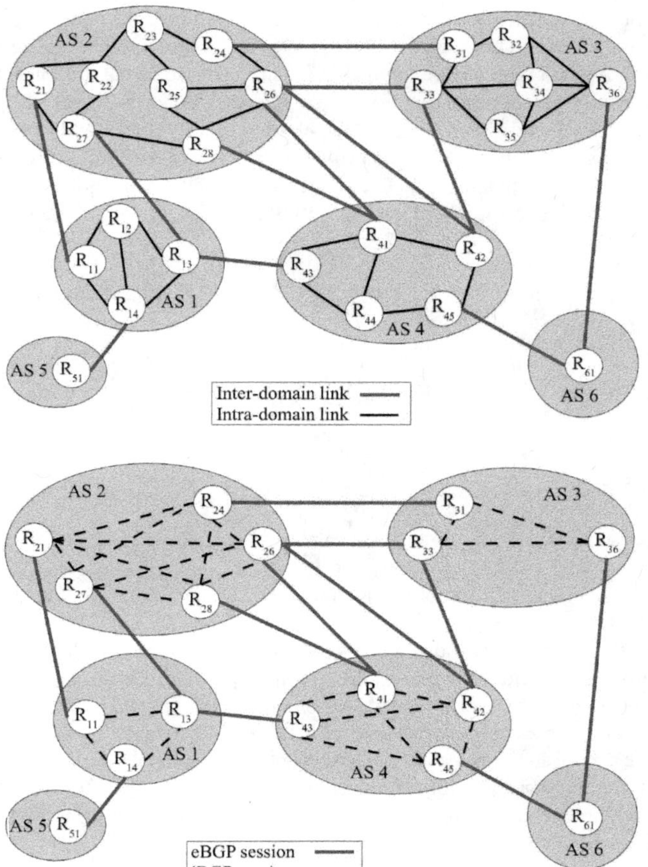

Figure 1.1: Simple Internet topology (top) and corresponding eBGP and iBGP sessions (bottom).

1.3 Route filtering

Inside a single domain, all routers are considered equal by the IGP. All usable intradomain links and routers are visible and used to compute the shortest paths between any pair of routers. In contrast, in the global Internet, all ASes are not equal and an AS will rarely agree to provide a transit service for all its connected ASes toward all destinations. Therefore, BGP allows a router to be selective in the route advertisements that it sends to neighboring eBGP routers, and to prefer some neighbors over others. This route filtering process is the basis of what is called "policy routing" in the Internet. To better understand the operation of BGP, it is useful to consider a simplified view of a BGP router as shown in Figure 1.2.

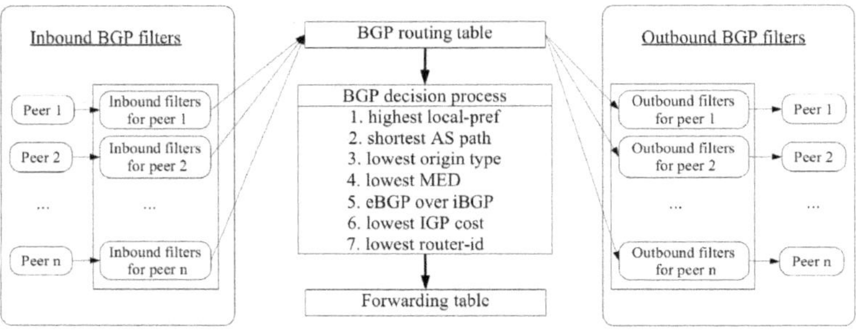

Figure 1.2: Simplified operation of a BGP router.

A BGP router processes and generates route advertisements as follows. First, the administrator specifies, for each BGP peer, an input filter (left of Figure 1.2) that is used to select the acceptable advertisements. For example, a BGP router could only select the advertisements with an `AS-Path` containing a set of trusted ASes. Once a route advertisement has been accepted by the input filter, it is placed in the BGP routing table, possibly after having updated some of its attributes. The BGP routing table thus contains all the acceptable routes received from the BGP neighbors.

Second, on the basis of the BGP routing table, the BGP decision process (center of Figure 1.2) will select the best route toward each known network. Based on the `BGP next-hop` of this best route and on the intradomain routing table, the router will install a route toward this network inside its forwarding table. This table is then looked up for each received packet and indicates the outgoing interface that must be used to reach the packets' destination.

Third, the BGP router will use its output filters (right of Figure 1.2) to select

among the best routes in the BGP routing table the routes that will be advertised to each BGP peer. At most one route will be advertised for each reachable destination. The BGP router will assemble and send the corresponding route advertisement messages after a possible update of some of their attributes.

The input and output filters used in combination with the BGP decision process are the key mechanisms that allow a network administrator to support within BGP the business relationships between two ASes. Many types of business relationships can be supported by BGP. Two of the most common relationships are the customer-to-provider and the peer-to-peer relationships [176, 90]. To understand how these two relationships are supported by BGP, consider the top part of Figure 1.1. If `AS5` is `AS1`'s customer, then `AS5` will configure its BGP router to announce its routes to `AS1`. `AS1` will accept these routes and announce them to its peer (`AS4`) and upstream provider (`AS2`). `AS1` will also announce to `AS5` all the routes it receives from `AS2` and `AS4`. If `AS1` and `AS4` have a peer-to-peer relationship on the link between R_{13} and R_{43}, then router R_{13} will only announce on this link the internal routes of `AS1` and the routes received from `AS1`'s customer (i.e. `AS5`). The routes received from `AS2` will be filtered and thus not announced on the $R_{13} - R_{43}$ link by router R_{13}. Due to this filtering, `AS1` will not carry traffic from `AS4` toward `AS2`.

1.4 The BGP decision process

A BGP router receives one route toward each destination from each of its peers. To select the best route among this set of routes, a BGP router relies on a set of criteria called the decision process. Most BGP routers apply a decision process similar in principle to the one shown in Figure 1.2 [156]. The set of routes with the same destination are analyzed by the criteria in the sequence indicated in Figure 1.2. These criteria act as filters and the nth criterion is only evaluated if more than one route has passed the $n - 1^{th}$ criterion. Most BGP implementations allow the network administrator to optionally disable some of the criteria of the BGP decision process.

The first rule of the BGP decision process uses the `local-pref` attribute. The `local-pref` attribute is an administrative cost specifying the preference among the different routes towards a given destination. The route with the highest value of the `local-pref` attribute is considered as the best one. This attribute can be considered as the one that allows to enforce the different types of peering relationships by indicating which route is to be preferred over another. The `local-pref` attribute is local to the AS receiving the route. It is set upon receival of the BGP route and not advertized outside the local domain.

The second rule of the BGP decision process uses the length of the `AS path`

of the route. Even though the original purpose of this attribute was to detect loops in the propagation of the BGP routes, this rule also constitutes a distance metric for the interdomain path followed by packets whose destination IP address matches the prefix of the route. The best route is the one having the shortest `AS path` length. The information contained in the `AS path` attribute concerns only the traffic that goes from the local AS towards the prefix of the route. The actual path used in the reverse direction ought not to be the same due to local policies enforced along the path between the local AS and the destination prefix [178, 91].

The third rule concerns the `origin type` of the route. This attribute identifies how the originating AS learned about the route, where IGP is preferred to EGP (a now defunct distance-vector protocol) which is preferred to INCOMPLETE. This attribute is becoming obsolete although it is still currently used in the BGP decision process [54].

The fourth rule is related to the `MULTI_EXIT_DISC` (for multi exit discriminator) attribute. Comparing routes with respect to the value of the MED attribute requires that these routes be learned from the same neighboring AS. Routes which do not have the MED attribute are considered to have the lowest possible MED value. The route with the lowest MED value is preferred. The MED attribute serves at selecting a particular egress point in the local domain. The input filter might override the `origin type` and the value of the MED attribute so that the use of these attributes for route selection usually depend on a mutual understanding between the two neighboring ASes.

The fifth rule then prefers the routes that have been learned through eBGP over those learned from iBGP.

The sixth rule prefers the route with the lowest IGP path cost to the next hop. This rule (along with the previous one) allows for what is commonly termed "hot-potato" routing, which means that the local AS tries to get rid of the IP packet as soon as possible. The logic behind "hot-potato" routing is to minimize the resource consumption implied by the forwarding of the IP packet that transits inside the local domain. In that way, IP packets are routed inside the local domain through the shortest path within the boundaries of the local domain according to the IGP metric.

The last rule of the BGP decision process chooses as the best route the one whose `router-id` (IP address of the BGP peer router) is the lowest. This last rule ensures that no two routes towards the same prefix can be chosen since at most one route towards any given prefix is advertised by a BGP peer.

According to the BGP-4 Internet draft [156], all rules other than the first one are called the "tie-breaking" rules of the BGP decision process. From our viewpoint, the choice of this word "tie-breaking" is unfortunate, if not misleading. The "official" metric for the BGP routes according to the Internet draft is thus the value of their `local-pref` attribute. Given that this attribute is not present in routes

exchanged between eBGP routers but its value is set through the input filters of the local BGP routers of the domain, rules 2 to 7 of the decision process shown on Figure 1.2 would then have no meaning concerning the quality of the routes. Our viewpoint in this book is rather to define the "tie-breaking" rules as "*the subset of the rules of the BGP decision process that allow an AS to choose a best route among those that cannot be differentiated according to the rules that are meaningful for the considered AS*". The drawback of the previous definition concerns its dependence on the AS being considered. This definition however admits the consideration of different subsets of the rules of the BGP decision process as the "tie-breaking" rules. Some ASes may simply not realize which rules constitute their actual "tie-breaking" rules. For instance, stubs rarely use rules beyond the second one to differentiate the quality of their routes so all rules after the second act as "tie-breaking" rules. Transit providers on the other hand often consider the rules up to the sixth one to assess a route's quality, only rule seven acts as a "tie-breaking" rule. Figure 1.3 illustrates the three different views of the "tie-breaking" rules according to the BGP-4 Internet draft [156], transit providers and stubs.

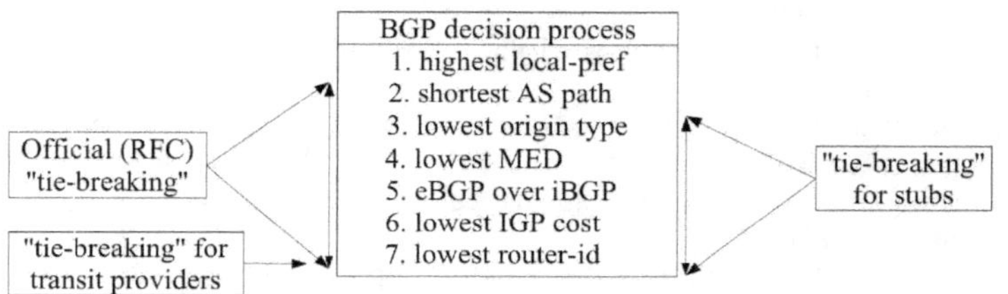

Figure 1.3: Various perspectives on the meaning of "tie-breaking" rules in the BGP decision process.

1.5 Interdomain Traffic Engineering

Interdomain traffic engineering requirements are diverse and often motivated by the need to balance the traffic on links with other ASes and to reduce the cost of carrying traffic on these links. These requirements depend on the connectivity of an AS with others but also on the type of business handled by this AS.

The connectivity between ASes is mainly composed of two types of relationships (also called peerings). The most frequent relationship between ASes is the

customer-to-provider relationship where a customer AS pays to use a link connected to its provider. The customer obtains a global connectivity to the Internet through its providers and becomes reachable by all other ASes in the Internet. A stub AS usually tries to maintain at least two of these links for performance and redundancy reasons [176]. In addition, ASes try to obtain *peer-to-peer* relationships with other ASes and then share the cost of the link with the other AS. *Customer-to-provider* peering relationships are typically private, as they involve a commercial business agreement. *Peer-to-peer* relationships on the other hand are increasingly relying on public exchange points, where hundreds of ASes are peering with each other. Negotiating the establishment of AS relationships is often a complicated process since technical and economical factors, as exposed in [24], need to be taken into account.

Moreover, an AS will want to optimize the way traffic enters or leaves its network, based on its business interests. For instance, content providers that host a lot of web or streaming servers and usually have several customer-to-provider relationships with transit ASes will try to optimize the way traffic leaves their networks. On the other hand, access providers that serve small and medium enterprises, dialup or xDSL users typically wish to optimize how traffic enters their network. Finally, a transit AS will try to balance the traffic on the multiple links it has with its peers.

Optimizing the way traffic enters or leaves a network means to favor one link over another to reach a given destination or to receive traffic from a given source. Interdomain traffic engineering of this kind can be performed by tweaking the BGP routes of the AS.

Interdomain traffic engineering can be used by operators having different requirements in mind. The first type of requirement originates from stub ASes that host a lot of content. Usually, these ASes have several customer-to-provider relationships with larger ASes and need to optimize how their content leaves their own network. If the amount of traffic of such ASes grows, they might have interest in negotiating peer-to-peer relationships with their providers and even become transit providers not to have to rely on customer-to-provider relationships at all. On the other hand, ASes that provide Internet access to dialup or broadband users typically wish to optimize how the Internet traffic enters their network, as they are content consumers. The traffic engineering goal of a content provider could thus differ much from the traffic engineering goal of an access provider. Finally, transit ASes that connect access and content providers also have their own interdomain traffic engineering requirements.

1.5.1 Control of the outgoing traffic

To control how the traffic leaves its network, an AS must be able to choose which route will be used to reach a particular destination through its peers. Since an AS has control on the decision process of its BGP routes, it can influence the selection of the best path. Two techniques are frequently used.

A first technique is to rely on the local-pref attribute. This optional attribute is only distributed inside an AS. It can be used to rank routes and is the first criteria of the BGP decision process (center of Figure 1.2). For example, consider a stub AS with two links toward one upstream provider : a primary link and a backup link. In this case, the BGP router of this AS could be configured to insert a low local-pref inside routes learned via the backup link and a higher value to routes learned via the primary link.

In practice, the manipulation of the local-pref attribute can also be based on passive or active measurements. Recently, a few companies have implemented solutions [33, 15] that allow multi-homed stub ASes and content-providers to engineer their interdomain traffic. These solutions usually measure the load on each interdomain link –some also rely on active measurements– to evaluate the performance of interdomain paths. Based on these measurements and some knowledge of the Internet topology, they attach appropriate values of the local-pref attribute to indicate which route should be considered as the best one by the BGP routers.

A second technique, often used by large transit providers, is to rely on the intradomain routing protocol to influence how a packet crosses the network of the transit provider. As shown on Figure 1.2, the BGP decision process will select the closest IGP neighbor (in terms of IGP cost) when comparing several equivalent routes received via an iBGP session. For example, consider in Figure 1.1 that router R_{27} receives one packet whose destination is R_{45}. The BGP decision process of router R_{27} will compare two routes towards R_{45}, one received via R_{28} and the other received via R_{26}. By selecting router R_{28} as the exit border router for this packet, AS2 ensures that this packet will consume as few resources as possible inside its own network. If a transit AS relies on the tuning of the weights of its intradomain routing protocol as described in [87], this tuning will influence its outgoing traffic.

1.5.2 Control of the incoming traffic

One method that can be used to control the traffic that enters an AS is to rely on selective advertisements and to announce different routes on different links. Note that advertising different routes to a particular neighboring AS can be considered as wrong [79]. Taking for example the topology given on the top part of Figure 1.1, if AS1 wants to balance the traffic coming from AS2 over the links

$R_{11}-R_{21}$ and $R_{13}-R_{27}$, it could announce only its internal routes on the $R_{11}-R_{21}$ link and only the routes learned from `AS5` on the $R_{13}-R_{27}$ link. Since `AS2` would only learn about `AS5` through router R_{27}, it would be forced to send the packets whose destination belongs to `AS5` via router R_{27}. Yet if the link $R_{13} - R_{27}$ fails, then `AS2` would not be able to reach `AS5` through `AS1`. This is not desirable and it should be possible to utilize link $R_{11} - R_{21}$ for the packets toward `AS5` at that time without being forced to change the routes that are advertised on this link.

A variant of the selective advertisements is the advertisement of more specific prefixes. This advertisement relies on the fact that an IP router will always select in its forwarding table the most specific route for each packet (i.e. the matching route with the longest prefix). For example, if a forwarding table contains both a route toward `16.0.0.0/8` and a route toward `16.1.2.0/24`, then a packet whose destination is `16.1.2.200` would be forwarded along the second route. The advertisement of more specific routes can also be used to control the incoming traffic. Assume for instance that prefix `16.0.0.0/8` belongs to `AS3` and that several important servers are part of the `16.1.2.0/24` subnetwork. If `AS3` prefers to receive the packets toward its servers on the R_{24}-R_{31} link, then it would advertise both `16.0.0.0/8` and `16.1.2.0/24` on this link and only `16.0.0.0/8` on its other external links. This solution has the advantage that if link R_{24}-R_{31} fails, then the subnetwork `16.1.2.0/24` would still be reachable through the other links.

Another method would be to allow an AS to indicate a ranking among the various route advertisements that it sends. Based on the use of the length of the `AS path` as the second criterion in the BGP decision process, a possible way to influence the selection of routes by distant ASes is to artificially increase the length of the `AS path` attribute. Coming back to the top part of Figure 1.1, assume that the primary interdomain link of `AS3` is link $R_{61} - R_{45}$ while link $R_{61} - R_{36}$ is only used as backup primary link. In this case, `AS6` would announce its routes normally on the primary link with an `AS path` of `AS6` while it would add its own AS number several times instead of once in the AS path attribute on the $R_{61} - R_{36}$ link. The route advertised on the primary link will be considered as the best route by all routers that do not rely on manually configured settings for the `local-pref` attribute. This technique can be combined with selective advertisements. For example, an AS could divide its address space in two prefixes $p1$ and $p2$ and advertise prefix $p1$ without prepending and prefix $p2$ with prepending on its first link and the opposite of its second link. Note that in practice predicting the actual effect of prepending on inbound traffic [152] or finding the optimal amount of prepending to balance traffic [92] is difficult.

The last method to allow an AS to control its incoming traffic is to rely on the MED attribute. This optional attribute can only be used by an AS multi-connected to another AS to influence the link that should be used by the other

AS to send packets toward a specific destination. The utilization of the MED attribute is usually subject to a negotiation between the two peering ASes [79] and some ASes do not accept to take the MED attribute into account in their decision process.

1.5.3 Community-based Traffic Engineering*

In addition to these techniques, several providers have been using the communities attribute to give their customers a finer control on the redistribution of their routes. The communities attribute is an optional attribute that can be attached to routes. This attribute can contain several 32 bits wide community values. Community values are often used to attach optional information to routes such as a code representing the city where the route was received or a code indicating whether the route was received from a peer or a customer. The community values can also be used for traffic engineering purposes. For instance, predefined community values can be attached to routes in order to request actions such as not announcing the route to a specified set of peers, prepending the AS path when announcing the route to a specified set of peers or setting the value of the local-pref attribute. However, this technique relies on an ad hoc definition of community values and on manual configurations of BGP filters which makes it difficult to use.

The IETF is currently considering the definition of a new standard type of extended communities that are called "redistribution communities" [30] to answer the shortcomings related to the use of classical communities for traffic engineering purposes [153]. Redistribution communities can be attached to routes to influence the redistribution of these routes by the upstream AS. The redistribution communities attached to a route contain both the traffic engineering action to be performed and the identification of the BGP peers that are affected by this action. One of the supported actions allows an AS to indicate to its upstream peer to not announce the attached route to some of its BGP peers.

Another type of action allows an AS to request its upstream peer to perform AS path prepending when redistributing a route to a specified peer. To understand the usefulness of such redistribution communities, let us consider again the top part of Figure 1.1, and assume that AS6 receives a lot of traffic from AS1 and AS2 and that it would like to receive the packets from AS1 (respectively AS2) on the R_{45}-R_{61} link (respectively the R_{36}-R_{61} link). AS6 cannot ensure such a traffic distribution by performing AS-Path prepending itself. However, this becomes possible with the redistribution communities by requesting AS4 to perform the prepending when announcing the AS6 routes to external peers. AS6 could thus advertise to AS4 its routes with a redistribution community that indicates that this route should be prepended two times when announced to AS2. With this redistribution community, AS4 would advertise AS path AS4:AS4:AS6 to AS2

and AS path AS4:AS6 to AS1. AS2 would hence receive two routes toward AS6, AS4:AS4:AS6 and AS3:AS6, and AS2 would select the route via AS3. AS1 on the other hand would select the AS4:AS6 route having a shorter AS path than the AS2:AS3:AS6 route.

1.5.4 Limitations

We have described several techniques that can be used by ISPs to engineer their interdomain traffic. However, there are limitations to be considered when deploying those techniques.

A first point to note is that the control of the outgoing traffic with BGP is based on the selection of the best route among the available ones. This selection can be performed on the basis of various parameters, but is limited by the diversity of routes received from neighboring ASes. The diversity of the routes received by an AS depends on the connectivity and the policy of these neighboring ASes [136, 137]. At most one route towards each prefix will be received from any BGP neighbor.

The control of the incoming traffic is based on a careful tuning of the advertisements sent by an AS. This tuning can cause several problems. First, an AS that advertises more specific prefixes or that divides its address space into distinct prefixes to announce them selectively will advertise a number of prefixes larger than required. These prefixes will be propagated throughout the global Internet and will increase the size of the BGP routing tables of potentially all ASes in the Internet. [36] reports that more specific routes constituted more than half of the entries in a BGP table. Faced with this increase of their BGP routing tables, several large ISPs have started to install filters to ignore the BGP advertisements corresponding to more specific prefixes [27]. The deployment of those filters implies that the more specific prefixes will not be announced by those large ISPs.

When considering the manipulation of the AS path attribute, we have mentioned that it can be used on backup links. It is sometimes also used to better balance the traffic. For instance, [36] reports that AS path prepending affected 6.5 percent of the BGP routes in November 2001. However, in practice it can be difficult to predict the outcome of performing AS path prepending on a given interdomain link [152]. Usually, providers that rely on AS path prepending try to find out the right amount of prepending on a trial and error basis.

The redistribution communities can provide a finer granularity than AS path prepending or selective announcements. In practice, it can be expected that those communities will be used to influence the redistribution of routes toward transit ISPs having a large number of customers. A recent study of the use of BGP communities [68] shows that more and more ISPs rely on them to perform traffic engineering.

An important point to bear in mind concerning the techniques that require changes to the attributes of the BGP advertisements is that any change to an attribute will force the route advertisement to be redistributed to potentially the entire Internet. Although it would be possible to define techniques relying on measurements to dynamically change the BGP advertisements of an AS for traffic engineering purposes, a widespread deployment of such techniques would increase the number of BGP messages exchanged and could lead to BGP instabilities [116, 99]. Any dynamic interdomain traffic engineering technique that involves frequent changes to the values of BGP attributes should be studied carefully before being deployed.

Finally, several features exist in BGP implementations to engineering the traffic. Cisco's BGP multipath [51] allows to load-balance outgoing traffic when several equal cost (only the router-id differs) eBGP routes are learned from a neighboring AS. Per-packet or per-destination load balancing is then performed by the router among the multiple paths. Unequal cost load-balancing is also possible by relying on the extended community attribute [165]. However, Cisco's BGP load-sharing techniques [52] are limited to multiple links on a single router and do not work well with multiple routers. In addition, these techniques are rather limited in their scope: outgoing load-balancing (packet-based or destination-based) and static incoming traffic distribution based on static relative available link bandwidth. The same features are also available on Juniper routers [110].

Table 1.1 provides a summary of BGP-based traffic engineering methods. Table 1.1 gives the scope (local or remote) of each method, the traffic direction concerned (inbound or outbound), and the conditions under which the traffic engineering method works.

Technique	**Scope**	**Traffic**	**Conditions**
`local-pref`	local AS	outbound	none
selective announcements	Internet	inbound	none
more specific prefixes	Internet	inbound	no prefix filtering
`AS-Path` prepending	Internet	inbound	none
`MED`	neighbor ASes	inbound	bilateral agreement
`communities`	variable	inbound	bilateral agreement

Table 1.1: Summary of BGP-based traffic engineering methods.

1.6 Conclusion

In this first part, we have described several techniques used today to control the flow of packets in the global Internet. We have described the current organization of the Internet and the key role played by BGP. We have explained how BGP is tuned today for interdomain traffic engineering purposes. We have shown that an AS has more control on its outgoing than on its incoming traffic. Several techniques can be used to control the incoming traffic, but they have limitations. The selective advertisements and the more specific prefixes have the drawback of increasing the size of the BGP routing tables. With AS path prepending, it can be difficult to select the appropriate value of prepending to achieve a given goal. Finally, we have explained how the redistribution communities could allow an AS to flexibly influence the redistribution of its routes toward non-directly connected ISPs.

Part II

Interdomain traffic characterization

In this second part, we analyze in details several interdomain traffic traces to develop a better understanding of the behavior of the traffic. Understanding the behavior of the traffic is important to realize how variable interdomain traffic is at different timescales. In the context of interdomain traffic engineering, the topological distribution of the traffic is particularly important. Indeed, it concerns the number of interdomain sources and destinations that will have to be influenced by the traffic engineering technique. This is not to say that traffic dynamics is unimportant. The dynamics of the traffic has deep implications on its prediction and control. Before designing a traffic engineering method, one should first properly understand over which timescales the traffic is stable enough to be controlled.

In Chapter 2, we analyze the behavior of the traffic dynamics that enters or leaves a stub AS over timescales ranging from seconds to several days. We then go to study the topological distribution of the interdomain traffic in Chapter 3. Finally, Chapter 4 analyzes the dynamics of the traffic on the interdomain topology.

Chapter 2

Interdomain traffic dynamics

2.1 Introduction

It could be argued that the aspect of the dynamics of the interdomain traffic is not fundamentally relevant to interdomain traffic engineering. The purpose of this chapter is to show evidence of the contrary. Many problems in engineering require the knowledge of the timescales over which controlling a system is possible. For most control applications, too large fluctuations in the systems dynamics mean that another timescale should be used for controlling it. In this chapter, we analyze the behavior of the interdomain traffic to understand on which timescales its behavior is smooth enough for traffic engineering be possible over these timescales. Note that parts of this chapter have appeared in [194, 198].

This chapter is structured as follows. Section 2.2 first introduces the main concepts concerning scaling processes. Section 2.2.1 goes on by discussing the practical issues of detecting scaling in real processes. Section 2.3 deals with the total traffic long-term self-similarity. Section 2.4 then considers the scaling in the process of the TCP flows arrivals. Finally, section 2.5 discusses the implications of this chapter.

2.2 Self-similarity and scaling processes

2.2.1 Scaling, self-similarity, and LRD*

Scaling behavior is a general term connected with the absence of a particular characteristic scale controlling the process under study. For an introduction to scaling in network traffic, we refer to [4]. [145, 71] provide detailed treatments of various aspects concerning scaling in network traffic. For more general discussions of scaling in natural phenomena we refer to [171, 102].

Self-similarity on the real line, or more precisely self-affinity [125], with parameter H $(0 < H < 1)$, is defined by

$$\{X(t), t \in \mathbb{R}\} \stackrel{d}{=} \{c^H X(t/c), t \in \mathbb{R}\}, \forall c > 0, \tag{2.1}$$

where $\stackrel{d}{=}$ denotes equivalence in distributions. Equation 2.1 means that the distribution of the original process X and the distribution of the rescaled process $c^H X(t/c)$ are the same. For $c > 1$, rescaling corresponds to zooming in (finer resolution), equivalent to viewing the process X at smaller timescales. $c < 1$ on the other hand corresponds to a zooming out of the process or a larger timescales perspective of the dynamics of the process X. The reason for the name "self-similarity" is that all timescales of the process are related to one another through this relationship between the distributions of the rescaled process, hence a similarity of the process with itself. The distributions of the process X seen at any timescale are connected through the affine relation defined by equation 2.1. The parameter H controls the behavior of the process X at all timescales.

The following property of self-similar processes concerns the scaling of their moments

$$E|X(t)|^q = E|X(1)|^q \, |t|^{qH} . \tag{2.2}$$

This property expresses the behavior of moment q (if it exists) of the process as a power law of time, defining a high-order scaling which is linear in H. A self-similar process will exhibit the same fluctuations (after rescaling) independent of the considered (time)scale, hence cannot be stationary due to the dependency of all its moments on time. Recall that for a process to be strictly stationary, its statistical properties must not depend on time.

In the networking literature, strict self-similarity as defined in equation 2.1 is rarely used. Rather, second-order stationarity is assumed, meaning that both the mean of the process and its variance are finite and time-invariant. This second-order stationarity assumption allows for an easier statistical analysis of the process' properties [29].

Statistical *self-similarity* in the context of discrete-time processes is defined through the following procedure. Let $X = \{X_i, i \geq 1\}$ be a second-order stationary sequence and build from it the following sequence

$$X^{(m)}(k) = \frac{1}{m} \sum_{i=(k-1)m+1}^{km} X_i, \;\; k = 1, 2, ... \tag{2.3}$$

Formula 2.3 defines the *m-aggregated* sequence obtained by partitioning the original sequence X into non-overlapping blocks of length m and viewing the block average as a new point of the sequence $X^{(m)}$. The series $X^{(m)} = \{X_k^{(m)} : k =$

$1, 2, ...\}$ is said to be *asymptotically self-similar* if

$$X \stackrel{d}{=} m^{1-H} X^{(m)} \quad as \ m \to \infty \tag{2.4}$$

where $\stackrel{d}{=}$ denotes equivalence in distribution. The previous property can be seen as the discrete-time equivalent of self-similarity on the real line of equation 2.1. The statistical self-similarity property states that all *m-aggregated* sequences ($X^{(m)}$) rescaled by m^{1-H} have the same distribution as the original process X. If the sequence is not second-order stationary, then the distribution of the normalized *m-aggregated* sequence might not converge. Asymptotically, the first two moments of the process might even be infinite, in which case the procedure used for building the *m-aggregated* sequence is meaningless. Deviations about second-order stationarity might bias the convergence of the distributions, even though finite discrete-time processes have finite moments. Note that we use the term "sequence" to refer to the succession of samples in time while the term "series" denotes the set of the *m-aggregated* sequences for $m = 1, 2, \ldots$

When one restricts the study of the process to its second-order properties, long-range dependence is often used. Long-range dependence (LRD), or long-memory, is associated with second-order stationary processes. A second-order stationary process displays LRD [29, 209] if its autocovariance $r(k)$ behaves as

$$r(k) \sim c_r |k|^{2H-2} \quad as \ |k| \to \infty \tag{2.5}$$

when $1/2 < H < 1$. In such a case, the correlations decay so slowly that $\sum_{k=-\infty}^{\infty} r(k) = \infty$. When $H = 1/2$, the process is uncorrelated ($r(k) \sim 0$). On the other hand, when $0 < H < 1/2$, we speak of short-range dependence (SRD), or anti-persistence, due to negative dependence ($r(k) < 0$ for $k \neq 0$) between time lags. The divergence of the correlations when $1/2 < H < 1$ also means that the spectral density behaves as

$$f(\lambda) \sim c_f |\lambda|^{1-2H} \quad as \ |\lambda| \to 0, \tag{2.6}$$

hence diverges at the origin ($f(0) = \infty$). Because of stationarity, LRD processes cannot exhibit fluctuations at any arbitrary scale, as do strictly self-similar processes.

An often mistaken relationship between LRD and self-similarity concerns the implication of LRD by strict self-similarity for $H > 1/2$. There are actually two types of LRD behaviors: 1) one that is limited to the strict definition in 2.5 and 2.6 for a second-order stationary process for which strict self-similarity does not hold, and 2) the consequence of LRD that arises for a strictly self-similar process for $H > 1/2$. When considering self-similar processes for which second-order stationarity does not hold, the second-order analysis is biased by the non-stationarity

of the process. Spectral and correlation analysis thus become tricky because the results of these analyses might be caused by the second-order properties of the process as well as bias from non-stationarity [148]. This distinction is not merely formal, but has important practical consequences on the type of process considered. For example, it is not always possible to distinguish in discrete signals between pure statistical correlations that span more than the length of the studied sample, pure LRD and LRD produced by a self-similar process.

Scaling detection in practice

The previous section dealt with theoretical properties of signals. Now, we will sensitize the reader to the difficulty of scaling detection and the bias caused by non-stationarity. Because in real-life signals neither LRD nor self-similarity occur in an asymptotic sense, these two properties point out to two different types of signals. LRD is related to statistical dependence within the time-series. In practice dependence in a time series has an upper cutoff value corresponding to the largest correlation length between two time instants. By "largest correlation length between two time instants", we mean for example the smallest distance between two time instants that makes the autocorrelation fall below the 0.95 confidence limit of $2/\sqrt{N}$ where N is the length of the sample. The autocorrelation of a time-series is the correlation between the original time-series and time-shifted versions of the time-series. The autocorrelation for time lag t measures the correlation within the time-series between the samples separated by t time intervals. The upper cutoff value is the smallest value of the time lag after which the autocorrelation of the time-series falls below this $2/\sqrt{N}$ confidence limit. LRD in this sense manifests itself from the smallest timescale up to this upper cutoff value.

Equation 2.4 defining statistical self-similarity does not specify any cut-off value. As it is asymptotic, the definition only defines a relationship between the original process and its *m-aggregated* series for m going to infinity. The relation in Equation 2.4 holds for large values of m, starting at some value of m. Practical constraints however require that the length of the sample be finite, so only the *m-aggregated* series up to a given value of m can be computed. The time-series analyst is thus faced with the double issue of finding the value of m from which statistical self-similarity seems to hold, and of relying on a sufficiently long time-series to be able to estimate the value of H in Equation 2.4.

Self-similarity is a global dependence arising through the consistent relationship between the original time series and its *m-aggregated* series. While this self-similarity implies LRD for $H > 1/2$ in the second-order stationary case, an upper cut-off value for the correlation length of a LRD signal will make a non-negligible difference between self-similarity and LRD in practice. Because LRD stops at this cutoff value (call it C), which can be smaller than the length of the time-series,

the *m-aggregated* series of the LRD signal will not be self-similar any more for $m > C$.

To illustrate this point, Figure 2.1 compares the *m-aggregated* series of a LRD signal with an upper correlation length cutoff value C of one thousand time units with a self-similar process (fractional Brownian motion) with Gaussian increments (fractional Gaussian noise). The units have no meaning on Figure 2.1. The LRD signal has been generated by inducing a correlation length of one thousand by generating a stationary Poisson process of mean equal to ten, going over the whole length of the generated series and adding the current value of the Poisson process to every time interval for the next one thousand time units, creating a dependence of lag one thousand in the time series. The self-similar process has been obtained by integrating the Gaussian noise[1]. Both processes start at a value of ten thousand units and end close to each other, but the LRD process is known to be second-order stationary while the self-similar one is non-stationary.

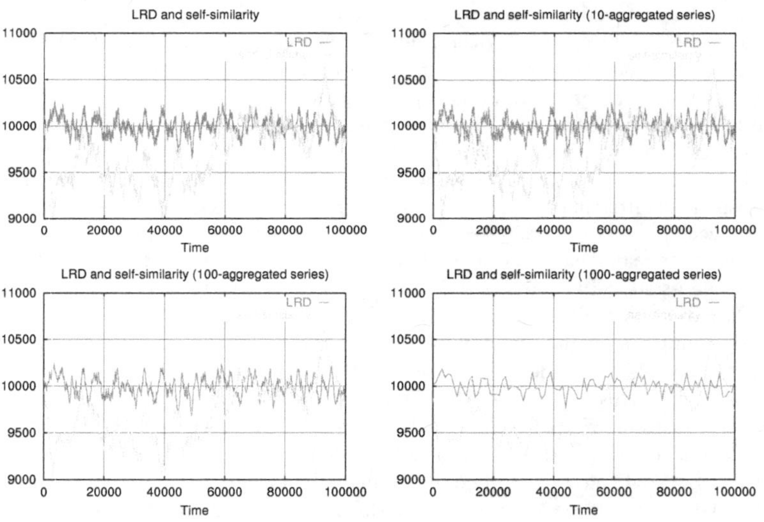

Figure 2.1: Difference between LRD and self-similarity.

The two signals compared in Figure 2.1 can be said to be qualitatively different because we know how they were generated. Suppose on the other hand that one has to analyze these two signals without knowing beforehand that one is LRD and

[1] More sophisticated techniques to generate self-similar or LRD processes exist [23, 61].

the other self-similar. For example, suppose that we want to check for LRD. We thus apply the definitions of LRD (from formula 2.6), considering that both processes are second-order stationary. The reader might contend that the self-similar signal of Figure 2.2 is *obviously* non-stationary. This is true indeed, because we chose the length of the time-series to be sufficiently long so that non-stationarity be visible on Figure 2.1. If we had samples corresponding to the signals within the [70000,80000] time interval only, non-stationarity would not be obvious. Many processes are physically constrained in the extent of their variation. For example, a finite number of users may connect to a web server at a given instant, or the size of the traffic bursts a single machine can send is limited by its network access capacity. Distinguishing between stationarity and non-stationarity is thus difficult [49, 135].

When comparing the spectrum of the two signals (left of Figure 2.2), we do not observe a major difference. The graphs of the two signals however (Figure 2.1) show that the LRD process seems stationary while the self-similar one is not. The autocorrelation function of the two processes (right of Figure 2.2) on the other hand reveals the difference between them. The stationary signal does not exhibit correlations beyond a time lag of one thousand since its correlations fall below the $2/\sqrt{N}$ confidence limit after a time lag of one thousand. The self-similar signal biases the autocorrelation analysis so much that the statistics tells that correlations are still close to one even for a time lag of one thousand.

Even though self-similarity and LRD are strongly related, these two concepts correspond to different processes. It is easy to show in synthetic signals that a process is second-order stationary while another is non-stationary. In practice many signals are non-stationary, in which case the time-series analyst has to identify the type of non-stationarity and how to de-trend. When different types of signals mix within the different timescales, then this de-trending is not possible without a-priori assumptions on the nature of the signal. For instance, in the case of the two synthetic signals we provided in this section, the right part of Figure 2.2 showing the autocorrelation allowed to see the difference between the two signals because we knew that one of them had limited correlations. If we had not known that correlations allow to see the difference between the two signals shown on Figure 2.1, we would have had to find the estimator that analyzes the right property of the signal to be able to know exactly in which way non-stationarity biases other estimators.

2.2.2 Classical scaling detection

In this section, we review the classical (non wavelet-based) methods for the analysis of scaling processes (and the Hurst parameter H). We only consider the basic ones, not their refinements, because it is mainly the aim and basic assumptions

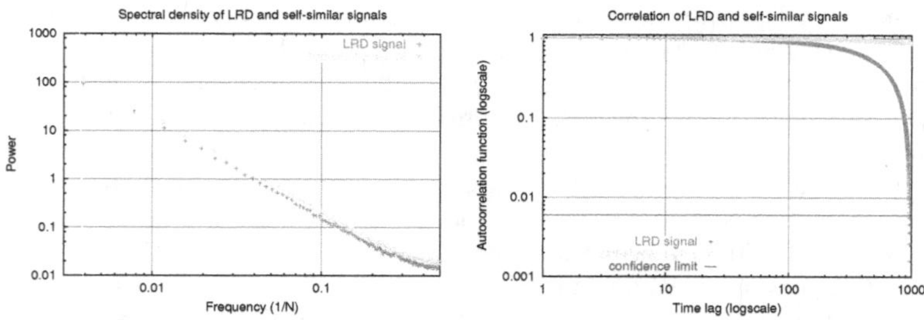

Figure 2.2: LRD from statistical correlation and self-similarity.

of a method that guides what can be inferred from it, refinements mostly affect the statistical bias of the analysis technique [180]. All of them are graphical, and consist in plotting a statistics on a log-log plot and determining a region where a linear behavior appears. The value of H is then inferred based on a least-square regression over the region of the graph where scaling is visually identified [29].

The first estimator is the well-known R/S statistic [29, 125]. Plotting the R/S statistic for large k on a log-log scale should be scattered around a straight line of slope H. A nice property of the R/S statistic is its robustness with respect to long-tailed marginal distributions, namely departure from Gaussianity [122]. See part VII of [125] for an in-depth treatment of the R/S statistic.

The second method for measuring H is the *aggregated variance* method [29], where one computes the sample variance of each *m-aggregated* sequence $X^{(m)}$ around its sample mean $\bar{X}^{(m)}$

$$s^2(m) = \frac{1}{m-1} \sum_{i=1}^{m} (X^{(m)}(k) - \bar{X}^{(m)})^2. \quad (2.7)$$

Then, the plot of $s^2(m)$ as a function of m on a log-log scale should follow a straight line with negative slope $2H - 2$ for large values of m. Because this *aggregated variance* method computes the variance of each complete *m-aggregated* sequence and not a local one (for subsets of the sequence), it can be affected by non-stationarity.

The previous two estimators mainly deal with the process variability (second order statistics), hence characterize the extent of the variations in the signal and their persistence with time aggregation.

The third estimator for H, the correlogram, relies on the time dependence within the samples. Long-range dependent data should exhibit very slowly decaying sample correlations proportional to k^{2H-2} for $1/2 < H < 1$, k representing

the time lag in the series. In order to properly interpret the autocorrelation function $r(k)$, the assumption of second order stationarity should hold. Variations in the mean and variance in the signal limit the use of this statistic, hence de-trending and variance normalization should be performed before computing the autocorrelation.

Finally, while the three previous methods were based on the time-domain, the final one is the periodogram, in the frequency-domain. If the series exhibits *long-memory* (LRD) then the estimated spectral density at the origin should behave as in equation 2.6. Plotting the periodogram near the origin with a log-log scale should roughly follow a straight line of slope $1 - 2H$. The biggest problem with frequency-domain statistics concerns the physical meaning of "frequency" components [148, 28]. If the signal is due to contain well-defined periodicities, then a smoothed periodogram might provide useful information about the signal's properties. In the case of statistical self-similarity on the other hand, there are not always good reasons to think of meaningful frequency components in the process, hence focusing on too small subsets of the periodogram might pose problems. Anyway, a frequency-domain analysis might show that global scaling exists in the signal through a $1/f$ noise-like behavior in the spectrum [124, 125]. See also [7] for a discussion of the estimation of the spectrum of $1/f$ processes in the wavelet domain.

Using estimators that rely on a stationarity assumption with non-stationary data might bias the estimation [60, 160]. A self-similar signal ideally requires an analysis tool capable of disentangling the scaling properties at each timescale. In the case of a signal containing both dependence and self-similarity, one has to check the cause of this dependence. A common practice in time-series analysis is to model non-stationarity by a trend and remove it from the data before analyzing the signal. In the case of scaling processes however, it is difficult to distinguish between actual periodic components, non-periodic shifts in the mean, or large variations in the signal due to a large variance. Since scaling processes have no particular localization in the frequency-domain (no characteristic timescale), it is often impossible to decide whether one should model the large timescales with a seasonal trend or not model it at all and let self-similarity explain non-stationarity that looks like a periodicity.

2.2.3 Wavelet-based scaling detection*

The estimators introduced in the previous section are the most widely used ones in the time-series literature. Nevertheless, their use for scaling processes is highly questionable because of the bias from non-stationarity. Wavelet-based tools constitute a large step forward in scaling detection thanks to a limited bias from non-stationarity in the estimation. A detailed treatment of wavelet analysis is outside

the scope of this thesis. For a complete treatment of wavelet theory and analysis, we refer the reader respectively to [62] and [5, 4, 6].

Wavelet signal decomposition consists in analyzing a signal $X(t)$ through a bandpass oscillating function $\psi_{j,k}$ where j represents the timescale and k the time instant. Throughout this chapter, small values of j represent the smallest timescales and large values represent the larger timescales. By scaling and shifting this function ψ, it is possible to break the signal into its timescale parts (at timescale j) and within each timescale along the time axis (at time k):

$$\psi_{j,k}(t) = 2^{-j/2}\psi(2^{-j}t - k), \quad j \in \mathbb{Z}, \quad k \in \mathbb{Z} \tag{2.8}$$

where the $\psi_{j,k}(t)$ form an orthonormal basis of L_2, the space of twice integrable functions. Any square integrable signal can hence be approximated by a finite linear combination of the $\psi_{j,k}(t)$.

A dilation and shifting operation by a factor (j,k) moves the original frequency f_0 of the function ψ to frequency $2^{-j}f_0$ while shifts it in time by a factor of $2^j k$. Recall that larger values of j relate to larger timescales, hence smaller frequencies ($2^{-j}f_0$). Figure 2.3 shows the tiling of the time-frequency plane when using dyadic dilation and shifting operators. On Figure 2.3, we show the original frequency f_0 and the corresponding time resolution. A dilation operation of factor j moves the original frequency f_0 to $2^{-j}f_0$, hence going to a coarser time resolution of a factor 2^j.

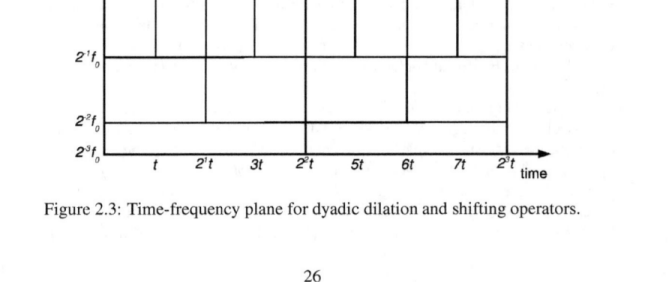

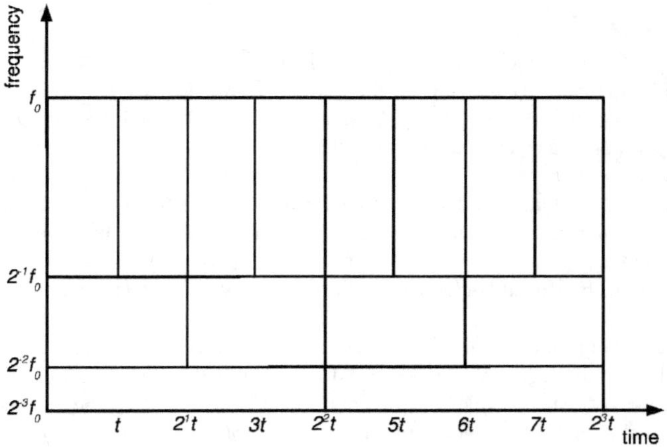

Figure 2.3: Time-frequency plane for dyadic dilation and shifting operators.

The discrete wavelet transform algorithm performs a dyadic tree decomposition (multiresolution analysis) of the signal

$$X(t) = \sum_{k} c_X(j_0, k)\phi_{j_0,k} + \sum_{j \leq j_0}^{J} \sum_{k} d_X(j, k)\psi_{j,k}(t) \qquad (2.9)$$

where the first term on the right-hand side represents an approximation of the signal while the second term represents the details, j_0 being the resolution depth, i.e. the coarsest resolution at which the signal is analyzed. The $c_X(j_0, k)$ are called the scaling coefficients and ϕ the scaling function. The $d_X(j, k)$ are called the wavelet coefficients and are of special interest because they are used as a substitute for the increments of the process X over the dyadic tree of the time-timescale plane. The orthogonal basis property of ϕ and ψ allow the easy computation of the coefficients by simple inner products with the signal. The nice properties of the wavelet transform make the wavelet coefficients better suited for statistical analysis in comparison to the original increments of the process. In addition, the wavelet transform has a $O(n)$ time complexity, making it computationally efficient. From now on, we shall use the $d(j, k)$ as a substitute for the process increments $X_\delta(k)$.

Now we shortly describe how the wavelet coefficients $d(j, k)$ are computed. For this, we rely on Figure 2.4. During the successive iterations of the wavelet decomposition, a discrete time-domain filter corresponding to the wavelet function is applied to the signal at the successive timescales. This filter is a sequence of weights applied on the discrete values of the signal. This filter does two things. First it generates the wavelet coefficients representing the variations in the signal that could be matched by the filter at the current timescale j ($d(j, k)$). Second, the filter produces the new signal ($c(j, k)$ on Figure 2.4) to be analyzed at the next timescale $j + 1$ (twice as large as the current timescale). This new signal contains everything in the sequence at the previous iteration that could not be matched by the wavelet function. The wavelet transform actually relies on two filters, a high-pass filter that generates the wavelet coefficients, and a low-pass filter that generates the smoothed version of the signal to be analyzed by the next iteration of the algorithm. For further details about the discrete wavelet transform, we refer to [208, 203].

Assume on Figure 2.4 that we start with a sequence containing 16 points on which we apply the discrete wavelet transform. After the first iteration of the algorithm, we get the 8 wavelet coefficients ($d(1, k)$) representing the variations in the original signal at a timescale equal to twice the precision of the original sequence. The effect of any iteration of the algorithm is to replace the original sequence by the wavelet coefficients in one half of the sequence length, and by the new sequence ($c(j, k)$) to be analyzed during the next iteration in the other

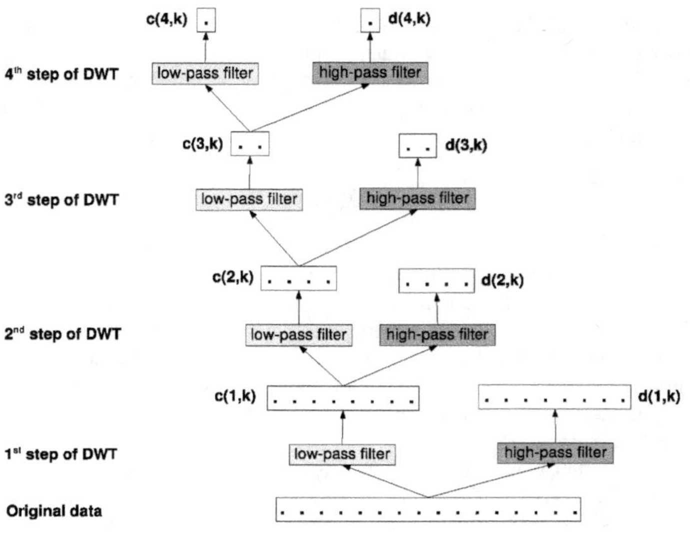

Figure 2.4: Discrete (forward) wavelet transform.

half of the original sequence. This allows the computation of the discrete wavelet transform algorithm in-line, i.e. in the memory locations used by the original sequence. The original sequence is hence lost but the algorithm has both a space and time complexity of $O(n)$ where n is the length of the original sequence.

The properties that make wavelet analysis suitable for studying scaling processes are the following:

1. scale invariance: the wavelet basis being built through the dilation operator, the analyzing functions family has a built-in scaling property that reproduces scaling present in the data;

2. robustness against non-stationarity: the mother wavelet (ψ_0) has a number $N \geq 1$ of vanishing moments $\int t^k \psi_0(t)dt \equiv 0,\ k = 0, ..., N-1$, allowing the wavelet to remove any polynomial trend of order up to $N - 1$;

3. almost decorrelation: under the assumption of $N \geq H + 1/2$, global LRD among the increments of the process can be turned into short-range dependence [184].

These properties ensure that the wavelet transform captures the scaling properties of the signal (point 1) and that a sound analysis of the signalcan be performed at

each timescale independently of other timescales (point 2), with a controlled bias coming from non-stationarity in the signal (point 3).

The most widely used wavelet-based estimator of scaling, the *logscale diagram* (LD) [5] consists in plotting $y_j = log_2(\mu_j)$ against j together with the confidence intervals, where

$$\mu_j = \frac{1}{n_j} \sum_{k=1}^{n_j} |d_X(j,k)|^2. \tag{2.10}$$

n_j in Equation 2.10 denotes the number of wavelet coefficients at octave j and $d_X(j,.)$ represent the wavelet coefficients at octave j. The *logscale diagram* is a second-order statistic of the wavelet coefficients that measures the variance of the wavelet coefficients at each octave j. Scaling detection can be performed through observation of linear alignment of the confidence intervals of the y_j within some octave range. If, and only if alignment is detected, can the estimation of scaling parameters be performed [5].

The slope of the LD indicates qualitative information about the signal's variability. Two situations are interesting concerning the slope of the LD: around zero, and positive. A slope of the LD about zero for some octave range points to a process whose variability is constant over the corresponding timescales. Such a process is uncorrelated, like white noise or a Poisson process. To understand the implication of correlations on scaling relationships in a signal, it is necessary to explain what correlations actually are. Suppose a zero mean process. Positive correlations in such a signal mean that whenever the value of the signal deviates (positively or negatively) from zero, then the probability that successive values of the signal will take values that deviate from the mean in the same way is larger than $1/2$. Successive positive or negative values are more likely than a positive value followed by a negative one on average. Negative correlations on the other hand mean that the probability is larger than $1/2$ for deviations about the mean in opposite directions for successive samples. Hence a positive value followed by a negative one is more likely than successive values of the same sign. Now let us consider the autocorrelation function of the zero-mean process. The autocorrelation function indicates how, on average, values separated by t time intervals are correlated over the whole process. A positive (negative) value of the autocorrelation function for time lag t means that samples of the process separated by t time units are positively (negatively) correlated. If any time lag of the autocorrelation function exhibits a different value compared to other time lags, there will be a shift in the LD for the corresponding timescale. The reason for this phenomenon is that large correlations (positive or negative) at time lag 4 for instance imply a higher likelihood of experiencing large deviations about the mean in the process every four samples, hence more energy in the LD for octave two. So far, we have

explained how correlations at some lag may imply changes in the LD at some octave. Scaling on the other hand concerns relationships among timescales in a process.

To understand how relationships among timescales can give us information about a process, it is best to come back at the procedure of the discrete wavelet transform. Assume that we actually apply the successive iterations of the algorithm on a zero-mean sequence. The wavelet coefficients at the first step will serve for computing the value of the LD at octave 1. Each successive iteration of the algorithm removes the variability matched at the current timescale analyzed by the algorithm. Getting a zero slope of the LD means that all timescales contain the same amount of variability. Now recall the third nice property of wavelets mentioned in section 2.2.3, the decorrelation of the wavelet coefficients at each timescale. This property ensures that the variance computed by the LD is the variance of an uncorrelated process, the wavelet coefficients at the particular timescale considered. In addition this process is very close to be zero-mean since the integral of the wavelet is zero (the first vanishing moment, moment zero). Hence the LD measures the variance of a zero-mean uncorrelated process. Note however that this process is not second-order stationary because although almost uncorrelated, the wavelet coefficients might have a non-stationary variance.

Now we come back to the problem of understanding the behavior of the LD for different types of signals. Each iteration of the discrete wavelet transform algorithm removes the contribution to the variability of the signal at the current timescale. What does mean that the contribution to the global variability at each timescale is the same? It means that the variability of the signal is not explained more by some timescale than some other. Another way to put it is to say that each timescale is responsible for the variations in the signal in exactly the same way. No variation in the signal occurs for some time lag preferentially: the variations in the process are random across timescales. If some correlation exists for some time lag in the process, the variations in the signal at this timescale will be larger, so would be the value of the LD for this timescale only. This explains that a slope of zero in the LD requires either no correlations or a constant autocorrelation for all time lags. For stationary signals, a constant autocorrelation function only occurs for white noise (zero correlations) or a constant signal (see [28] p. 135). A zero slope for the LD therefore implies a scaling process with $H = 1/2$.

The case of a positive slope of the LD on the other hand means that successive timescales contribute to increasingly large fractions of the total variability of the signal. The successive iterations of the discrete wavelet transform find wavelet coefficients with larger and larger variances. If the process is second-order stationary, the increasing variance for the successive timescales cannot come from non-stationarity but only from homogeneous variability in the signal corresponding to large scale correlations. In the case of LRD signals, the positive slope

extends to the timescale corresponding to the upper cutoff time lag, the largest correlation length allowed by the physics of the process. LRD is defined in equation 2.5 through an autocorrelation function that decreases geometrically as the time lag goes to infinity. Even though the discrete wavelet transform removes the correlation contained in the timescale analyzed by the current iteration of the algorithm, the successive timescales analyzed by the wavelet transform deal with increasingly large fractions of the autocorrelation's time lags. This increasing span of the time lags in the correlations of the process allow the variability to build up for increasing timescales, explaining the larger variances of the wavelet coefficients for larger timescales. In the case of strict self-similarity on the other hand, the positive slope of the LD is not caused by physical correlations in the signal but by the larger variability that arises with the scaling of the moments with time of equation 2.2. Even if the extent of the variations in the self-similar signal are limited by the physics of the process, LRD arising for $H > 1/2$ is enough to explain the positive slope of the LD. It was shown in [5] however that the slope of the LD for a strictly self-similar signal will be different from the slope arising due to a LRD one. A slope of the LD larger than 1 might point to self-similarity. In the case of network traffic on the other hand, it is more likely to be related to non-stationarity or missing data [161].

2.3 Long-term scaling of interdomain traffic

The previous section has introduced the necessary notions to understand the nature of scaling in network traffic. In this section, we analyze the long-term properties of the traffic. By long-term, we mean timescales of minutes up to days.

2.3.1 Measurements

Many researchers have chosen to study the behavior of network traffic by relying on packet level traces captured at a single link (e.g. [119, 147, 128]). Such traces allow the analysis of the traffic at the packet scale, but require a large storage space. For this reason, most of the used traces either correspond to a low traffic volume or a short period of time. We have chosen a different approach. In order to better understand the long-term behavior of the traffic, we consider a less precise trace that spans several physical links over six complete days. For this, we rely on a `Netflow` [53] trace collected at the border routers of the Belgian research network BELNET in December 1999. The trace covers all the interdomain traffic received by the BELNET on all its external links. It corresponds to 2.1 TBytes of traffic. BELNET provides access to the Internet as well as to high speed European research networks to universities, government and research

institutions in Belgium. At the time of the traffic capture, its national network was based on a 34 Mbps backbone linking major Belgian universities. Its users are mainly researchers or students with high speed connections (at least Ethernet) to the backbone, although some institutions also provide dial-up service to their users. It was also at that time the Internet service provider with the largest external capacity in Belgium.

The BELNET network is connected to a few tens of peers with high bandwidth links. It maintains high bandwidth peerings with two transit providers, with the Dutch SURFNET network and was connected to the TEN-155 European research network, without providing any transit service to its peers. In addition, BELNET is present with high bandwidth links at the Belgian (BNIX) and Dutch (AMS-IX) national interconnection points with a total of about 40 peering agreements in operation. The `Netflow` trace we used aggregated the information from all upstream links, so that we have the incoming traffic information as if it was received through a single external link.

`Netflow` provides the aggregated information of the layer-4 flows, by recording the starting time, the ending time and the total volume in bytes for each unidirectional TCP and UDP flow. `Netflow` raises some issues concerning the layer-4 flow-level measurements, because it uses a timeout policy for flushing out its flow cache. Depending on the configuration of the router, some flows could be considered as terminated while this need not be true from the TCP connection viewpoint. Fortunately, this problem mainly concerns fine-grained measurements for flow arrivals which we do not consider here. Furthermore, we use a coarse time scale and only study the total traffic for aggregated traffic sources (not layer-4). Hence, the bias from the flow timeout policy should not interfere too much with our results. The finest notion of a flow we use here is the traffic evolution every minute for some external IP address. Many layer-4 flows may thus be aggregated into a single flow.

The utilization of `Netflow` forces us to approximate the layer-4 flows as equivalent to fluid flows. More precisely, a flow transmitting *M* bytes between T_{start} and T_{stop} is modeled as a fluid flow transmitting $M/(T_{stop} - T_{start})$ bytes every second between T_{start} and T_{stop}. This approximation leads to an inaccurate estimation of the short-term burstiness of the traffic but allows for traffic traces of longer durations to be collected. All traffic volume information is summarized over equally-spaced one minute intervals throughout the six days. The time granularity of the trace is consequently of one minute.

The part of the `Netflow` statistics on which we rely is the information of the total traffic volume received from each external IP address for every minute of the measurements. For each one minute interval of the measurements, we have the number of IP addresses that are sending traffic during that particular minute as well as the corresponding traffic volume for each of them. This choice of one

minute for the time granularity was driven by our intuition that these timescales are sufficiently coarse not to be too much biased by the traffic dynamics due to TCP flows, but rather IP-level characteristics. The actual time granularity of the original trace was one second.

Even though the time granularity of the trace is one minute, it accounts for more than 42 million values recorded over more than 8600 samples. On average, 4912 external IP addresses were sending traffic each minute. This trace hence provides a fine measurement of the traffic dynamics for time-scales spanning minutes to hours.

2.3.2 Total traffic scaling

In this section, we focus on detecting the scaling of the total traffic time-series. We look at the total traffic received at the incoming access links for every one-minute time interval during the 6 days of the measurements. Figure 2.5 shows the evolution of the total traffic during the period of the measurements. While the global evolution of total traffic exhibits a stable daily periodicity with peak hours located during the day, there are important deviations around the average traffic evolution throughout the day that give self-similar traffic this highly bursty look. The mean traffic over the six days was about 32 Mbps, with a one-minute maximum peak at 126 Mbps and a standard deviation of 21 Mbps. The trace begins around 1 AM and finishes six days later around 1 AM also.

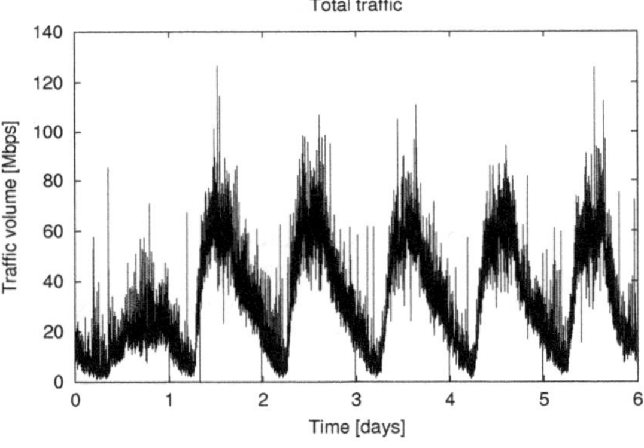

Figure 2.5: Total traffic evolution.

Figure 2.6 presents the estimation of the Hurst parameter using the four estimators introduced in section 2.2.1. The R/S (top left of Figure 2.6) and the ag-

gregated variance (top right of Figure 2.6) estimators indicate a value of H close to 1. The correlogram (bottom left of Figure 2.6) for its part indicates a large correlation for timescales up to one hour. Finally, the periodogram shows that a bias from non-stationarity occurs with a highly negative slope at low frequencies characteristic of a non-stationary $1/f$ noise [123]. At such large timescales, non-stationarity in the mean and variance is obvious from Figure 2.5.

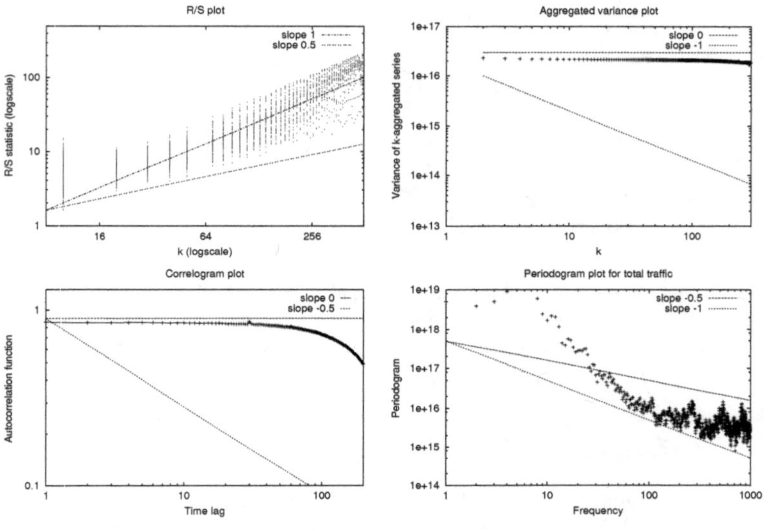

Figure 2.6: Estimation of H for total traffic.

Most of the estimators for H are defined in an asymptotic way. They provide a good estimate for large time-series only. However, our time-series containing more than 8600 samples, we can be quite confident with the previous estimators at least in the value range they provided, with H closer to 1 than $1/2$ for all of them. Non-stationarity in the signal might however bias these estimators. To confirm the presence of scaling while minimizing the bias from non-stationarity, Figure 2.7 provides the *logscale diagram* for the total traffic. Figure 2.7 shows the *logscale diagram* of the total traffic trace together with the 0.95 Gaussian confidence intervals. The *logscale diagram* shows that only octaves 5 to 9, timescales ranging from 30 minutes to 8 hours, contain scaling with a measured H of 0.84. The variance of the estimation of the value of H is 0.04. This variance of the estimation of H depends partly on the data length. The short timescales ($1 \leq j \leq 5$, up to 15 minutes) show a non-correlated signal with a value of H close to $1/2$,

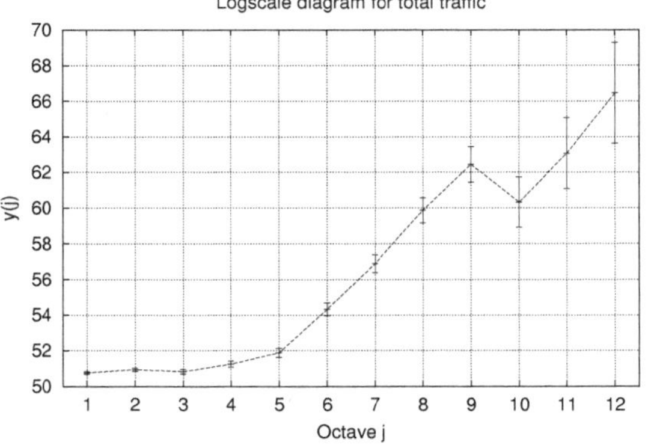

Figure 2.7: Logscale Diagram for total traffic.

indicating that short-term variations can be considered as "random" on such a coarse-grained trace. Large timescales ($j \geq 10$, half a day and more) show a large standard deviation of the estimator y_j partly caused by the too short length of the data.

2.4 Scaling in TCP flow arrivals

The previous section showed that the total traffic seen on access links contained scaling for timescales between minutes and up to one hour. It showed that scaling was present at the interdomain level, but not why. Scaling in the total traffic can have several roots. Three main roads have been explored in the literature to explain the presence of scaling in network traffic.

The first explanation of scling relies on the distributional properties of the flow activity periods that were shown to be heavy-tailed [59, 147]. The complementary proof of Taqqu, Willinger and Sherman [182] then provided a formal justification for the presence of scaling through the superposition of a large number of independent ON/OFF sources with heavy-tailed ON and/or OFF periods. Park, Kim and Crovella [143] made the connection between distributional properties of the file sizes and the modulating effect of the TCP/IP stack and showed that heavy-tails in the applicative flows were mapped to heavy-tailed activity periods at the network layer.

A second explanation was investigated by Feldmann, Gilbert and Willinger in [83] that studied the short time-scales behavior of TCP traffic. They showed

that multiplicative processes described well its working, adding multifractals and cascades among the scaling behaviors present in network traffic. These two first roads, heavy-tails and the TCP dynamics, relate to traffic volume. The first is related to the size of the files to be carried over the network while the second concerns the way flows are broken by TCP to be sent as IP packets. While the previous two aspects tell that the information to be exchanged as well as the behavior of the network and its protocol can be considered as partly responsible for scaling in the traffic, one may wonder whether users of the network have anything to do with scaling in the traffic, allowing us to introduce the third road.

The third road must be traced back in 1995, when Paxson and Floyd [147] found that the connection arrival process of some applications failed to be Poisson. This finding was unexpected due to the belief that users initiating actions independently should lead to a Poisson process. With the growth of the web, the situation was due to change even more, because many flows are initiated for every user session. The first step to understand this aspect was [84] that explained self-similarity in the TCP connections arrival process by heavy-tails in the number of TCP connections within user sessions. Another remarkable effort in the modeling of the TCP connection arrivals is Feldmann in [81] that showed that TCP connections arrivals are self-similar, through a second-order analysis. The aim of [81] was not to study the time-scale dependent properties of this process, but rather to fit a probability distribution to model TCP connection arrivals. The first thorough study devoted only to the scaling properties of the TCP flows arrivals was [198] that showed that this process is by no means simply self-similar. [198] showed that the TCP flow arrivals is actually a non-stationary process with different scaling components at different timescales. [198] showed that there exists a non-stationary statistical correlation at the seconds to minutes timescales, stable self-similarity at the minutes to hours timescales and that the "time of the day" trend occurs for timescales larger than about half an hour. Due to second-order properties analysis, [198] was not able reveal the exact nature of the components at each timescale. The goal of the present section is to provide a detailed study of the fine scaling properties of the process of the flow arrivals. Our aim is to determine the nature of the components of this process, for timescales ranging between less than seconds to days.

2.4.1 TCP Traffic Traces

In this section, we describe the traffic traces on which we rely to analyze the behavior of the TCP flow arrivals.

LBL-CONN-7

The "lbl-conn-7" trace, available on the Internet Traffic Archive [3] contains thirty days' worth of all wide-area TCP connections (606, 497 connections) between the Lawrence Berkeley Laboratory (LBL) and the rest of the world. It was gathered with `tcpdump`[2] on a Sun Sparcstation using the BPF kernel packet filter between September 16 1993 through October 15 1993. The time precision of the trace is one microsecond but we used a one hundred milliseconds precision, by not having taken into account the digits corresponding to these smaller timescales of the timestamps. Only the TCP connections that ended normally (with both SYN and FIN) were used. We call this trace the "LBL" trace. More details about this trace can be found in [146, 147, 3].

UC Berkeley Home IP Web

The "UC Berkeley Home IP Web" trace is also available from the Internet Traffic Archive [3]. This dataset consists of 18 days of `http` traces gathered from the Home IP service offered by UC Berkeley to its students, faculty, and staff. All clients were using modems. This client trace was collected with a packet sniffing machine placed at the head-end of the Home IP modem bank. Only traffic destined for port 80 was traced, all non-*http* protocols and *http* connections for other ports were excluded from this trace. It contains 9, 244, 728 flows spanning the period from November 1st 1996 through November 19th 1996. Although a better precision was available, we used a one hundred milliseconds time precision for the timestamps by removing further digits in the timestamp. More information on this trace can be found on the ITA website [98]. We call this trace "UCB".

AUCK IV

The Auckland-IV data set [140] is a continuous 45 days GPS-synchronized IP header trace captured with a DAG3 system [138] at the University of Auckland Internet uplink by the WAND research group between February and April 2001. The University of Auckland ITSS department is operating an OC3 ATM link via Clear Communications, which is carrying a variety of services off the main campus. This trace contains six and a half weeks of GPS-synchronized IP headers taken with a pair of DAG3 cards at the University of Auckland Internet access link. The tap was installed at an OC3 link carrying a number of Classical-IP-over-ATM, LANE and POTS services. The trace contains all Classical-IP headers of a pair of redundant VPI/VCI's, which connects the university to their local service provider. We had a different trace for both directions, incoming and outgoing.

[2]ftp://ftp.ee.lbl.gov/tcpdump.tar.Z

We selected a contiguous 7-days sample in this trace for our analysis. There were $11,919,853$ incoming flows ($12,335,605$ outgoing) during this sample 7-days period. In this trace, we solely rely on packets that had the SYN bit set and relied on a ten milliseconds precision mainly due to trace size limitations. We call this trace "AUCK".

Yucom

The Yucom trace consists of all flows between modem clients of a small commercial Belgian Internet service provider, Yucom. Yucom provides Internet access to dialup users through regular modem pools. The trace spans less than five days between April 17th and April 21st 2001. It is a Netflow trace [53], not a packet-level one. The collecting machine was located within the premises of the service provider and was running OSU *flow-tools* [159]. Due to the flow timeout policy used by Netflow, we did not rely on the raw flow information because it overestimates the number of flows. Instead we considered the incoming flows that had the SYN flag set, hence TCP traffic only. This means that even if we do not have the correct TCP flows information because of the Netflow timeout, we have the correct information concerning the new TCP flows (SYN). During the whole period of the trace, $59,581,814$ incoming TCP flows that had the SYN flag set were recorded. Among the incoming TCP flows, `http` and `https` represent about 72 percent of the traffic volume and 77 percent of the flows. Port numbers corresponding to "chat" applications (mainly 6667) represent about 16 percent of the traffic volume and 8 percent of the flows. Finally, e-mail applications represent about 5 percent of the traffic volume and 6 percent of the flows. We call this trace the "Yucom" trace. We used a 100 milliseconds time precision for this trace and only study its incoming flows.

PSC

The PSC trace consists of all outgoing flows seen on two monitoring points internal to the Pittsburgh Supercomputing Center, an American commodity and Internet2 network provider to organizations in western Pennsylvania. The flows were collected continuously over 24 hours starting from March 12th 2002 16:40 GMT at both monitoring points [198]. These points covered all outbound commodity traffic from non-dormitory hosts at a large university, in addition to all outbound commodity traffic at several smaller organizations. The collection was made using the CoralReef package from CAIDA [112] using typical Netflow settings for flow expiration. During the flow period the network served by one monitoring point experienced considerable congestion, so that rate limits during peak usage times were enforced. We consider the TCP traffic only, but because we had no infor-

mation about the TCP flags all terminated flows (TCP FIN or CoralReef timeout) were treated as complete flows. We used a 10 milliseconds time precision for this trace. There were $39,786,219$ TCP flows in the trace. The important applications we identified were `Gnutella` with about 22 percent of the bytes and about 12 percent of the flows, `kazaa` with about 20 percent of the bytes and 8 percent of the flows, and `http` with about 10 percent of the bytes and 60 of the flows. A large fraction of the bytes were thus generated by P2P applications. We call this trace "PSC".

2.4.2 TCP flow arrivals

This section studies the process of the flow arrivals defined as the number of arriving flows seen for every time bin, for the five traces. We first study the second-order properties of this process, globally over the whole time-span of the traces, as well as in a time-dependent way. After, we deal with the higher-order properties. Finally, we study the behavior of the largest clients of the PSC trace to understand the difference found between the traces. Each analysis is preceded by a presentation of the estimators on which we rely in the analysis.

In this section we rely solely on wavelet-based scaling estimators, hence reading the introductory section on wavelet analysis (section 2.2.3) is highly recommended. Because of the different time precision of the traces, the same octaves on different graphs may represent different timescales. The Auckland trace and the PSC trace use a 10 milliseconds precision, hence octave one corresponds to a time unit of of twenty milliseconds. For all other traces, we use a one hundred milliseconds time precision so that octave one represents two hundred milliseconds. Recall that each subsequent octave reduces the amount of data by two, corresponding to a timescale two times larger than the previous octave. Having relied on a common timescale for all traces would have prevented us to show some interesting behaviors, so we preferred working with different time granularities for the traces although it certainly complexifies the task of the reader.

Second-order scaling

Because of the importance of non-stationarity for understanding the flow arrivals, we rely on a method introduced in [187, 198] that consists in computing the LD over fixed-length time intervals. In this method, one plots on a three dimensional graph the evolution over time of the LDs computed over fixed size time intervals, hence its name "3D-LD". Correctly interpreting the 3D-LD requires some care. The first property of the analyzed trace seen on the 3D-LD is the evolution in time of the general level of each LD. Since the LD measures the variance of the wavelet coefficients of the signal, the value of the LD for octave one provides

an indication of the total energy contained in the signal. The higher the value of the LD at octave one, the higher the variations in the signal. A second-order stationary signal should thus have a 3D-LD whose first octave does not change with time. Non-stationarity in the variance can thus be detected by studying the time variations in the first octave of the successive LDs. Because of the vanishing moments property of the wavelet, non-stationarity in the mean is removed since trends matched by a polynomial of order up to $N - 1$, with N is the number of vanishing moments of the wavelet, are automatically removed. Nevertheless, non-stationarity in the mean can still be indirectly detected in the 3D-LD with the value of the last octave that represents the energy contained in the largest timescales of the analyzed block. However, the value of the LD for the last octave represents the energy of the signal that could not be matched by the wavelet analysis for smaller timescales, so it does not always represent an accurate picture of the variations over time of the whole block analyzed by the LD, but depends on the analyzing wavelet used. Depending on the type of non-stationarity in the mean that the signal contains, different wavelet functions will match such non-stationarities differently. Non-stationarity in the variance of the wavelet coefficients will thus appear in the 3D-LD through shifts in the level of the surface that follow the non-stationarity of the signal over timescales larger than the size of a block over which any LD is computed.

The main information provided by the 3D-LD concerns how the variability of the process decomposed among the different timescales changes over time. As already discussed in section 2.2.3, the slope of the LD over some octave range gives qualitative information about the signal. A slope close to zero points to an uncorrelated signal. A positive slope points to LRD, self-similarity or non-stationarity. The interest of the 3D-LD is to help visualize the changes of the LD in time and ease the interpretation of the LD over time durations where non-stationarity might bias the LD. It has been shown in [187, 198] that whenever several components exist at different timescales and whose scaling behavior changes with time, the LD might be largely biased. If the non-stationarity in the signal is significant or the different components of the signal at the different timescales are complex, then the LD might point to a process whose scaling properties are far from those of the real process. The 3D-LD is then useful to validate and correct the information provided by the LD over the whole analyzed sequence. Nevertheless, we insist that the nature of scaling cannot be automatically inferred simply from the 3D-LD, it is merely helps to better understand the time-varying nature of the LD. The nature of scaling has to be found in the physical understanding of the process or its high-order properties, as will be shown later.

Second-order analysis

We study the second-order scaling properties of the flow arrivals as defined in section 2.4.1. We study their global scaling properties with the LD as well as their time-dependent second-order scaling properties with the 3D-LD.

First we analyze the LD of the TCP flow arrivals. The interest of the LD is at identifying the presence of scaling over a range of timescales in the process taken as a whole. If a straight line of positive slope can be fitted to a portion of the LD, then scaling (be it self-similarity, LRD or even statistical dependence) has been detected.

Figure 2.8 shows the logscale diagram for all traces. As it appears on the graphs, no strict scaling can be found in any trace, but only some limited octave ranges that could be fitted a straight line when taking into account the confidence intervals. Based on these figures only, it is difficult to conclude much about scaling. WE can only say that there are roughly three different regions in each of the graphs. Here we can say that timescales below seconds (octaves < 8 for Auckland and PSC, octaves < 5 for UCB, Yucom and LBL) display low correlations (or uncorrelated for Auckland outgoing, UCB and LBL), timescales between seconds and minutes contain statistical dependence or LRD (slope < 1) while the timescales between minutes and hours might be self-similar or non-stationary (slope > 1). The issue with the largest timescales is that the confidence intervals tend to be so large that the size of the samples are not sufficient to conclude anything. We have studied these large timescales in [198] and found that the "time of the day" behavior (non-stationarity) starts at a timescale of about half an hour.

Given that flow arrivals have been shown to have non-stationary second-order properties [198], we know that we should not take the results of the LD at face value. Henceforth, we now turn to the 3D-LD of the flow arrivals. Figure 2.9 shows the evolution of the LD computed for time blocks of a little less than three hours for Auckland, one hundred and ten minutes for UCB, three and a half hours for Yucom, forty-five minutes for PSC and twenty-nine hours for LBL. First, the "time of the day" pattern is apparent from the largest octaves of the graph. Second, all graphs provide a 3D-LD that has roughly the same look as the LD although the "time of the day" non-stationarity is apparent for Auckland and Yucom as well as the "day of the week" pattern for LBL. This daily pattern constitutes the first source of non-stationarity. The second type of non-stationarity appears for the incoming flows of Auckland (top right of Figure 2.9), where days two and three exhibit sharp peaks explaining the larger values of the LD for the Auckland incoming flows over octaves [8,15] (top right of Figure 2.8). This shows that a single short-term event in the traffic is able to significantly bias the LD. Finally, the PSC trace is known to contain the effect of rate limitation for the largest client. Its effect is obvious (bottom left of Figure 2.9) with a constant positive slope for

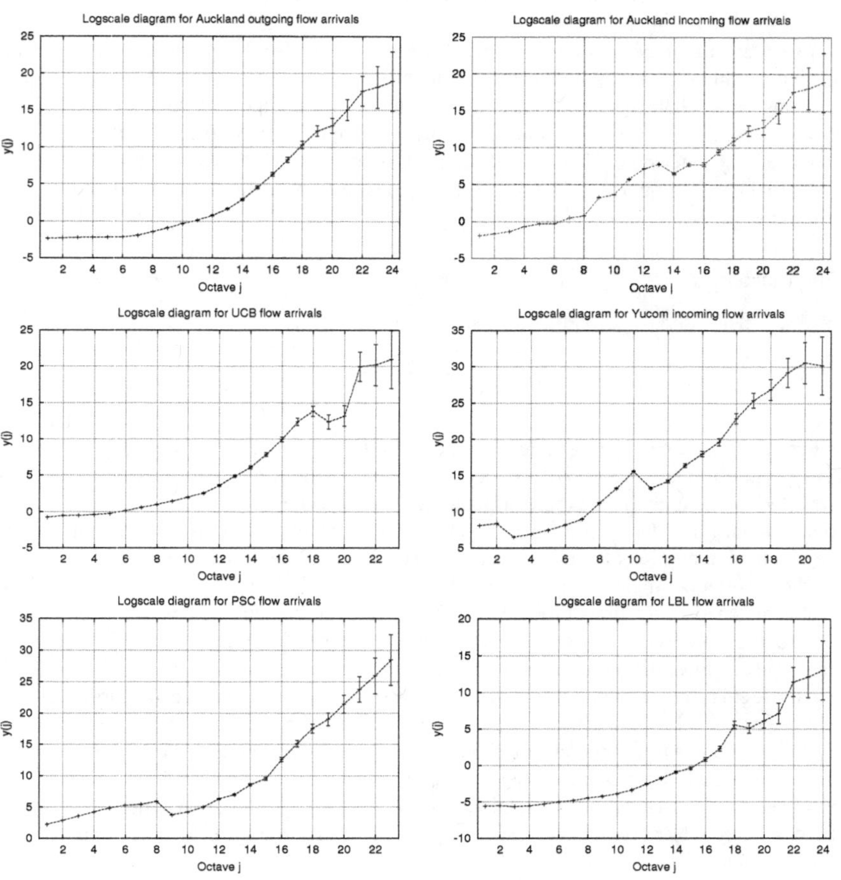

Figure 2.8: Logscale diagrams.

octaves [1,8]. One can see that this component smoothes the surface that would be there if no rate limitation had been performed. Any mechanisms that spread the flows and induces statistical dependence among the flow arrivals, like shaping, policing, or even flow-control as performed by TCP, will have such an effect on the LD.

Because the logscale diagram only deals with second-order properties, it is difficult to infer the qualitative nature of the process at each timescale. This section has shown that we should expect to see at least three components in the traffic below seconds, from second to minutes, and over minutes. However, we cannot yet precisely determine what type of scaling occurs at which timescale: correlation, LRD, self-similarity? The next section tries to answer this question, based on higher-order properties of the signal.

High-order scaling*

The wavelet-based partition function is defined as

$$S(q,j) = \sum_{k} |2^{-j/2}\, d_X(j,k)|^q \tag{2.11}$$

where $2^{-j/2}$ is a normalizing factor to remove the L_2 normalization of the wavelet. The partition function captures the local scaling behavior of a signal. Raising the wavelet coefficients to an exponent magnifies the importance of the largest coefficients that arise due to a local irregularity, while it reduces the importance of small coefficients. Recall that a wavelet is an oscillating function, hence the value of the wavelet coefficient is proportional to the size of the variation in the signal that is matched by the inner product with the wavelet. The smaller the wavelet coefficients at some scale j, the smaller the value of the partition function for $S(q,j)$ for large q. The partition function makes it possible to study the importance of the local irregularities at some scale j. A relatively smooth process that has no particularly large local irregularities will have small values of the $S(q,j)$ for $q > 2$, we call such a process a "second-order" process. An irregular process on the other hand will tend to have larger values of $S(q,j)$ for large values of q, we call such a process a "high-order" process. Note that we only study the positive moments here because we are only interested in the degree of irregularity of a process, not its smoothness.

Multiscaling is a relaxation of strict scaling where the linearity relation among moments does not hold anymore but can vary for each moment q [171, 202]:

- self-similarity : $E|d(j,k)|^q \propto exp(qH\ \ln(2^j))$,
- multiscaling : $E|d(j,k)|^q \propto exp(H(q)\ \ln(2^j))$.

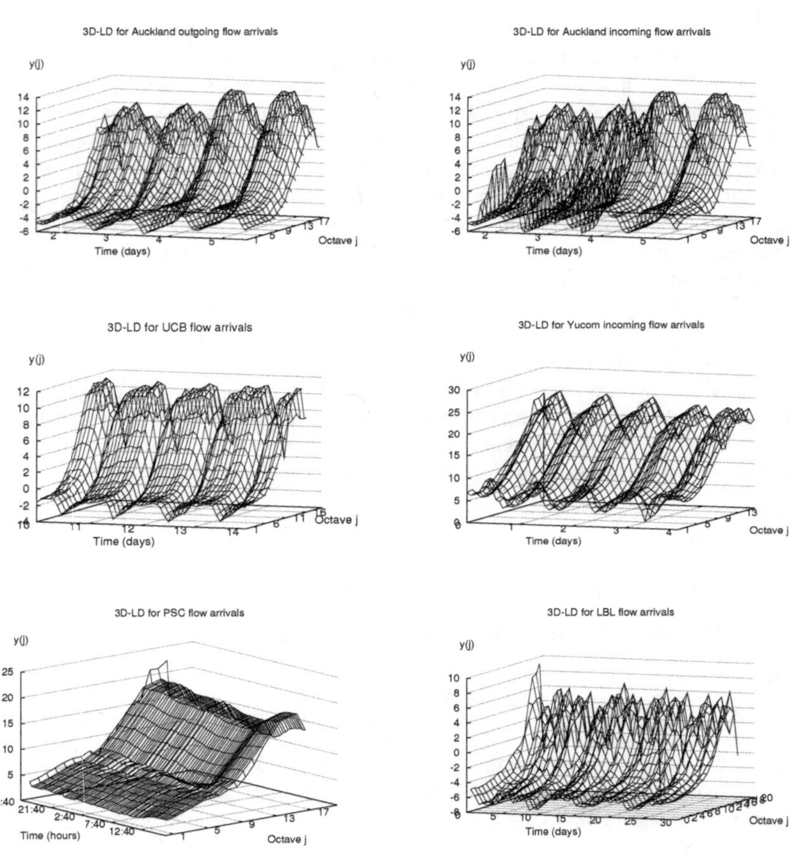

Figure 2.9: 3D-logscale diagrams.

Strict self-similarity implies that all moments behave like a power-law of the timescale, where the power-law is linear with respect to H. Multiscaling on the other hand allows the power-law relationship among moments to depend on the moment q. Self-similarity imposes that a single parameter H controls the behavior of the process. Multiscaling on the other hand permits a non-linear behavior of the moments with respect to one another. An example of multiscaling is multifractality [109, 158] where there exists a whole spectrum of local scaling exponents related to the local (ir)regularity of the sample path of the process. Let $\alpha(t_0)$ denote the local scaling exponent of the process at time t_0, then a self-similar process has $\alpha(t_0) = H$ while a multifractal process has a non-constant scaling exponent $\alpha(t_0) = H(t_0)$. This exponent $\alpha(t_0)$ is related to the local geometrical properties of the function[3], with large values of $(\alpha(t_0) > 1)$ mean small local variations while $\alpha(t_0) < 1$ arises when the local irregularities are large. Multifractality occurs when $H(q)$ depends non-linearly on q but linearly with respect to j: the moments $S(q, j)$ then tend to diverge relative to one another for the smallest timescales.

Besides power-laws of the moments **in time**, it is also possible to study the power-law relationship among moments (q) at a given timescale. The partition function provides gives hints about the type of process that arises at each timescale through the multiscaling paradigm. Multiscaling allows for three possible behaviors related to the linearity of high-order moments relative to low-order ones:

1. Sub-linearity of a LRD, statistically dependent or non-scaling process implies that $M(j) < 1$,

2. Linearity of a self-similar process implies that $M(j) \sim 1$,

3. Sup-linearity of a multiscaling process implies that $M(j) > 1$,

with

$$M(j) = \left\langle \left| \frac{log(S(q+1, j)) - log(S(q, j))}{log(S(2, j)) - log(S(1, j))} \right| \right\rangle_q, \qquad (2.12)$$

where $\langle\rangle_q$ represents the averaging over the values of q. Care must be taken when arguing about "the nature of the process", in the same way as with the logscale diagram. The reason is that summing or averaging leads to a potential bias of large values occurring in some part of the data. For example, the LD is largely biased against second- and higher-order non-stationarity. Apparent multiscaling in the process could hide self-similarity or LRD if some irregularities are large enough compared to the other components of the process at a given timescale. Hence, care must be taken not to consider too seriously what the $M(j)$ says without carrying a

[3]Multifractal theory deals with mmathematical functions. The sample path of a process can then be seen as a mathematical function.

time-dependent analysis in order to check whether non-stationarity in the process does not bias the estimation. Instead of the $M(j)$, we shall rather use $log(M(j))$ that maps the three types of scaling processes, LRD (or statistical dependence), self-similarity and multiscaling respectively to the negative, zero and positive values of $log(M(j))$. It is also important to recall that scaling among moments cannot be used without considering a sufficiently large number of timescales in order to the qualitative analysis to be meaningful. It is not possible to analyze one single timescale through the $M(j)$ because scaling has still to be present in the low order properties for the high-order analysis to make sense. The multiscaling paradigm is useful as a complement to second-order analysis, not as a tool by itself. Once more, because of the averaging process the value 0 should not be taken too strictly.

High-order analysis*

The counterpart of the LD for high-order properties is the partition function described in section 2.4.2. The additional information it provides about a signal concerns the degree of homogeneity in time of the wavelet coefficients at each timescale. While the LD gives information about the variance of the irregularities in the process, the partition function tells about the distribution of large and small wavelet coefficients in the signal for each timescale. For this, we plot $log(S(q, j))$ as a function of the scale j, for a range of values of the moments q. If the low order moments ($q \leq 2$) have a larger value of $log(S(q, j))$ than high order moments, we conclude that the process is a second-order one, well described by the low order properties of the wavelet coefficients (the variance of its increments). This process in turn must be relatively smooth, without too many large-valued coefficients. Figure 2.10 presents the partition function for all traffic traces.

Two important things must be looked at when interpreting the partition function. First, discriminating between second-order and high-order processes can be done by determining for which values of q the $log(S(q, j))$ is the largest. A second-order process will have larger $log(S(q, j))$ for $q \leq 2$ while a high-order process for $q > 2$. Figure 2.10 shows that Auckland outgoing flows, UCB and LBL are second-order processes for the timescales larger than a few seconds because of their larger values of the partition function for $q \leq 2$. The Auckland incoming flows on the other hand are a second-order process for timescales larger than 5 minutes only. All other traces are high-order processes for all the considered timescales with their larger value of the partition function for large values of q. This implies that most of the traces contain large wavelet coefficients at most of the considered timescales.

A recurrent problem with wavelet analysis, as well as all the analysis techniques known so far, concerns non-stationarity. Wavelet estimators of scaling rely on some sufficiently large sample of a time series where the wavelet coefficients

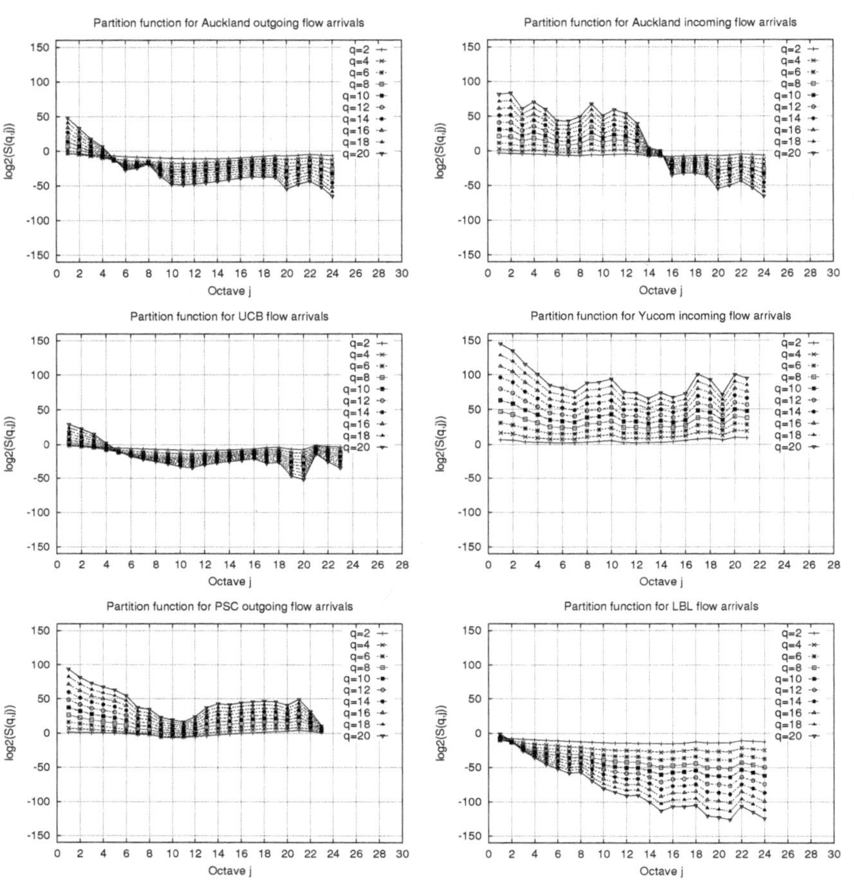

Figure 2.10: Partition functions.

on the fixed dyadic grid are assumed to be stationary, at least for what concerns the studied properties. The wavelet coefficients then can be used to compute a meaningful statistic of some property of the process. The problem in practice is that one would like to infer properties without having the certainty that this stationarity assumption holds. A way to deal with this problem consists in computing the statistics over many subsamples of the time-series and compare the subsamples to determine what is a true property of the process and what is an "outlier".

Now we turn to the time-dependent partition function, to determine whether the high-order behavior identified in the partition function arises for limited periods of the process or is a true property of the process that lasts for its whole duration. Figure 2.11 shows the distribution of the partition function for subsets of the process for moments two and six. Note that the size of these subsets is at least 2^{16} in order to rely on a sufficiently large sample for each subsequence. Figure 2.11 shows how frequently the high-order properties ($log(S(6,j))$) are larger than the low order ones ($log(S(2,j))$) on the constant-size subsamples. The plot of the Auckland outgoing trace (top left of Figure 2.11) reveals a single high-order subsample that does not bias the partition function (top left of Figure 2.10). The plot of the Auckland incoming trace (top right of Figure 2.11) on the other hand exhibits a few subsamples having a high-order behavior for octaves [8,14]. This time these events are sufficiently strong to bias the partition function so as to consider these timescales as high-order while they are low-order most of the time. The same bias occurs for the PSC trace for octaves [9,14]. The other traces do not seem to contain such bias.

Up to now, we have shown that the process of the flow arrivals can be second-order or high-order depending on the considered trace. We did not address the question of the "nature" of the process that arises at any given timescale, through the use of the $M(j)$ estimator defined in Formula 2.12. In the same way as for the partition function, we compute for subsamples of the whole traces the $M(j)$ for each j, and plot $log(M(j))$ as a function of j. Figure 2.12 presents the values of the $log(M(j))$ for each j and each trace.

Negative values $log(M(j))$ on some graph of Figure 2.12 indicate LRD, statistical dependence or no scaling at all depending on the results of the second-order analysis. A LRD process, being second-order stationary, requires a positive slope of the LD for the largest timescales, a stationary 3D-LD, no high-order scaling and a negative $log(M(j))$. Statistical dependence is like LRD except that the positive slope of the LD stops before the largest timescales, at the upper cutoff value of the correlation length. A self-similar process requires a positive slope of the LD for all timescales, a stationary 3D-LD, high-order scaling, and a $log(M(j))$ of zero. Multiscaling appears for positive values of the $log(M(j))$.

The Auckland traces and UCB have mostly negative values of the $log(M(j))$, with the largest timescales being close to self-similar with an almost zero $log(M(j))$.

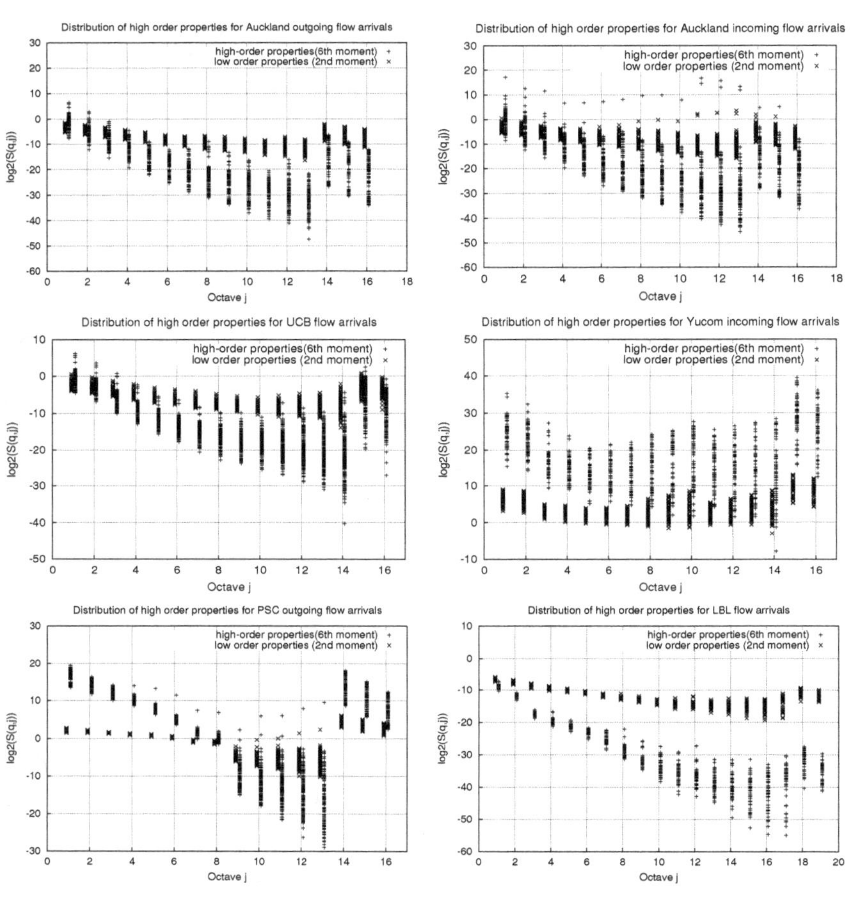

Figure 2.11: Stationarity check of partition function.

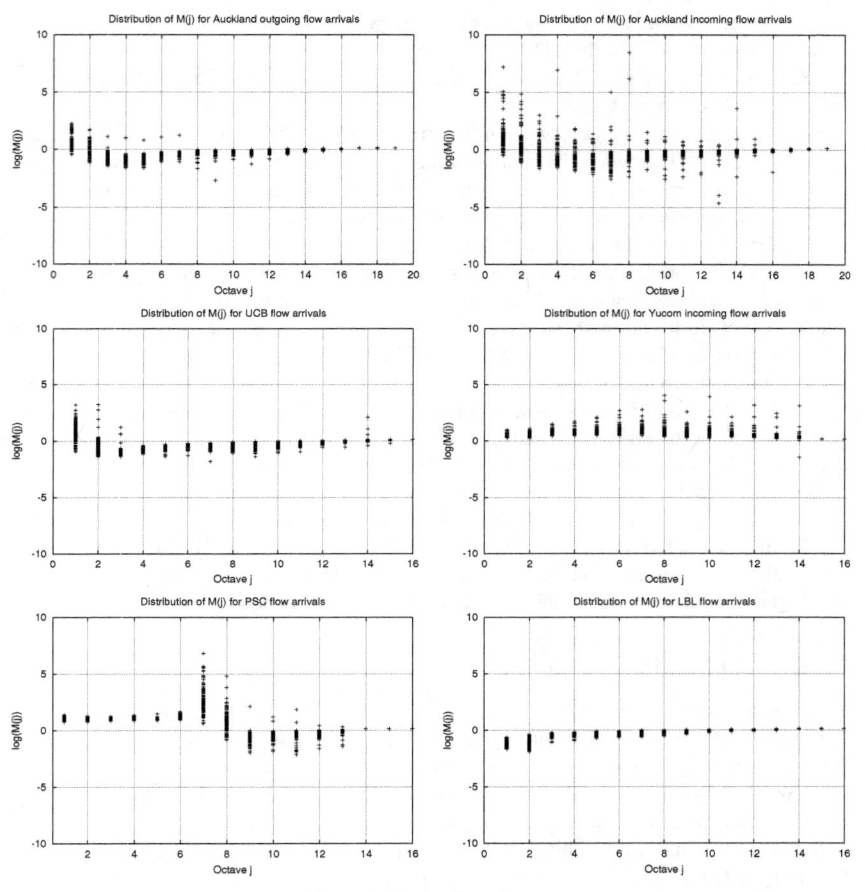

Figure 2.12: M(j) estimator.

Going towards smaller timescales shows decreasing values of the $log(M(j))$ until octave five (less than five hundred milliseconds) where it increases in a multifractal-like and time-dependent way, the values of the $log(M(j))$ increasing more or less linearly. Note that this multifractal behavior is probably due to a Poisson process since the second-order analysis indicated an almost uncorrelated process at these timescales with a slope of the LD close to zero.

Yucom on the other hand has all values of the $log(M(j))$ well above zero all the time, with sharp peaks quite often. This process is not multiscaling but contains some kind of high-order intermittency due to the instability with time of the $M(j)$ estimator. The would-be multiscaling behavior of this trace looks like a noisy high-order process, with sharp peaks occurring without much consistency. The second-order properties of this process however do not differ from the ones of the other flow arrival processes, the only difference with being the unidentified non-stationarity through its second-order properties. The flows of this trace could suffer from congestion, which creates intermittency in the flow arrivals. Unfortunatey, lack of information about the context of this trace leaves the cause of the phenomenon unexplained. All `http` traffic of the Yucom trace, incoming as well as outgoing, behave in that way.

PSC contains a mix of stationarity for the smallest and the largest timescales, with stable multiscaling at the smallest timescales (octaves [1,5]), LRD, statistical dependence or no scaling at all at octaves [9,13] while self-similarity at octaves larger 13 where second-order self-similarity was found in [198]. Octaves [6,9] exhibit a spikiness which we shall explain in the next section.

Finally, the LBL trace shows a neat graph with negative values of the $log(M(j))$ for almost all timescales and an increase of the $log(M(j))$ when going towards the largest timescales, corresponding to LRD or statistical dependence.

2.4.3 Applications behavior

The previous section has left unanswered the reasons for the differences in behaviors we found among the different traces. In this section, we partly address this issue by studying in more details the PSC trace. We focus on this trace for three reasons. First, we have the information about the internal clients of PSC. We can analyze each client's traffic separately. Second, we know that the largest client undergoes rate limitation. Third, we know that a large fraction of the traffic is P2P (kazaa, Gnutella,...). We can thus compare the difference in scaling behavior between P2P applications and `http` traffic.

Table 2.1 first compares the flows number for the two largest clients, for `http` and P2P (only kazaa and Gnutella, ports 1214 and 6346). The number of P2P flows is quite large, representing 18 percent of the flows for the largest clients, 57 percent of the flows for the second largest client. P2P traffic is becoming more

Table 2.1: Comparison of flows mix for largest 2 PSC clients.

Client	http	P2P	Total
1	9,767,651	2,265,086	12,032,737
2	2,038,776	2,769,587	4,808,363

and more popular, so a careful study of its behavior is appropriate.

Figure 2.13 presents the LD and 3D-LD for these two largest clients of PSC. The difference between the second-order properties of the two clients is obvious: client 1 has the mark of its rate limitation while client 2 has not. The LD of client 1 indicates an increase in the variance of the wavelet coefficients for the small timescales (up to a few seconds) while client 2 has a more classical LD. The 3D-LD confirms this behavior with an increased variance of the wavelet coefficients for client 1 throughout the day. Rate limitation affects the flows arrivals of client 1 all the time.

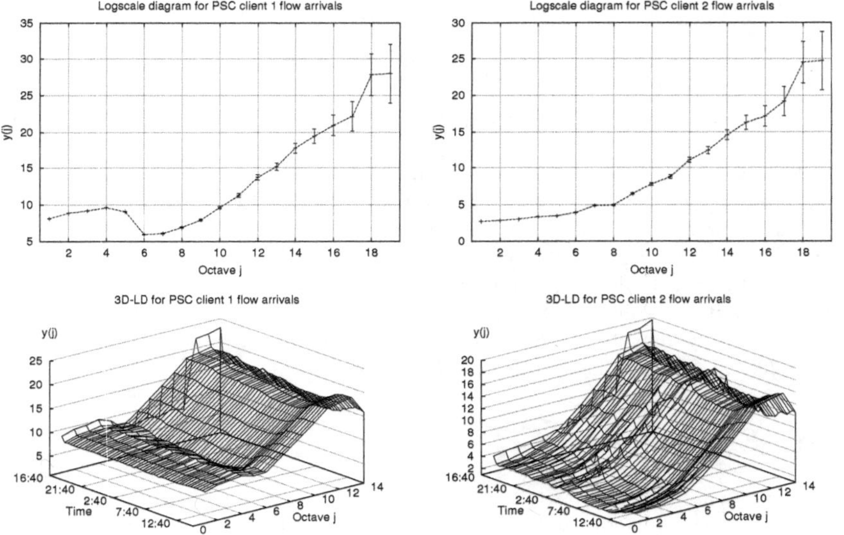

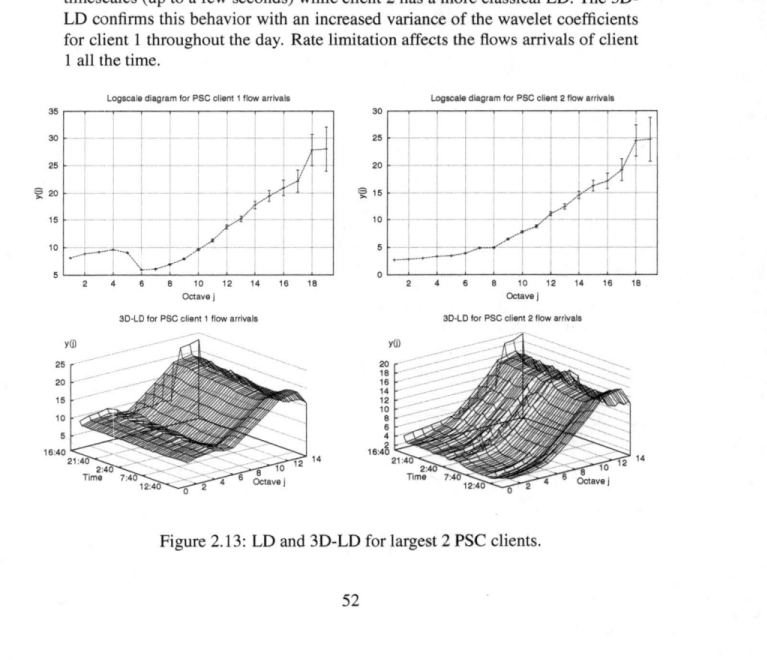

Figure 2.13: LD and 3D-LD for largest 2 PSC clients.

Figure 2.14 presents the evolution of the $log(M(j))$ over the day of the PSC trace also for the largest two clients. The effect of rate limitation (through a token bucket) for client 1 is to induce a linear high-order moments scaling (left of Figure 2.11) for timescales up to a few as well as stable multiscaling (bottom left of Figure 2.11) for the small timescales (up to a few seconds in our case). The effect of rate limitation is to shift the properties of the flows towards larger timescales, and to create time-homogeneous multiscaling over the small timescales. [82] did show that the small timescales of congested TCP traffic was multifractal. Here we show that such models as multiscaling and multifractals are even applicable for flow arrivals, not only for traffic volume at the packet-level.

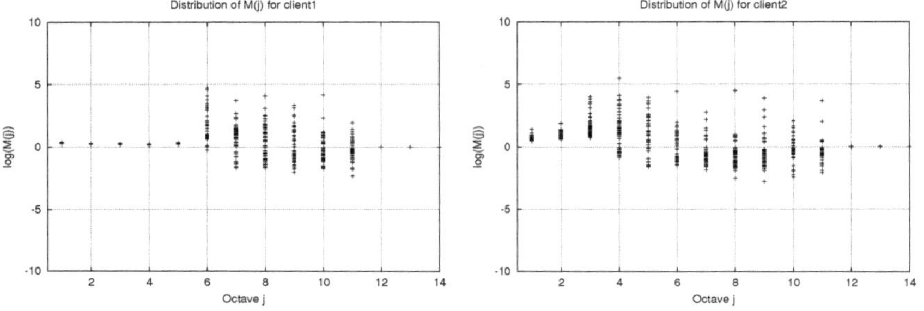

Figure 2.14: M(j) for largest PSC clients.

Finally, Figure 2.15 compares the behavior of the flows of the two largest clients, by splitting their flows into `http` (port 80) and P2P (ports 1214 and 6346). We can see on this figure that P2P traffic does not suffer from rate limitation, with clients 1 and 2 having a similar behavior of the $log(M(j))$, while `http` flows show a different behavior for them. `http` flows for client 1 exhibit a graph of the $log(M(j))$ similar to the one of all flows, while client 2 seems to be less impacted by client 1's rate limitation, although its behavior is different from normal `http` traffic like UCB's for instance (middle left of Figure 2.12).

2.5 Evaluation

In this section, we discuss the implications of the findings of this chapter on the roots of scaling in Internet traffic. Firstly, [59, 143] have shown the influence of the distributional properties of web traffic and file sizes on the statistical self-similarity of the traffic. This self-similarity was a second-order one, not strict self-similarity, and it involved a single control parameter H for all timescales. [182]

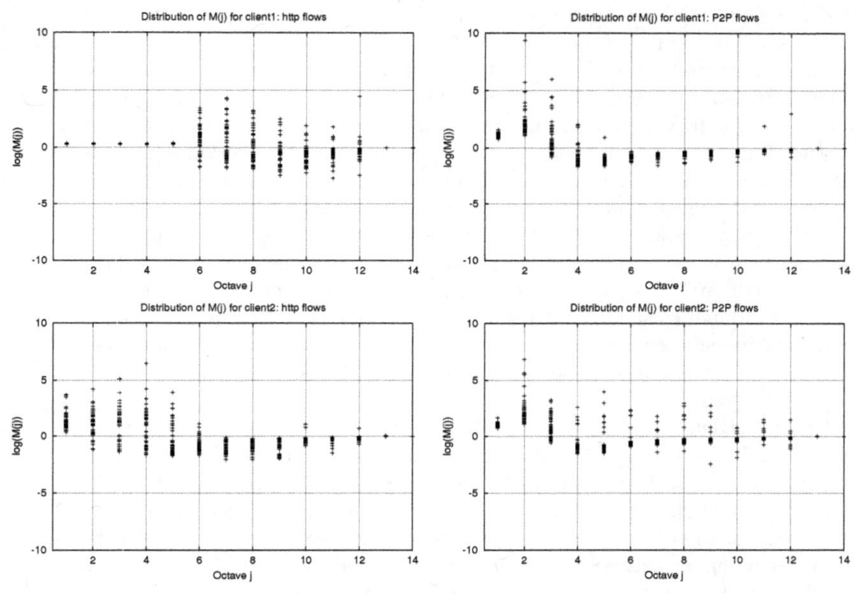

Figure 2.15: Comparison of scaling for PSC clients.

also proved that an aggregation process with independent ON/OFF sources and heavy-tailed ON/OFF distribution times gives rise to a second-order self-similar process. Secondly, papers dealing with scaling in Internet traffic have convincingly shown the influence of the TCP protocol and its dynamics on the traffic scaling properties [144, 83, 94, 82].

In section 2.4, we used a third perspective: the flow arrivals. We studied the process of the flow arrivals, and have shown not only that there is second-order scaling in this process, confirming [81, 198]; but that in addition higher-order scaling was also necessary to properly describe its dynamics. We showed that a wide range of scaling behaviors are present in this process, from multifractality at the small timescales of the `http` flows, to LRD, statistical dependence or no scaling at the medium timescales and finally exact self-similarity or LRD at the largest timescales. We point out the importance of the user's behavior through the process of the flow arrivals as another possible cause for the scaling in Internet traffic. While flow size distribution [59] in web traffic and network mechanisms [83, 94, 82] are undoubtedly involved in the scaling of network traffic, flow arrivals also to contain almost all types of scaling behaviors.

This chapter asks for more investigations in the relationships between scaling in network traffic, users and applications behavior, through the process of the flow arrivals. We showed that our understanding of the process of the flow arrivals is still limited and that interesting information might be extracted from the relationships between the network conditions and user's behavior. The high-order moments of the flow arrivals provide unexploited information about the dynamics of network traffic.

In this chapter we studied the total traffic over timescales larger than minutes and the TCP flows arrivals for timescales ranging from milliseconds to days. Another interesting scaling property of the traffic is caused by the way TCP breaks the traffic of the flows into IP packets. This process is well modeled by a *conservative cascade* [94]. A *conservative cascade* is a cascade that accommodates the two competing objectives of deterministic and random cascades: 1) preservation of the total mass of the process at each step of the cascade and 2) randomness of the distribution of the mass among the subintervals. These two properties are intuitively valid for TCP traffic, where there is a random distribution of the mass of the traffic among the different traffic flows aggregated in the total traffic and a deterministic breaking of the IP packets within a flow. The distribution of the packets is a mix between the deterministic way with which the TCP protocol distributes the mass of the traffic within a flow, and the randomness induced by the behavior of the network and its users. We studied packet-level traffic traces and showed in [189] that TCP traffic is consistent with a conservative cascade whose parameters' values are time-invariant, for timescales between a few milliseconds and a few hundreds of milliseconds. [189] shows that while the cascade model

seems to be an invariant of the traffic, the time-varying network conditions affect the second-order properties of the process as well as its multifractality.

On the one hand, conservative cascades are an invariant of the traffic for what concerns the timescales where TCP really matters, from milliseconds up to a few hundreds of milliseconds. Timescales of milliseconds correspond to the smallest timetick for the computer to send IP packets. With gigabit local area networks, these smaller timescales where the cascade model applies will go to microseconds. A few hundreds of milliseconds correspond more or less to a few round trip times, or the timescale at which TCP computes its estimate of the available bandwidth on the end-to-end path. The cascade model will therefore become more and more prevalent as the throughput of TCP flows increases, as the range of timescales over which TCP will drive the traffic dynamics will increase in turn.

On the other hand, the cascade model cannot capture what lies beyond the behavior of TCP: timescales larger than a few RTT's. The time-varying network capacity that influences how TCP chooses to send data is not captured by the parameters of the cascade model. This is why [189] shows that the changes in network conditions imply second-order non-stationarity that changes the variability of the traffic differently on different timescales, to which the cascade model is insensitive. [189] shows that only traffic *received* by a stub network is affected by this second-order non-stationarity, not traffic *emitted* by the stub network. Only the time-varying network conditions like available bandwidth and end-to-end delays can explain this phenomenon. Much remains to be done on the traffic dynamics side before we can state that we understand Internet traffic dynamics. Non-stationarity must be taken into account to fill the gap between mathematical models and the real traffic dynamics.

2.6 Conclusion

Besides the various technical implications of network traffic self-similarity on the understanding of the traffic dynamics, we conclude this chapter by discussing the implications on interdomain traffic engineering.

Self-similarity by itself is not an issue. The main issue concerns the implications of the properties of a self-similar signal on traffic prediction. From a statistical viewpoint, self-similarity implies that the extent of the variations in the signal scale linearly with the considered timescale. The aspect related to long-range dependence is secondary for what concerns traffic engineering. Correlations in the signal are of interest for applications leveraging the sample path properties of the process. For traffic engineering on the other hand, we are rather interested in knowing whether the process's variability is "stable" for control purposes. The analysis of the scaling properties of the traffic are important in this book because

we need to know at which timescales variability in the interdomain traffic becomes sufficiently limited not to be an issue for traffic engineering. We showed in this chapter that from the viewpoint of a stub AS, interdomain traffic variability becomes limited over timescales larger than one hour or so. Hence from a control perspective, interdomain traffic engineering should preferably work on timescales larger than hours.

Chapter 3

Topological distribution of interdomain traffic

3.1 Introduction

The behavior of Internet traffic has been analyzed by researchers almost since the creation of the first computer networks [113]. Many studies have tried to better understand the microscopic or packet-level behavior of the Internet. In this case, the analysis is usually performed on the basis of a capture of all the packets (or the packet headers) that flow through a given link for some period of time. This type of analysis has been popular thanks to the availability of powerful packet capture tools [43, 108, 58]. By analyzing the packets in the recorded trace, researchers have identified scaling (see section 2.2.1) in network traffic [119, 147] and studied aspects like the packet size distribution, the application mix or the number of concurrent TCP connections [56, 185].

Fewer studies have analyzed the macroscopic behavior of the Internet. By macroscopic behavior, we mean the topological distribution of the traffic throughout the global Internet. One of the early papers on this topic is [113] that analyzed the traffic distribution on the ARPANET in 1974. An important finding from this paper was that a few sites were responsible for a large part of the traffic. Similar studies were conducted in 1993 on the NSFNet backbone [55] on the basis of packet-level traces and SNMP statistics from backbone routers. They confirmed the findings of [113]. More recently, [76] analyzed several one hour packet level traces from universities and a commercial backbone to evaluate the impact of aggregating flows at the Autonomous System level.

In parallel to the measurements discussed in the literature, many network providers have deployed measurement systems to better understand the dynamics of the IP traffic inside their network notably for traffic engineering purposes.

Several papers have described the tools deployed by these network providers [129, 41, 85, 18, 89], but few papers have used the output of such tools to analyze the macroscopic behavior of Internet traffic.

In this chapter, we present a detailed analysis of the characteristics of interdomain traffic by studying two one week long traces of the entire interdomain traffic received by two non-transit providers. Note that this chapter is a largely extended version of [195].

The structure of this chapter is the following. Section 3.2 first describes the two considered providers. Section 3.3 studies the aggregation provided by different notions of the traffic flows. Then, section 3.4 analyzes the topological distribution of the traffic. This topological analysis is refined in section 3.5 that studies the distribution of the active traffic sources.

3.2 Measurement environment

To obtain a better understanding of the characteristics of interdomain traffic, we have relied on Netflow [53] traces of two different non-transit providers. Netflow is a traffic monitoring facility supported by Cisco routers. When enabled, the router regularly transmits information about all unidirectional layer-4 flows that passed through it to a monitoring station. With Netflow, the monitoring station knows the starting and ending timestamps of the layer-4 flows (TCP connections and UDP flows), the flow volume (in bytes and packets), and the transport protocol and port numbers. Netflow is often used for billing purposes or by providers that need to better understand the traffic inside their network. Compared to the traditional packet-level traces that are often analyzed in the literature, Netflow has the advantage of being able to monitor multiple links during long periods of time. The main drawback of Netflow is that it does not capture the short-term variations of the traffic, but this is not a problem in our case since we focus on medium to long-term traffic variations.

3.2.1 The studied providers

The only characteristic common to both providers is that they do not offer transit service. Besides, they serve different customers and it can be expected that these customers have different requirements on the network. Due to technical reasons, it was unfortunately impossible to obtain traces from the two studied providers covering the same period of time.

The first trace was collected in December 1999 and covers six successive days of all the interdomain traffic received by BELNET. BELNET provides connectivity for the research and education institutions located in Belgium. At that time,

the BELNET network was composed of a 34 Mbps star-shaped backbone linking the major universities. Its interdomain connectivity was mainly provided through 34 and 45 Mbps links with two large tier-1 providers. In addition, BELNET had a 45 Mbps link to the European research network, TEN-155, and was present at the BNIX and AMS-IX interconnection points with a total of 63 peering agreements in operation. Although some universities provided dialup access for their students, the typical BELNET user had a 10 Mbps access link to the BELNET network through their university LAN. During the six days period, BELNET received 2.1 TBytes of data. BELNET is representative of research networks and could also be representative of a provider providing services to high bandwidth users with cable modem or ADSL.

The left part of Figure 3.1 shows the evolution of total traffic for BELNET during the period of the measurements. While the global evolution of the total traffic exhibits a daily periodicity, with peak hours located during the day, there are important deviations around the average traffic evolution throughout the day. The mean traffic over the six days period was slightly larger than 32 Mbps, with a one-minute maximum peak at 126 Mbps and a standard deviation of 21 Mbps. The trace begins around 1 AM on a Sunday and finishes six days later around 1 AM also.

The second trace was collected in April 2001 and covers a little less than five consecutive days of all the interdomain traffic received by Yucom. Yucom is a commercial provider that provides Internet access to dialup users through regular modem pools. At that time, the interdomain connectivity of Yucom was mainly provided through high bandwidth links to two transit providers. In addition to this transit service, Yucom was also present at the BNIX interconnection point with 15 peering agreements in operation. During the five days of the trace, Yucom received 1.1 TBytes of data. Yucom is representative of providers composed of low bandwidth users.

The right part of Figure 3.1 presents the total traffic evolution for Yucom during the measurements. The trace starts around 8:30 AM on a Tuesday and finishes almost 5 days later at midnight. The total traffic also exhibits a daily periodicity with peak hours located during the evening, in accordance with the typical user profile, a dialup user. It had an average total traffic of about 23 Mbps over the measurements, with a one-minute maximum peak at 64 Mbps and a standard deviation of 12 Mbps.

3.2.2 Interdomain Topology

Before analyzing in details the collected traffic statistics, it is useful to have a first look at the BGP table of the studied providers. In this chapter, we assume that the BGP table of both providers was stable during the period of the measurements and

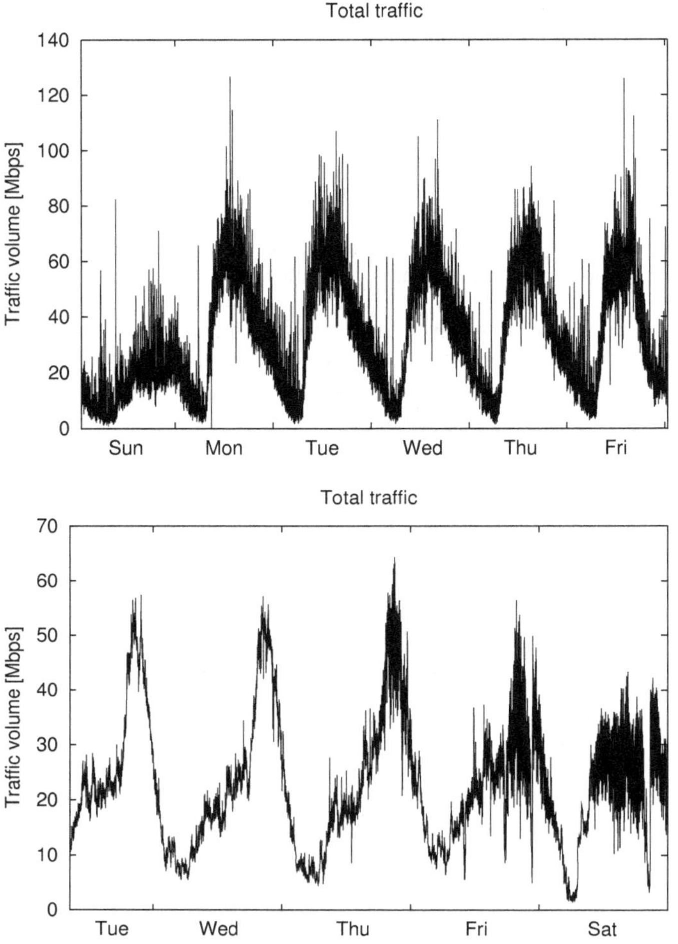

Figure 3.1: Total traffic evolution, BELNET (top) and Yucom (bottom).

perform all our analysis based on a single BGP table for each provider. We assume in this chapter that the influence of the changes in the BGP routes is negligible. Chapter 4 will confirm that relying on a single BGP routing table is not a bad approximation.

The routing table of Yucom contained $102,345$ prefixes, covering about 26 percent of the total IPv4 address space. This coverage of the total IPv4 address space is similar for BELNET, with about 24 percent but for $68,609$ prefixes only. Form late 1999 to mid-2001, 30 percent of additional prefixes are necessary to cover a similar percentage of the IPv4 address space. Although having different numbers of prefixes in their BGP routing table, the two providers cover a similar percentage of the IPv4 address space. The average address span per prefix for each provider, about $15,200$ IP addresses for BELNET and about $11,000$ IP addresses for Yucom, explains this increase in the number of prefixes necessary to cover a similar percentage of the IPv4 space. Yucom knew $10,560$ distinct ASes while BELNET 6298. This difference is mainly due to the large increase in the number of multi-homed sites during the last few years [107].

Figure 3.2 compares the distribution of the potentially reachable IP addresses at each AS hop distance for the BGP routing tables of the two providers. For each BGP prefix, we attributed the number of IP addresses according to its prefix length to the AS hop distance equal to its unprepended AS path length. While this does not measure the actual number of IP addresses at each AS hop distance, it reflects the potential reachability of IP hosts for each AS hop distance seen by BGP routes. The main difference between the two ISPs is the more compact distribution for Yucom around a distance of three AS hops. BELNET has its reachable address space more spread over distances of three and four AS hops. The first three AS hops for Yucom provide almost eighty percent of the reachable address space while only about sixty percent for the research ISP. The difference between the distribution of the reachable IP prefixes seen from the two providers are likely to be due to the way the two providers are connected to the Internet as well as the sixteen months delay between the two traces.

3.3 Aggregation of interdomain flows

To understand the topological variability of interdomain traffic and the possible levels of aggregation, we consider in this chapter two different types of interdomain flows. Generally, a flow is defined as a set of IP packets that share a common characteristic. For example, a micro-flow is usually defined as the set of IP packets that belong to the same TCP connection, namely the IP packets that share the same source address, destination address, IP protocol field, source and destination ports. In this chapter, we consider two different types of network-layer flows. A

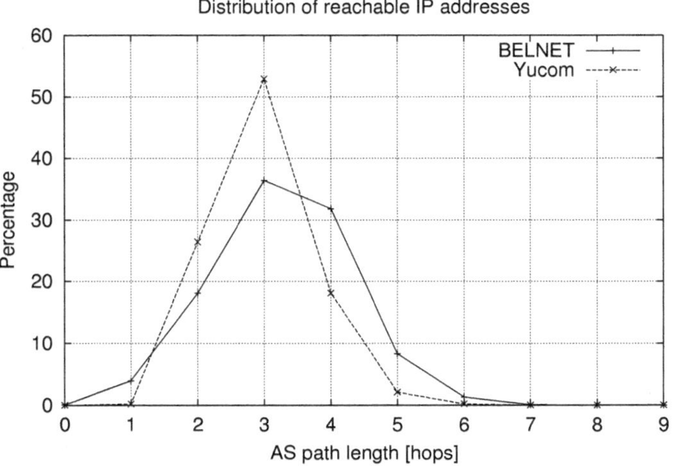

Figure 3.2: Distribution of reachable IP addresses.

prefix flow is the set of IP packets whose source addresses belong to a given network prefix as seen from the BGP table of the studied provider. An *AS flow* is defined as the set of IP packets whose source addresses belong to a given AS as seen from the BGP table of the studied provider. We do not use explicitly throughout this chapter the term "flow" to designate traffic coming from a traffic source, but mostly the terms "prefix" and "AS" (or "source AS") to denote a *prefix flow* and *AS flow* respectively.

We study in this section the amount of aggregation provided by the AS and prefix flows. Figure 3.3 shows the cumulative percentage of the traffic or *order statistics*, for AS and prefix flows. On this figure, we ordered the prefixes and ASes by decreasing order of the total traffic volume sent by them to the studied network over the whole measurements, and we computed their cumulative contribution to the total traffic over the measurements. The x-axis uses a logarithmic scale to emphasize the low *order statistics*. Both providers have a similar cumulative traffic distribution for their most important interdomain traffic sources. The largest one hundred AS flows capture about 72 percent of the total traffic for Yucom and a little less than 60 percent for BELNET. The largest one hundred prefix flows capture about 52 percent of the total traffic for Yucom and a little more than 40 percent for BELNET. 90 percent of the total traffic volume is captured by only 4.7 percent of the ASes and by 4.1 percent of the prefixes present in the

BGP table of Yucom. BELNET required 9.8 percent of the ASes and 4.5 percent of the prefixes present in its BGP routing table to capture 90 percent of the total traffic volume. These results are similar to the findings of earlier studies on the Internet of the 1970's [113] and the early 1990's [55]. On the other hand, some ASes and prefixes contributed to a small fraction of the total traffic. For Yucom, more than 4000 different ASes sent each less than one megabyte of data during the measurement period and some ASes only sent a single packet during this period. For BELNET, 719 ASes sent less than one megabyte of data during the six days measurement period.

Over the measurement period, BELNET received IP packets from 5,606 ASes and 35,688 prefixes, corresponding to 89 percent of the ASes present in its BGP table. Yucom received IP packets from 7668 ASes and 35,693 prefixes, corresponding to 72 percent of the ASes present in its BGP table. These figures show that even relatively small providers receive traffic from a large portion of the Internet during a one week period, even though some sources only send a few packets.

3.4 Topological distribution of the traffic

The amount of traffic aggregation provided by the interdomain flows is not the only important aspect to be considered when studying the macroscopic behavior of Internet traffic. Another important aspect concerns the topological distribution of the traffic. By topological distribution, we mean the distance between the traffic sources and the studied ISP. This distance is important for two reasons. First, [131] has shown that the performance of an Internet path decreases with the distance between the source and destination AS, even if [106] showed more mitigated results. Second, if the distance between the source and the destination AS is large, it will be difficult for either the source or the destination to apply mechanisms to control the traffic flow in order to perform interdomain traffic engineering [20, 155] (see also Chapter 1).

Figure 3.4 presents the cumulative traffic distribution for the order statistics of the ASes for each AS hop distance. Considering AS flows does not show the traffic aggregation occurring on the AS-level topology. We thus computed the traffic volume sent by each prefix of the BGP routing table. Then, based on the `AS path` attribute of the BGP routes, we attributed to each AS appearing in the `AS path` attribute of the route the total traffic received from the prefix of the BGP route. Doing this amounts to consider each AS not as a traffic source but as a transit. The objective of this procedure is to attribute to each AS of the Internet the traffic volume it has seen for the two studied providers during the period of the measurements. An AS located at an AS hop distance of n is seen as the source of the traffic it generates as well as of all the traffic it forwards when considering

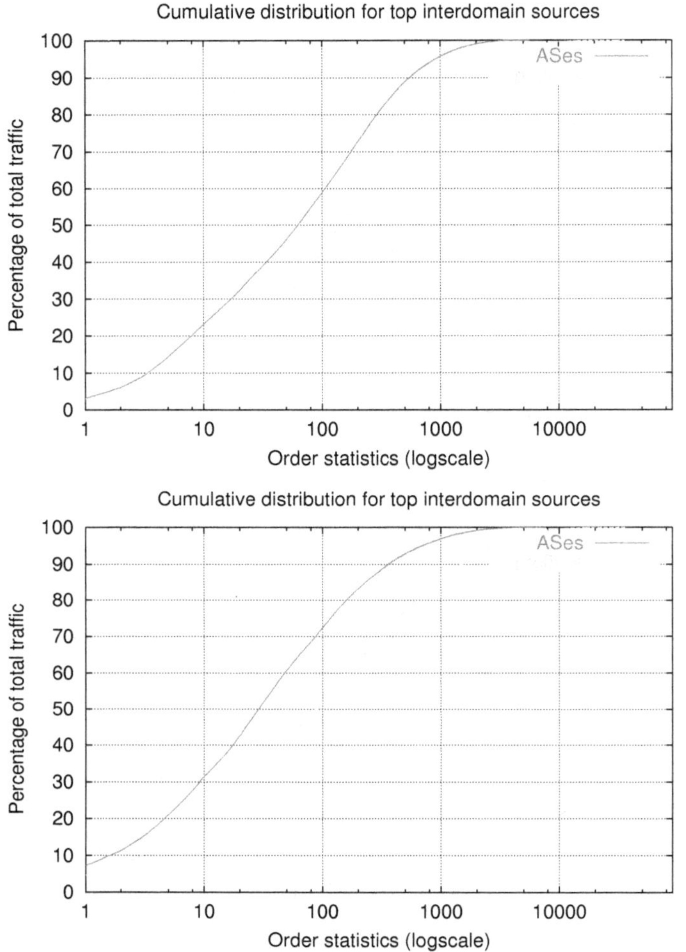

Figure 3.3: Traffic aggregation for interdomain flows, BELNET (top) and Yucom (bottom).

the unprepended `AS path` information of the BGP routing table. Because each AS hop distance does not contribute evenly to the total traffic, we have plotted the cumulative traffic percentage for every AS hop distance with respect to the total traffic seen during the measurements.

Each curve of Figure 3.4 shows the cumulative traffic percentage seen by the largest ASes ordered by decreasing amount of the traffic volume seen by them over the whole measurement period. A point of a curve shows the traffic percentage seen by the largest ASes in traffic volume at some AS hop distance. For example, the first curves on top of both graphs of Figure 3.4 show the cumulative percentage of traffic captured by the ASes located at one AS hop form the two considered providers. The AS located at one AS hop are the BGP peers of the two studied providers. Both curves end up at one hundred percent of the total traffic, indicating that all traffic was seen by the ASes located at one AS hop. The curves for larger AS hop distances on the other hand do not attain one hundred percent of the total traffic. The difference between the total percentage of traffic seen at two successive AS hop distances, say n and $n+1$, shows what percentage of the traffic was generated by ASes located at AS hop distance n. For instance, the difference between the largest traffic percentages for the one AS hop distance and two AS hop distances on Figure 3.4 give the percentage of traffic generated by ASes located at one AS hop. In this particular case, the ASes at one AS hop generated 11 percent of the total traffic volume for BELNET and a little more than 6 percent for Yucom. ASes at two AS hops generated a little more than 12 percent of the total traffic volume for BELNET and about 42 percent for Yucom. Then, ASes at three AS hops generated about 46 percent of the traffic for BELNET and about 44 for Yucom. The last AS hop distance for which a significant percentage of the traffic was received is for four AS hops with about 25 percent for BELNET and about 7 percent for Yucom. Subsequent AS hop distances did not send a significant percentage of the total traffic for neither of the two considered providers.

Now if we focus our attention on one particular curve of Figure 3.4, we can study how many ASes must be considered to capture some percentage of the total traffic. Take for instance the curve for a distance of one AS hop. On this curve, we see that the AS located at one AS hop that sees the largest amount of traffic for BELNET sees 64 percent of the traffic (42 percent for Yucom). The three ASes at an AS hop distance of one that see the largest amount of traffic for BELNET carry 83 percent of the BELNET traffic, while for Yucom the largest three ASes at one AS hop carry 87 percent of the traffic. Since ASes located at one AS hop are either transit providers used by the local AS for its Internet connectivity or peer ASes with whom local traffic is exchanged, the expected situation is the one where a very small number of ASes at one AS hop see a large fraction of the total traffic sent or received by a non-transit provider.

At a distance of two AS hops, a significant difference between BELNET and

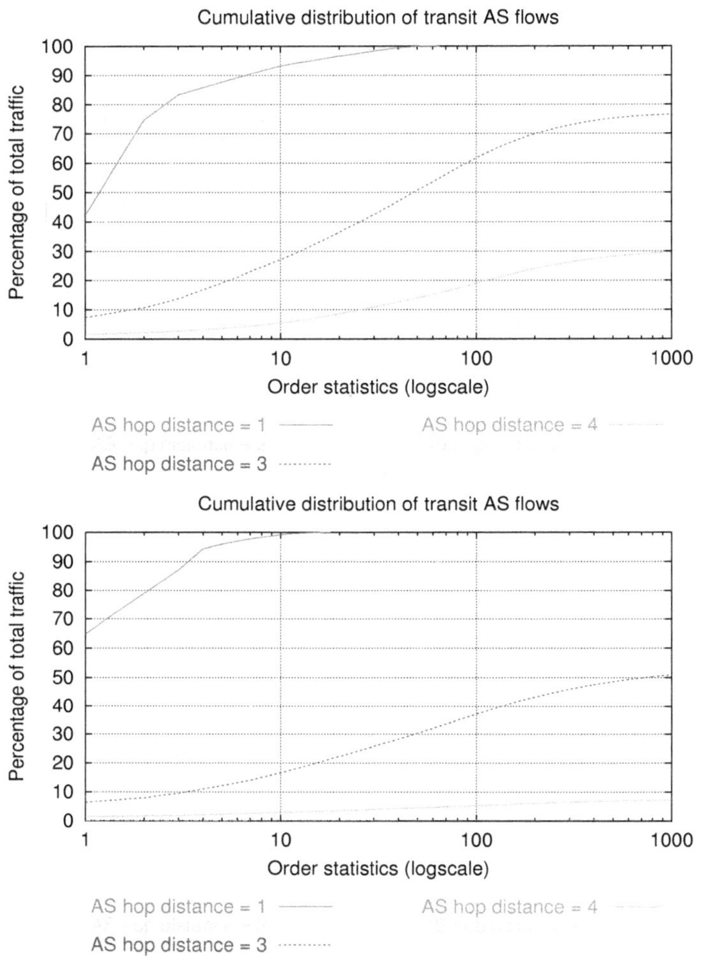

Figure 3.4: Topological traffic distribution, BELNET (top) and Yucom (bottom).

Yucom appears. At two AS hops, the largest ten ASes for BELNET see about 75 percent of the traffic, while for Yucom the largest ten ASes at two AS hops see only about 50 percent of the traffic. However, the largest one hundred ASes at two AS hops capture a similar percentage of the total traffic of BELNET and Yucom, about 90 percent. The number of ASes required to capture a given percentage of the total traffic at some AS hop distance indicates how the incoming traffic is distributed on the AS-level topology. Suppose that we wanted to know for each AS hop distance what is the smallest number of ASes that see 40 percent of the total traffic. Then we take Figure 3.4 and find when the curve for each AS hop distance intersects the horizontal line indicating 40 percent of the total traffic. For BELNET, it gives one AS at one AS hop, two ASes at two AS hops, and less than 30 at three AS hops. For Yucom, it gives one AS at one AS hop, eight ASes at two AS hops, and more than one hundred ASes at three AS hops. The incoming traffic of Yucom is hence distributed over more ASes on the interdomain topology than the one of BELNET.

3.5 Activity of the Traffic Sources

This section studies the *activity* of the interdomain traffic sources, i.e. the behavior over time of the ASes that generate traffic. We say that a traffic source is *active* during a given time interval whenever at least one byte of traffic is originated from it during the considered time interval. Note that this notion of *activity* gives no respect to the variations in traffic volume over time. We look at the number of AS flows that are *active* over some time interval as well as the number of *active* AS flows at different timescales.

We first study the activity of the AS flows in Figure 3.5. Figure 3.5 shows, over the whole measurement period, the average number of *active* AS flows at each AS hop count, for timescales of fifteen minutes, one hour, four hours and twelve hours. The number of *active* AS flows is compared with the total number of ASes known from the BGP table. If we compare the number of ASes known from the BGP table for each AS hop distance, we realize that it differs from the distribution of the reachable address space for each AS hop distance shown on Figure 3.2. For BELNET, most ASes are located at four, three and then five AS hops. The distribution of the reachable space of Figure 3.2 on the other hand tells that there are more potential sources at three, two, then four AS hops. For Yucom, most ASes are located at three, four and then five AS hops for a distribution of the reachable space with more potential sources at three, two and four AS hops. This shows that the distribution of the reachable address space for each AS hop distance is a poor predictor for the distribution of the ASes with whom traffic is exchanged.

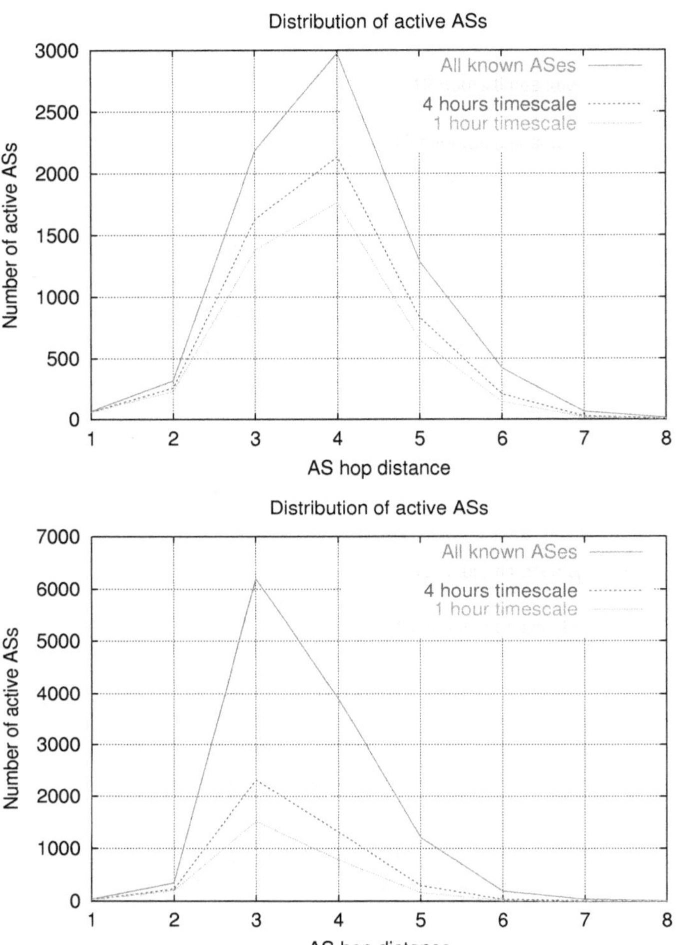

Figure 3.5: Number of *active* ASes, BELNET (top) and Yucom (bottom).

The different curves of Figure 3.5 also show that the distribution of the average number of *active* ASes does not change fundamentally across the different timescales considered. The look of the distribution is similar for all timescales, what changes is the total number of *active* ASes that varies by a factor of less than ten between the smallest and the largest timescales considered. During a given fifteen minutes period, BELNET receives packets from 3088 different ASes on average. For the twelve hours timescale, the number increases to 5711 ASes. Yucom for its part receives on average packets from 1632 ASes during a fifteen minutes interval and from 5721 ASes for a twelve hours interval. These numbers should be compared to the 6298 ASes (resp. $10,560$) that appear in the BGP table of BELNET (resp. Yucom).

The ASes that are direct peers of the studied providers are the most *active* source ASes. During any fifteen minutes period, 94 percent of the ASes known at a distance of one AS hop are *active* for Yucom. During a period twelve hours, all ASes known at one AS hop are *active* for Yucom. For BELNET, 90 percent of the ASes at one AS hop are *active* on average during any fifteen minutes interval and this percentage increases to 96 percent when considering twelve hours intervals. The percentage of *active* ASes at a distance of two AS hops ranges between 58 percent for the fifteen minutes intervals and 85 percent for the twelve hours intervals for BELNET. The same numbers for Yucom are of respectively of 62 percent and 90 percent.

Let us turn to *active* prefixes. Figure 3.6 shows the activity of prefixes, which exhibits a similar behavior to the one of the ASes. For BELNET, there are on average $7,029$ different *active* prefixes during a fifteen minutes interval, about 10 percent of the prefixes contained in its BGP table. For Yucom, the average number of *active* prefixes during a fifteen minutes interval is $3,719$, representing only about 4 percent of the prefixes of its BGP table. The average number of *active* prefixes for Yucom increases to $18,948$ for twelve hours intervals, to be compared with the $102,345$ prefixes known from its BGP table.

The distribution of the traffic sources (in ASes or prefixes) in terms of AS hop distance does not have to be consistent with how many reachable prefixes or ASes are known in the BGP table at a given AS hop distance. Furthermore, considering AS flows or prefix flows does not significantly change the number of *active* sources on average at a given timescale.

3.6 Conclusion

In this chapter, we have analyzed the topological distribution of the traffic on the basis of two one week traces covering all the interdomain traffic of two different medium size non-transit providers. We expect that similar results would be found

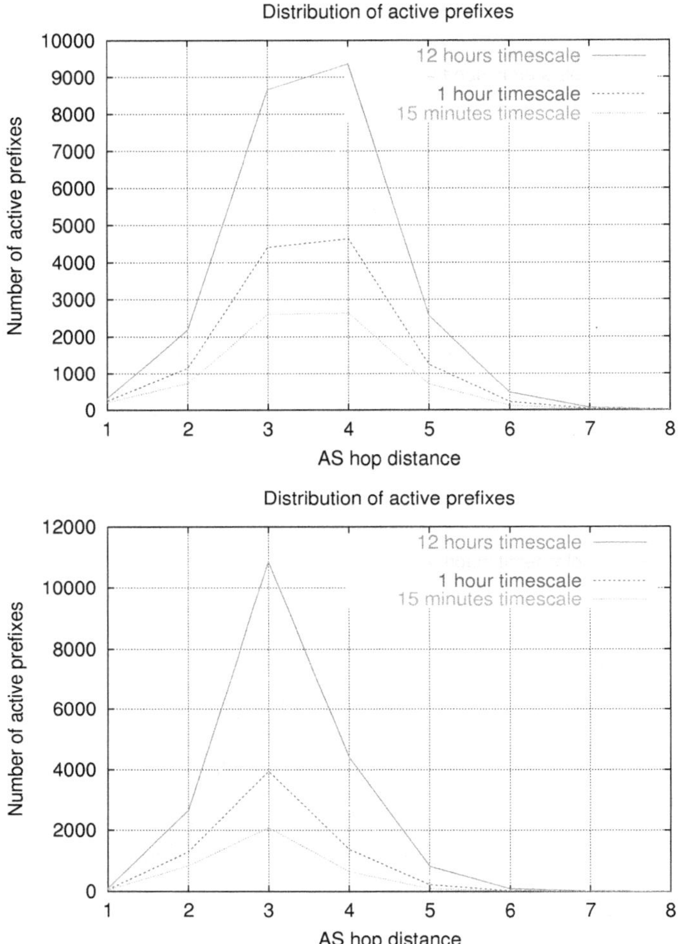

Figure 3.6: Number of *active* prefixes, BELNET (left) and Yucom (right).

for most small to medium ISPs that constitute the majority of the ISPs in the Internet. The topological distribution of the interdomain traffic however mainly depends on the way the considered provider is connected to the Internet as well as the most popular ASes with whom traffic is exchanged.

The main results are the following. First, the distribution of the reachable address space spans few AS hops. Second, we confirm that a small number of interdomain traffic sources account for an important part of the total traffic received by the studied providers, with a relatively small difference between the prefix and the AS aggregation levels. Third, the most important part of the traffic is produced by sources located at a distance of a few AS hops, but the distribution of the traffic differs from the distribution of the reachable address space. Fourth, the distribution of the average number of interdomain traffic sources is also limited in terms of the AS hop distance and can also differ from the distribution of the reachable address space as well as of the distribution of the traffic. On the other hand, the distribution (not the total number of sources) of the average number of interdomain traffic sources does not change much with the considered timescale. Fifth, the studied providers, despite their size, receive traffic from a large fraction of the global Internet.

Chapter 4

Topological dynamics of interdomain traffic

4.1 Introduction

The common message of papers dealing with interdomain traffic [113, 55, 76, 80, 195, 155] is that a limited number of ASes are responsible for a large fraction of the interdomain traffic. The main limitation of these studies is that they consider the traffic distribution for a given time period. Variations of the topological traffic distribution within the whole measurements period are not considered.

The aim of this chapter is to study the dynamics of the traffic on the interdomain topology, as seen by two stub ASes. Compared to [157, 168, 114, 166] that focus on a subset of the applications seen in the traffic, our goal is to study the dynamics of all the traffic sent by stub ASes, independently of the underlying applications. This chapter focuses exclusively on these stub ASes, as they constitute the vast majority of all ASes [176, 10]. During the last years, the fraction of stub ASes has remained stable around 85 percent [170].

In section 4.2 we explain the context of this chapter. Section 4.3 presents the traffic traces as well as the procedure used to merge the traffic and BGP routing information. Section 4.4 studies the stability of the BGP routing for the interdomain traffic. Section 4.5 analyzes the topological properties of the interdomain traffic over the whole traces while section 4.6 looks at the dynamics of the properties of the interdomain topology for the traffic. Note that most of the content of this chapter also appeared as [199].

4.2 Studying the interdomain topology

In order to understand the topological distribution of interdomain traffic we rely on an AS-level graph. This graph is built from the AS path information contained in the BGP advertisements received by the studied stub ASes. More precisely, each AS is one node of the interdomain graph and an edge correspond to the peerings between two ASes. A node of our AS-level graph may correspond to several distinct prefixes. For example, the node corresponding to a large provider will correspond to the domain of this provider and to several of its smaller clients that do not have an AS number. Furthermore, an edge in the interdomain graph may correspond to several distinct physical links or peerings. Since BGP only advertises one path to each destination, the considered interdomain graph does not show all interconnections between domains.

On this interdomain graph, we are interested in the paths followed by IP packets sent by the monitored stub AS to the global Internet. Those paths are selected by the BGP decision process [175] based on the advertisements received from each of its BGP peers. Since BGP is a dynamic routing protocol, these paths may change over time. Furthermore, since an AS may advertise several prefixes and the AS Path information is associated with one prefix, there might be several distinct paths between the studied stub AS and a given destination AS.

In this chapter, we call an "edge" an AS pair appearing as two consecutive and distinct ASes in the AS path of the *best BGP route* for some traffic destination. With the BGP routes, we know how the traffic is forwarded across the AS-level topology, without respect to physical links or multiple peering points between two ASes. With the AS path information of the BGP routes, only "edges" of the AS-level topology are known. An example will better illustrate this notion of an "edge". In the left part of Figure 4.1, the stub AS (AS A) has two BGP peers (AS B and AS C) and each BGP peer announces one route towards each of the two destination ASes (AS F and AS G). In this topology, the stub AS receives two routes towards destination AS F from its two peers. The route advertised by peer 1 has AS path *B-F* while the route advertised by peer 2 has AS path *C-B-F*. Assuming that the stub chooses as its best route towards AS F the route advertised by AS B having the shortest AS path, we have two edges in the interdomain topology for this AS path, namely edge <A,B> and edge <B,F>. Our stub AS receives two routes for destination AS 2 having the same AS path length of 3 from its two peers, *B-D-G* for peer 1 and *C-E-G* from peer 2. If we also assume that our stub chooses the route advertised by peer 2 we have three new edges; <A,C>, <C,E> and <E,G>.

In the AS-level graph, we are interested in studying and understanding the topological distribution of the interdomain traffic. For this, we need to associate each "edge" with the amount of traffic it carries. We thus count all network traf-

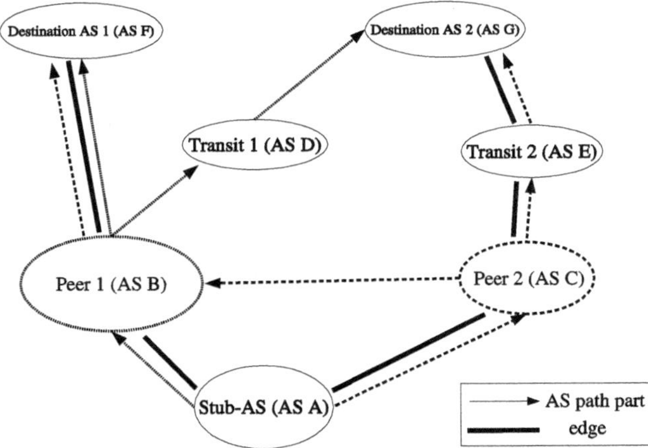

Figure 4.1: Example AS-level topology.

fic that transits between two ASes connected by that edge. An edge does not represent a physical link or a single peering between two ASes, but all peerings between two ASes. From a topological characterization perspective, "edges" represent an aggregate of links or peerings. From a traffic engineering perspective on the other hand, "edges" represent a meaningful traffic aggregate that could be controlled through BGP tweaking like AS path prepending, MED, or redistribution communities [155].

4.3 Measurements

In this chapter, we rely on two traffic traces along with the associated BGP routing information. The first trace is a one entire month trace of the outbound traffic from the "Université catholique de Louvain" (UCL), Belgium. The BGP data was gathered from a session established with the associated network provider, BELNET. To the best of our knowledge, this trace is the longest and most recent interdomain traffic trace of all the interdomain traffic reported in the literature. The second trace, mainly used for validation and comparison purposes, was gathered over 24 hours from the outbound commodity traffic of the Pittsburgh Supercomputing Center (PSC), Pennsylvania, USA. Its associated BGP information was gathered from an on-site route server.

Capturing such interdomain traffic traces is particularly difficult because of

the anonymization issues involved in analyzing the whole interdomain traffic of some network. Most ISPs at least require that all IP addresses be anonymized so that most traces publicly available have all IP addresses related information lost [213], making them useless for studying the topological properties of the traffic that require that all flows or packets be routed with BGP routing tables to know where the sources and destinations of the traffic are located and to determine the actual path between them when the traffic crossed the Internet.

Given that the aim of this chapter is to study the macroscopic properties of interdomain traffic, the decision was made to use a minimal time granularity of one hour. The reason for this choice is mainly practical, in that for a trace of one month duration a finer granularity would prove to be cumbersome. In addition, Chapter 2 has shown that at timescales smaller than one hour, LRD and self-similarity will make the dynamics of the traffic difficult to understand. This choice of one hour is therefore the smallest timescale at which we can expect a relatively smooth behavior of the traffic on the AS-level topology, along with a reasonable time granularity of the study for practical purposes.

4.3.1 The UCL trace

The primary data set used is a month long trace of all the outbound traffic from the University of Louvain between March 19th 2003 0:00 and April 18th 2003 23:59 CET. During this period, 8.4 TBytes of traffic were observed with an average bit rate of 25.1 Mbps. The University's connectivity is a heavily loaded full duplex fast Ethernet link to the service provider BELNET (AS2611).

Inbound and outbound flows were tapped and sniffed using separate fast Ethernet NICs in a single host. The host used to capture the traffic was a P4 2.4 GHz with 256 MB of RAM running FreeBSD 4.7. Two fast Ethernet NICs were used to gather traffic information while a third was used to place the host on the University's internal network. To gather the traffic statistics, we relied on `nProbe` [67], a flow-level collection software that captures packets arriving at a NIC and emulates the behavior of `Netflow` by exporting flow-level statistics. The maximum flow lifetime and timeout period were forced to be 30 seconds. No packets were reported as dropped by `nProbe` during the one month period. An eBGP session was established between the sniffing machine and the BELNET router serving the university using the Zebra routing daemon [2].

The majority of the university's user base consists of about 25,000 students and 5,000 additional staff using the internal university network. The university also provides ADSL and cable modem access to some students and staff. The university's Internet connectivity is provided by the Belgian research network, BELNET, which is connected to two tier-1 providers and the European research network GEANT, in addition to more than 100 peers at several national intercon-

nection points (BNIX, AMS-IX, and SFINX). The internal network of BELNET consisted of all Belgian universities and research institutions, linked by a 2.5 Gbps star configuration between its main routers in Brussels and each university. To our knowledge, BELNET did not perform any sort of traffic engineering on its commercial providers but relied on the shortest AS path towards each destination. All academic traffic was sent and received through the GEANT network. A limit of 45 Mbps was enforced by BELNET for commodity traffic of the university in the incoming direction only, and no traffic limitation was applied for traffic among the BELNET sites nor for traffic exchanged with universities in Europe or elsewhere through the GEANT network.

4.3.2 The PSC trace

The Pittsburgh Supercomputing Center (PSC) is a regional aggregation point of presence located in western Pennsylvania, USA. PSC provides commodity and Internet2 access to local universities and organizations including Carnegie Mellon University, The University of Pittsburgh, Pennsylvania State University, West Virginia University, The City of Pittsburgh, and others. At the time of the study PSC had a maximum capacity of 395 Mbps of commodity access though AT&T at 145 Mbps and Verio with 250 Mbps via an OC12. Additionally PSC had a full OC48 of Internet2 connectivity through the Abilene network.

The PSC trace used in this study is composed of all the outbound commodity traffic from 8:47 AM 18 April 2003 to 8:45 AM 19 April 2003. During this period 1.7 TBytes of traffic were observed for an average rate of about 164 Mbps.

The trace was taken by inserting an optical splitter on the interior Gigabit Ethernet commodity interface of a Juniper M40 router. The optical split was then attached to a PIII 800 dual processor system with 1 GB RAM running FreeBSD 4.6 with 2 Syskonnect 9843 Gigabit Ethernet cards on a 64bit PCI bus. The flow data was captured using CoralReef's [16] `crl_flow` with all flows being expired in 30 second intervals. Each trace interval captured more than one million packets and one hundred thousand flows. Packet loss during the captures averaged under ten thousand packets per interval or less than one percent.

BGP data was collected by establishing a session with the local route server using the Zebra [2] BGP routing daemon. A full dump of the table was taken on 17 April 2003 and all BGP updates stored for later integration. This method should provide highly accurate BGP information for any arbitrary time during the capture session.

The user community consists of about 100,000 students and an additional 25,000 thousand faculty and staff members. While all university members do have access to the connectivity provided by PSC, some member universities have purchased independent connectivity for the residence halls.

4.3.3 Post-processing

Having relied on flow-level collection tools, it was necessary to route the flows using the BGP data so that the destination IP's were correctly attributed to the BGP best-matching prefix. To accomplish this, we read the traffic flow statistics and maintained the BGP routing information synchronized with the starting timestamps of the flows. The goal is to determine at each flow if the starting timestamp of the next flow is after the next BGP update. If the next BGP update happens before the commencement of that flow we apply this update and iterate until a BGP update is encountered that has occurred after the start of the next flow. Once the BGP information is updated, we find the best matching prefix in the BGP routing table and add this information to the flow data in addition to the corresponding AS path information.

Because the BGP routing information only deals with outbound traffic, we used the best-matching prefix and AS path information only for outbound traffic. Due to routing policy AS paths are often asymmetrical in that the path followed by incoming IP packets need not be the same as the AS path for this IP destination [178, 91].

4.4 Stability of the interdomain paths

Any study of the topological stability of Internet traffic depends on the assumption that the AS-level topology as seen from BGP routing information is also stable [157]. Therefore we first explore the stability of the AS paths of the BGP routes. To do this we rely exclusively on the traces and BGP data collected from UCL. The month long trace provides the necessary depth of information which the PSC trace of 24 hours simply cannot.

Figure 4.2 compares the number of days any distinct AS path was during the UCL trace for all AS paths and paths where at least one byte of data was seen over that month. During this period $103,853$ distinct AS paths were seen in the BGP advertisements but only $31,151$ carried any traffic at all. Due to BGP dynamics more than 50 percent of the distinct AS paths were present in the BGP routing table for less than 9 minutes. On the other hand, more than 42 percent of the AS paths over which traffic was sent were found in the BGP routing table for 99 percent of the duration of the month. These stable AS paths having traffic represented slightly more than 13 percent of all AS paths seen over the month. These results are similar to those from [157] which showed that most BGP update events occur for a small fraction of the prefixes. [169] studied the lifetime of triples <prefix,source AS, destination AS> also confirms our observations by demonstrating that the majority of these triples have a single AS path present in

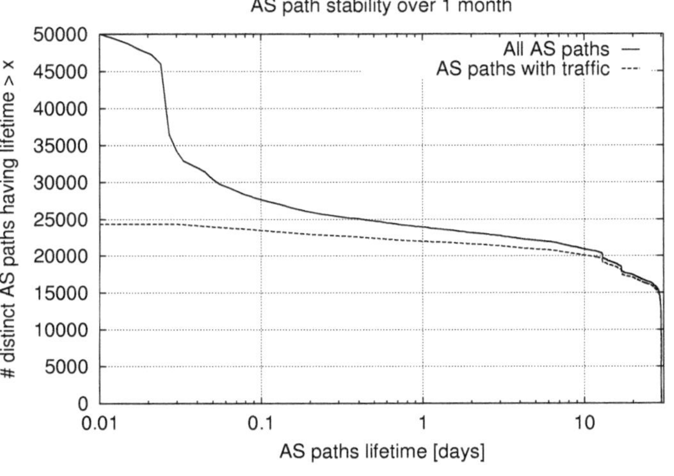

Figure 4.2: Stability of AS paths.

the routing tables over long periods. [205] also showed that the primary AS path for a DNS A root server lasted for long time periods.

Figure 4.2 provides an overview of the stability of the AS paths but without respect to the traffic they see. We are more interested in the stability of the AS paths that carry the majority of the traffic. Figure 4.3 compares the longevity in presence of the AS paths and their associated traffic over time. We define "presence" as the number of seconds during which an AS path was found among the BGP best routes over the one month of the trace. The curve labeled "AS paths" on Figure 4.3 gives the percentage of the total time AS paths were present in the BGP routing table. The curve labeled "traffic" shows the percentage of traffic for AS paths present for more than some percentage of the one month period. The latter curve shows that more than 95 percent of the total traffic was carried by AS paths that were present in the BGP routing table more than 99 percent of the time. In contrast, only 42 percent of the AS paths having traffic were present for more than 99 percent of the time in the BGP routing table. Our results are similar to those of [157] that showed that a few popular web sites for AT&T traffic had stable BGP routes. Our results however are more precise and detailed than [157] since we consider all prefixes in the traffic, not a subset of them. The prefixes for the largest fraction of the traffic of our stub AS have a single AS path that lasts for

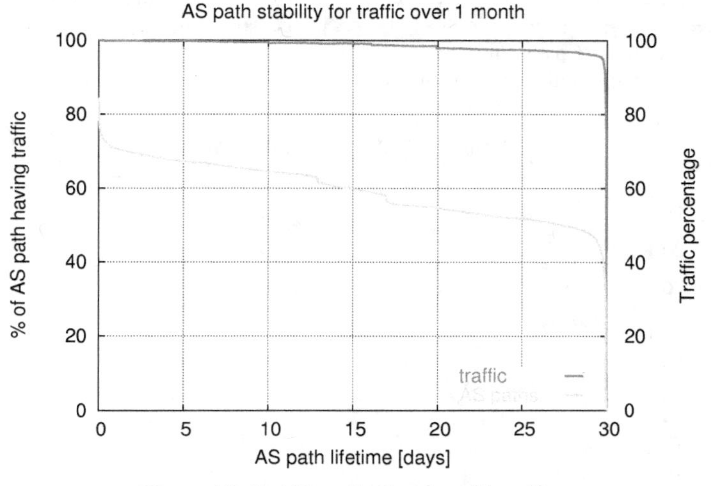

Figure 4.3: Stability of AS paths with traffic.

more than 99 percent of the one month period of our trace.

The view of the AS-level topology provided by BGP is thus relatively stable, even over time periods as long as one month. It is thus safe to rely on the AS paths to study the topological traffic distribution.

4.5 Topological distribution of interdomain traffic

Before looking at the dynamics of the traffic on the AS-level topology, we first want to build an understanding of how traffic gets aggregated on the AS-level topology. A primary concern in traffic engineering is how large is the interdomain topology that will be affected if a given fraction of the traffic is controlled [152]. The question we ask in this section is the following: how large is the smallest AS-level graph which represents an arbitrary percentage of the total traffic exchanged with remote ASes?

First, let us look at how much traffic traverses an AS path over the one month for UCL and the studied day for PSC. Figure 4.4 presents the cumulative traffic percentage carried by an increasing number of AS paths over the whole of the UCL and PSC traces. Figure 4.4 shows that the 20 largest AS paths carry slightly more than 50 percent of the total traffic for UCL and 40 percent for PSC. Addi-

tionally, we see that for the 90th percentile of the UCL traffic requires about 200 AS paths while PSC needs more than 400 AS paths. The cumulative distribution of the traffic for AS paths shown here is similar to the results of [195, 155] for interdomain traffic sources and destinations (prefixes and ASes).

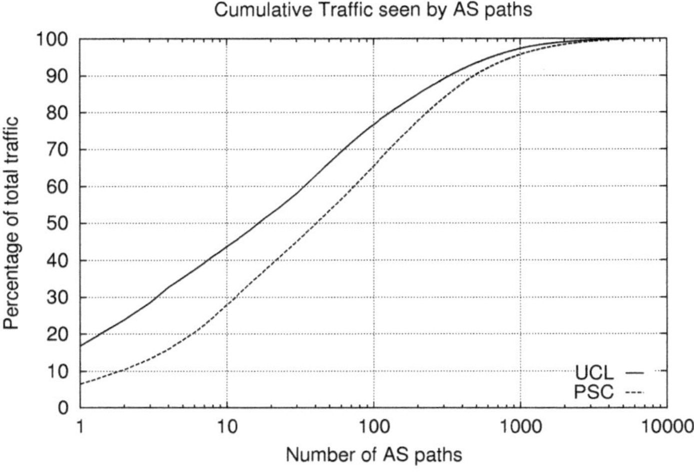

Figure 4.4: Cumulative traffic captured by AS paths.

Now let us determine how many interdomain "edges" carry the greatest portion of the observed traffic. For this, we start from the complete AS-level topology for all AS paths that carried traffic, and remove all edges that fall below as specified traffic percentage. Figure 4.5 provides the pseudo-code to compute the number of edges with more than x percent of the total traffic. In the pseudo-code of Figure 4.5, we first attribute the traffic to the appropriate edges for each AS path (lines 2 to 12) in the `Edge` array. We then sort the transit edges by byte count in ascending order (line 14). This result is stored in the array `Sorted_edges_traffic` and finally we exclude the edges having less traffic than the x percent threshold. The resulting value is the number of transit edges with more than x percent of the traffic and is assigned to `left_edges`.

Figure 4.6 gives the results of the above code for both the UCL and PSC datasets. The curves of Figure 4.6 show the number of edges and AS paths remaining in the AS-level topology, after imposing the constraint on the amount of traffic they carry. A total of $22,352$ edges and $31,151$ AS paths were seen for UCL

```
// Sum up traffic for transit edges
foreach AS_PATH k {
// Sum up traffic to know the total
Tot_traffic+=k.traffic
// Add the traffic for each edge of the AS path
i ← 1
while (i < (k.as_path_length-1)) {
   // Add AS path traffic for the current edge
   Edge[k.as[i]][k.as[i+1]]+= k.traffic
   i++
} // end while
} // end foreach
// Sort transit edges by increasing amount of traffic
Sorted_edges_traffic ← sort_increase(Edge)
// Determine how many edges are left for a trigger of x% of the traffic
// Initialize edge counter
left_edges ← size(Sorted_edges_traffic)
while ((Sorted_edges[left_edges]*100/Tot_traffic) < x) {
// This edge has less than x % of traffic, prune it
left_edges–
} // end while
// left_edges transit edges seeing at least x % of traffic are left
```

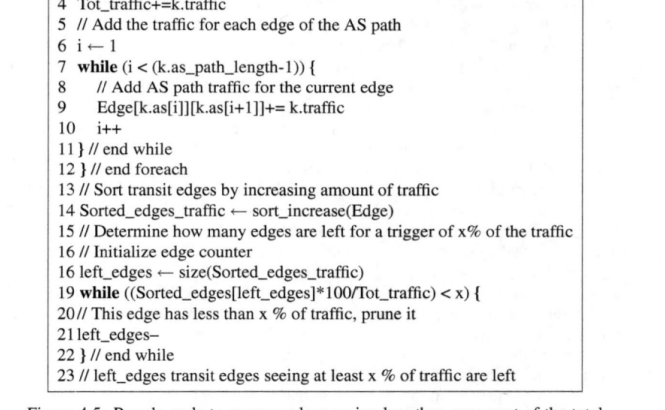

Figure 4.5: Pseudo-code to remove edges seeing less than x percent of the total traffic volume.

and $15,961$ edges and $17,263$ AS paths for PSC. The effect of increasing the traffic percentage threshold reduces the number of edges and AS paths in what seems to follow a long-tailed distribution, namely a distribution whose tail does not fall like an exponential for large values of the threshold. Only for the largest 50 edges and AS paths in terms of the carried traffic do we see a discontinuous behavior of the curves, for threshold values larger than about half a percent of the total traffic. For less than 50 transit edges and AS paths, the horizontal distance between the two curves increases. An increase in the horizontal distance between the curves of edges and AS paths means that for that particular number of remaining edges and AS paths, one of two traffic aggregates requires a larger threshold than the other. Because edges aggregate the traffic from potentially several AS paths, the "edges" curve is always on the right of the "AS paths" curve on Figure 4.6. We know from Figure 4.4 that these 50 largest AS paths are indeed carrying the majority of the traffic, 50 percent for PSC and 66 percent for UCL.

Looking closely at the tail of the curves of Figure 4.4 shows a marked difference between UCL and PSC in that the horizontal distance between the UCL's AS paths and edges curves does not grow as rapidly or as much as PSC's. The reason lies in the way traffic aggregation occurs for these largest edges for these two stubs. UCL has a significant percentage of its traffic with direct peers of its provider, while all the PSC traffic crosses tier-1 providers. A significant traffic aggregation occurs for the few largest edges of PSC. Figure 4.4 proves this point by providing the cumulative traffic percentage carried by edges at each AS hop distance. So we take all edges at some AS hop distance and count how many bytes have been carried by these edges. We then sort the edges at this AS hop distance by decreasing byte count and plot the cumulative percentage of traffic carried by the largest n edges at this AS hop distance from the local AS. The curves for an AS hop distance of 1 on Figure 4.7 illustrate the main difference between UCL and PSC. The largest direct peer of UCL's provider sees only a little more than 40 percent of the total traffic, while the largest peer of PSC sees a little less than 80 percent of the total traffic. UCL's provider has many direct peers with whom it exchanges traffic at local interconnection points. The PSC trace on the other hand shows that the two largest direct peers see all the traffic, simply because the PSC trace contains only commodity traffic that is sent through its commercial providers.

The curves of Figure 4.7 for an AS hop distance of 2 and 3 provide interesting information about traffic aggregation on the AS-level topology. Obviously, traffic aggregation occurs on the edges with the direct peers. However, traffic aggregation is desirable farther in the topology because traffic engineering techniques are due to perform finer operations than moving all the traffic from one peering to another [152, 92]. Traffic engineering will have to work on traffic aggregates to allow an AS to control the amount of traffic that crosses the links with its direct

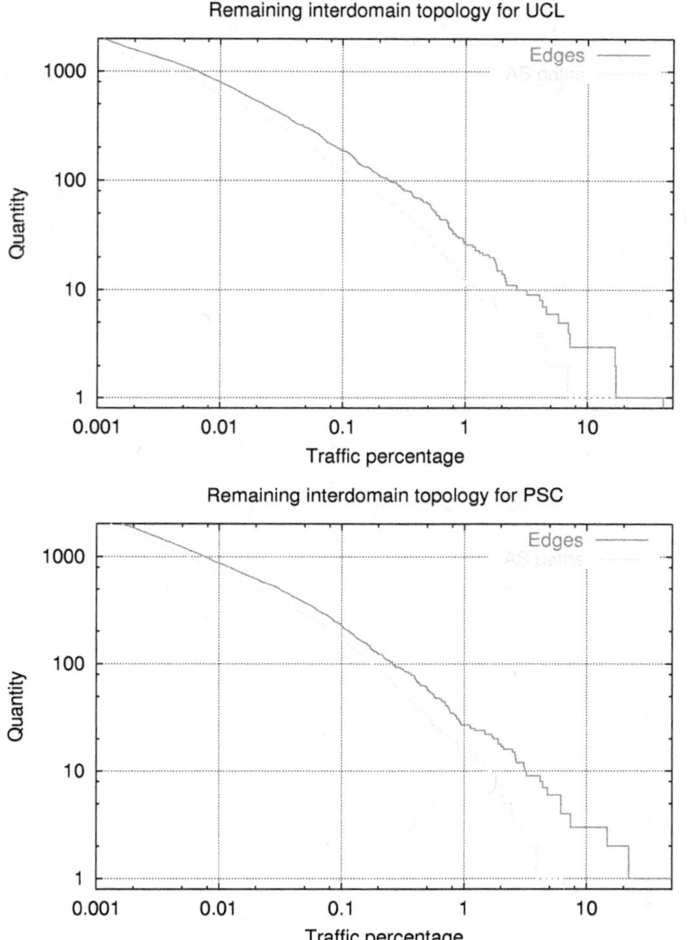

Figure 4.6: Effect of trigger for pruning the interdomain topology: UCL (top) and PSC (bottom).

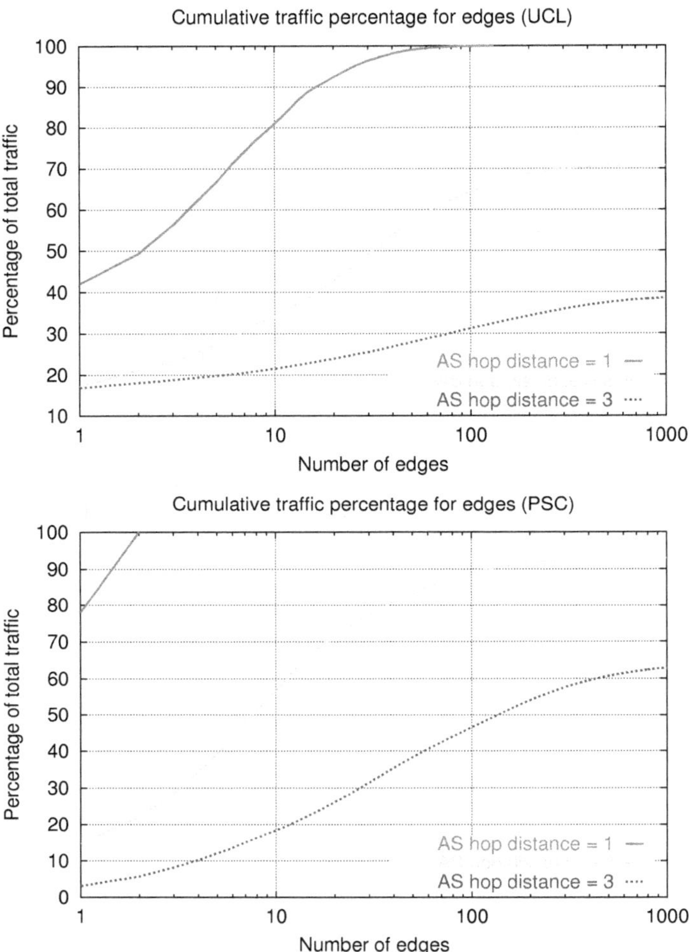

Figure 4.7: Cumulative traffic carried by edges.

peers. This is why it is important to know how traffic is split a few AS hops away in the AS-level topology. Figure 4.7 tells us that if we were to influence the largest 10 edges, then we would be able to control about 35 percent of the total traffic for UCL and a little less than 60 percent for PSC. This information is valuable for traffic engineering since we might be able to influence a large fraction of the total traffic by tweaking a few edges. If we need to control for instance 90 percent of the total traffic to perform fine-grained traffic engineering, then we need to rely on edges at 2 AS hops or more because working on direct peers will move too large amounts of traffic. This means that we would then need to tweak in the order of hundreds of edges due to the lack of traffic aggregation beyond direct peers. For PSC, controlling 90 percent of the traffic by tweaking edges at two AS hops requires to influence more than 100 edges. For UCL, there is a lot of traffic with direct peers so that edges two AS hops away see only about 70 percent of the total traffic. Had UCL not that much traffic with peers at public interconnection points, it would show a similar topological aggregation to the one of PSC. Traffic exchanged at public interconnection points is far cheaper than traffic exchanged with commercial peers, so that most stub ASes have strong incentives to peer at those public interconnection points.

4.6 Short-term stability of interdomain traffic

The last two sections examined the properties of the traffic on the AS-level topology over the entire duration of the measurements. Another important aspect is the stability of the traffic over smaller timescales. Understanding at which timescale the AS paths are stable is of interest for practical interdomain traffic engineering techniques [196, 152, 92]. Therefore in this section we outline the connection between the large timescale features of the traffic on the AS-level topology and smaller timescales. The analysis in this section is based solely on the UCL dataset. The PSC dataset is only used to confirm the results on the UCL data. Furthermore, we will not use edges in the rest of this chapter. The reason for using AS paths only is that there is no substantial difference between AS paths and edges as far as traffic aggregation and route control are concerned. We will therefore use the more familiar notion of AS path.

4.6.1 Short-term evolution of top AS paths

We start with an analysis of the dynamics of what we call the top AS paths. The "top AS paths" are nothing more than the AS paths sorted by decreasing amount of traffic over the considered time period (hour, day or month). For instance, the

top 50 percent AS paths for some hour is the smallest set of AS paths that carry 50 percent of the total traffic during that particular hour.

Figure 4.8 shows the hourly evolution of the number of AS paths present in the top x percent AS paths for four different values of x. To create this graph we computed the amount of traffic seen by each AS path for each hour, sorted them by decreasing amount of the total traffic, computed the cumulative traffic distribution and finally plotted the hourly evolution of the number of AS paths for various values of the traffic percentage every hour. Each point of the curves shows how many AS paths are necessary to carry a specified percentage of the total traffic during a given hour.

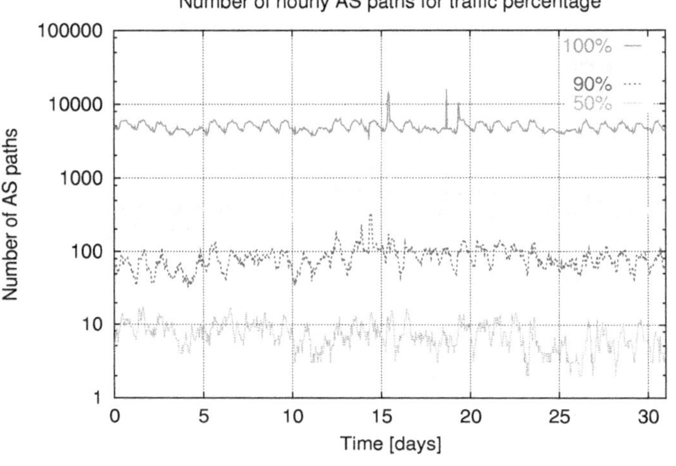

Figure 4.8: Hourly evolution of the number of top AS paths.

The use of a logarithmic scale for the y-axis reduces the visual importance of the dramatic increase in the number of AS paths needed as the percentage of traffic increases. Capturing 50 percent of the traffic every hour requires between 5 and 10 AS paths, while 90 percent requires 100 AS paths, and capturing all the traffic demands as many as 5000 AS paths. These numbers are smaller than those of Figure 4.4 for the cumulative percentage carried by the largest AS paths. On the timescale of hours, large AS paths thus see proportionately a larger fraction of the traffic than over one month, the latter needing 20 AS paths for 50 percent of the traffic and more than 200 AS paths for 90 percent of the traffic. AS paths

carrying much traffic over small timescales capture a larger fraction of the total traffic on small timescales than on large timescales.

The hourly evolution of the number of AS paths for the traffic percentage seems to follow a daily periodicity similar to the one of the total traffic evolution. The larger number of AS paths seen during the busiest hours of the day is likely to be due to a larger number of active IP addresses. Note however that for fifty percent of the hourly traffic, this daily periodicity is less evident because the periodicity is due principally to the total number of active IP addresses [194], not the largest AS paths in traffic.

4.6.2 Intersection of hourly and monthly top AS paths

The above results do not say whether the AS paths seen each hour are the same ones that are seen over the whole month. Without this information, we cannot determine the actual stability of the traffic on the AS-level topology. If traffic engineering can rely on a subset of the paths that capture a significant fraction of the traffic, such paths must also have relatively low churn for traffic engineering not to have to change too often the set of paths it uses to reach its solution. We must therefore compare the individual AS paths over both the monthly and hourly timescales to understand their churn on small timescales.

The overlap between the hourly and monthly AS path statistics can be seen in Figure 4.9. Figure 4.9 shows three curves. The first one, labeled "# top 90 % AS paths (over 1 month)", provides the number of top AS paths that carry 90 percent of the traffic over the whole month. 330 AS paths are required to capture 90 percent of the traffic over the whole month. The second curve of Figure 4.9, labeled "# top 90 % AS paths (hourly)", provides the hourly evolution of the number of AS paths representing ninety percent of the traffic every hour. This curve is the same as the curve labeled "90 %" on Figure 4.8. The third curve, labeled "intersection", shows the number of AS paths among the top 90 percent hourly AS paths that are also in the top 90 percent monthly AS paths. The absence of significant difference between the second and the third curves indicates that, at least in terms of the number of AS paths, there is a significant overlap between AS paths that carry much traffic in the short and the long term.

An significant overlap in number between the top AS paths over the month and the top AS paths over hours does not mean that the overlap in amount of traffic carried is significant. The top curve of Figure 4.10, labeled "hourly capture of monthly top 90 % AS paths", shows the hourly evolution of the largest AS paths that carry 90 percent of the total monthly traffic. This curve shows that if we could guess what are the 330 largest AS paths in traffic over the coming month, then we could control about 90 percent of the hourly traffic. The middle curve of Figure 4.10, labeled "hourly capture of 90 % intersection", shows the hourly

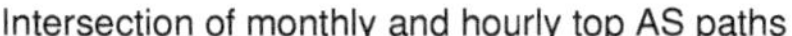

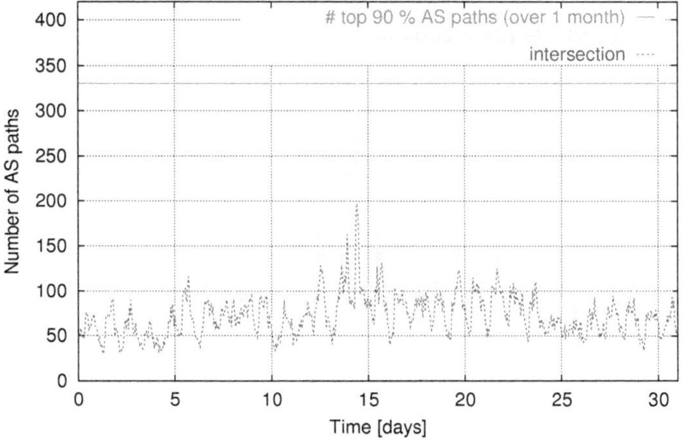

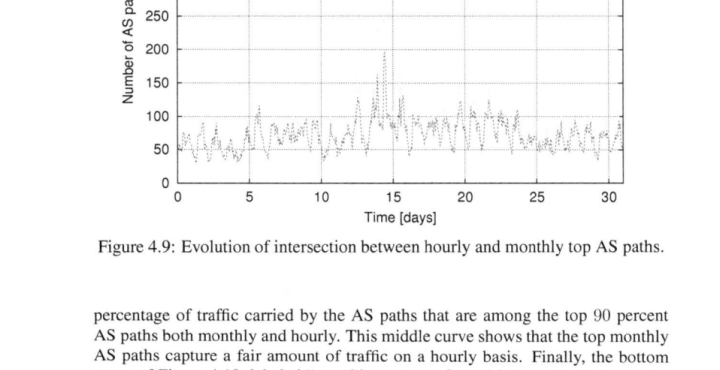

Figure 4.9: Evolution of intersection between hourly and monthly top AS paths.

percentage of traffic carried by the AS paths that are among the top 90 percent AS paths both monthly and hourly. This middle curve shows that the top monthly AS paths capture a fair amount of traffic on a hourly basis. Finally, the bottom curve of Figure 4.10, labeled "monthly capture of 90 % intersection", shows the percentage of the monthly traffic carried by the hourly top 90 % AS paths. This bottom curve shows that the AS paths that are among the hourly top 90 % AS paths capture only between fifty and seventy percent of the monthly traffic. Hence the AS paths that capture most of the traffic on a hourly basis are not the same as the AS paths that capture most of the traffic on a monthly basis.

Figure 4.10 tells us that assuming that we could know which AS paths are important over long time intervals in advance, we could include them and have a good certainty that this set of AS paths will capture a fair amount of the traffic in the future. However, we cannot make this assumption. Even though 330 AS paths capture 90 percent of the traffic over a one month period, less than 100 of them are active on a hourly basis. Therefore, to capture 90 percent of traffic with a large probability of success, we would have to take into account more than three times as many AS paths as required.

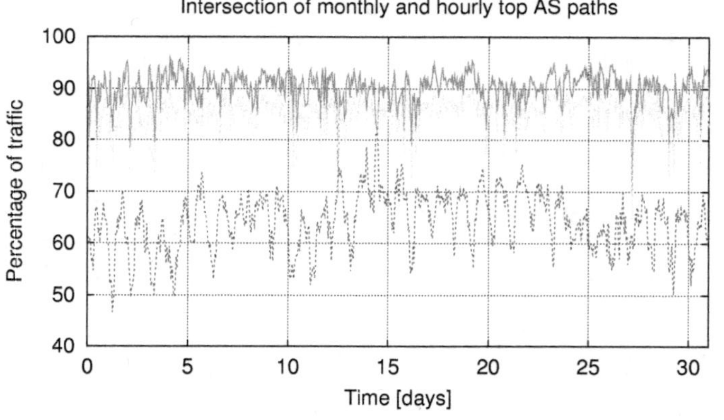

Figure 4.10: Evolution of traffic captured by intersection of top AS paths.

4.6.3 Presence of short-term top AS paths

The previous section showed that hourly AS paths that see much traffic have much in common with the monthly top AS paths. In practice we do not have such knowledge of which AS paths will carry large amounts of traffic in the future, we only know those from the past. This section studies for how much time most AS paths are among the hourly top AS paths. For this, we record for each hour the top AS paths that capture 90 percent of the traffic, and count for each of those paths in how many hourly bins this path appears in the top AS paths that capture 90 percent of the traffic during those hourly bin.

Figure 4.11 counts the number of hours over the month that each AS path appears in the hourly top 90 percent AS paths. We also compute for each of these AS paths the amount of traffic it captures during the hours it is among the top 90 percent AS paths. The curve labeled "AS paths percentage" (lower curve) gives the cumulative percentage of the AS paths which have ever appeared among the top 90 percent in terms of the number of hours it was present, hence the name "presence". The other curve labeled "traffic percentage" (higher curve), gives the corresponding cumulative percentage of the total traffic carried by the previous AS paths during a given number of hours over the month. A total of 2, 139 AS paths were seen among all the hourly top 90 percent AS paths. This is more than

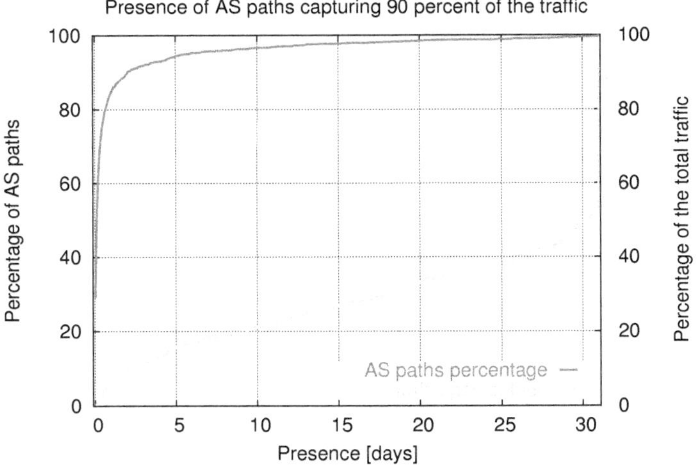

Figure 4.11: Presence of hourly top ninety percent AS paths.

6 times the minimal number of paths required to capture 90 percent of the traffic over the whole month. 626 of these paths were seen only during a single one hour bin while only 6 paths were seen during all hourly bins. The curve labeled "AS paths percentage" shows that most AS paths were seen for a small fraction of the time. More than 96 percent of the 2, 139 AS paths were seen for less than one third of the time and represent about 20 percent of the total traffic of the whole month. Additionally, 50 percent of the total traffic appears for AS paths seen among the hourly top 90 percent for about 80 percent of the time. The 6 AS paths that are present during all time periods carry about 36 percent of the total traffic. More than 20 percent of the traffic appears for AS paths that are seen in less than one third of the hourly bins during the month. This leaves about 25 percent of the traffic being carried by AS paths that are moderately but not entirely stable.

The relatively limited presence of the top 90 percent AS paths might be due to the fact that most of them carry a too limited amount of traffic. The presence of AS paths that carry more traffic might be better than what Figure 4.11 showed. Figure 4.12 shows the same information as Figure 4.11 but for the top 50 percent top AS paths. 360 AS paths were seen among the hourly top 50 percent AS paths. This number is pretty close to the 330 AS paths that capture 90 percent of the traffic over the whole month. One third of these AS paths were present during

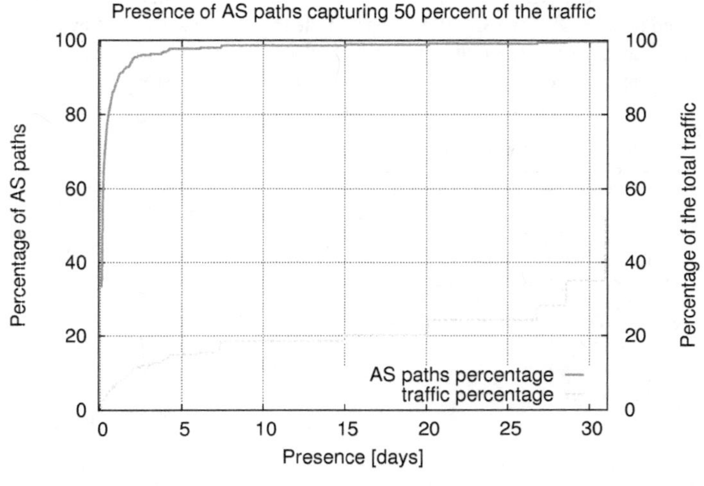

Figure 4.12: Presence of hourly top fifty percent AS paths.

only two hourly bins among the hourly top 50 percent top AS paths. Slightly less than 90 percent of these 360 AS paths were present during 24 hourly bins of the month. Among these 360 AS paths, those that are present for less than one third of the hourly bins carry about 20 percent of the total traffic. AS paths that are present more than half of the hourly intervals carry 30 percent of the total traffic. This shows that even the largest AS paths in terms of the hourly traffic have a high variability in presence. This variability in presence of the AS paths is not an artifact of selecting paths that do not carry enough traffic, but concerns even the most important paths in traffic.

This section showed that while a few AS paths can be considered as stable over the whole month, there is a significant churn in the AS paths that carry the largest amount of traffic over hourly timescales.

4.6.4 Traffic Capture for top AS paths

To this point we have assumed foreknowledge of which AS paths carry the largest portion of the traffic. Under this assumption, it is easy to show that a significant percentage of the hourly traffic travels AS paths that are dominant in the monthly traffic statistics. However, we have shown in Figure 4.11 that while some of the

hourly top 90 percent AS paths are stable, the majority are moderately to highly unstable and do account for a significant part of the traffic. This means that knowledge of AS paths with a large amount of traffic over short times scales will not necessarily make traffic engineering over longer timescales possible.

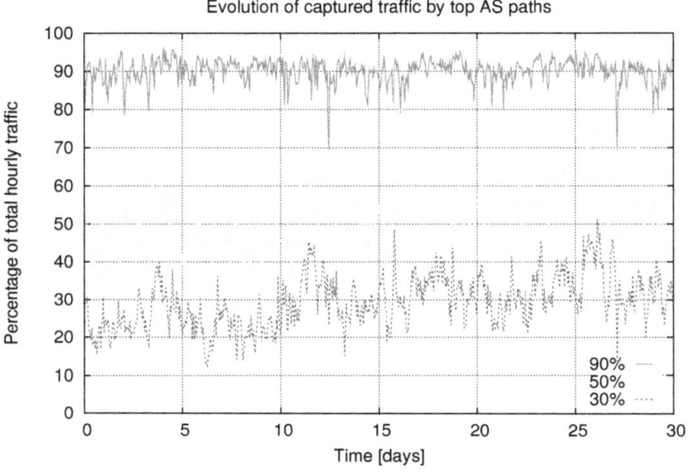

Figure 4.13: Evolution of percentage of traffic captured by top AS paths.

Figure 4.13 shows the hourly percentage of traffic captured by the top 30, 50, and 90 percent AS paths over the length of the one month of the UCL trace. We first determined the AS paths that carried the most traffic over the month and then plotted the time evolution of the hourly traffic that these top AS paths represent. If these paths were stable we would expect to see a relatively flat line with little variation. However, we see that the AS paths that carry 30 percent of the monthly traffic rarely carry 30 percent of the hourly traffic. Insteadwe observe large variations in the hourly traffic, generally between 20 and 40 percent. We see similar variations as we include more AS paths to increase the percentage of hourly traffic captured, although the absolute level of variation decreases as we include more paths. Figure 4.13 showed that even though the largest AS paths in traffic over the one month capture a large fraction of the traffic on a hourly basis, the traffic percentage these AS paths capture every hour varies a lot, particularly for the largest AS paths in traffic.

Using a time window to predict the largest AS paths in traffic might give some insight into the actual stability of AS paths over time. Therefore, we show in

Figure 4.14 how much traffic over the next time interval is captured on average when the AS paths that capture most of the traffic over the last time intervals are used as a predictor. For each hour we first determined the AS paths that carry 50 percent and 90 percent of the traffic during the previous time windows, and computed how much traffic these ASes paths carry over the next hour. Finally, the average over the length of the trace of the percentage of traffic seen for each time window length is computed and plotted on Figure 4.14. The same process was repeated for time windows of spanning days. The results are shown in the top graph of Figure 4.14 along with the standard deviation (bottom of Figure 4.14).

The top of Figure 4.14 shows that increasing the length of the the time window increases the average percentage of traffic captured. However, the gains for larger time windows are limited. This is especially true for increasing the time window to periods of multiple days. However, if we compare the standard deviations of the hours window and the days window we discover a notable difference. The standard deviation for all time windows and the AS paths capturing fifty percent over the time window is larger than for the ninety percent AS paths. The smaller number of AS paths for the fifty percent capture are responsible for this since the smaller set of AS paths accounts for a comparatively larger portion of the total traffic. The smaller sets of AS paths are thus more vulnerable to variations in the traffic they carry. However, the very simple prediction technique used gives good results, indicating that the popularity of the top AS paths does not change so much over hours or days. This is an encouraging result for interdomain traffic engineering that needs stability in the amount of traffic that can be controlled.

4.7 Implications on interdomain traffic engineering

Most studies on interdomain traffic insist that a few sources and destinations generate most of the interdomain traffic. This optimistic view is misleading because it hides an equally important facet of the network topology: influencing most of the interdomain traffic requires that a significant number of ASes be considered –, far more than previously believed. For instance, in section 4.6.1 it was shown that to control 50 percent of its traffic a small stub like UCL would have to tweak about 10 AS paths every hour, or about 100 AS paths every hour to control 90 percent of its traffic. While these numbers might not seem particularly large, making the necessary adjustments to BGP on a nearly continuous basis is not a trivial task, see Part III.

This is not to say it is impossible though. For outbound traffic, tweaking BGP is much easier and does not impact BGP routing outside the local domain of the stub AS. For inbound traffic things are a little more intricate. First, the AS path information contained in the BGP routing table of the local stub does not provide the

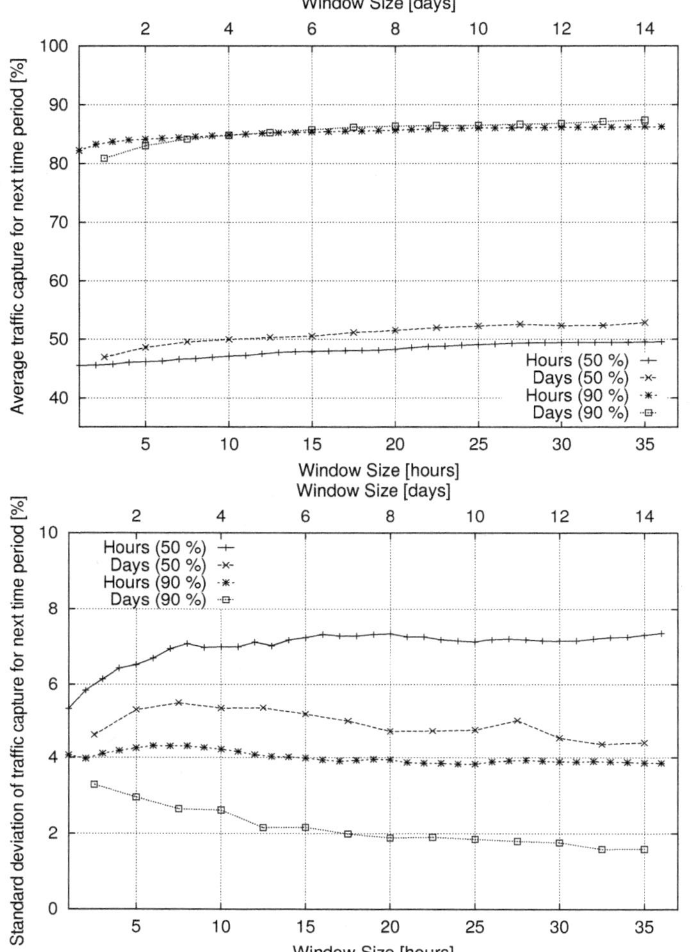

Figure 4.14: Traffic capture for windowed 50 % and 90 % top AS paths: average (top) and standard deviation (bottom).

right information concerning the inbound path followed by IP packets [178, 91]. Second, even if the local stub had the whole map of the AS-level Internet, which is a very strong assumption [126, 136], there is no guarantee that influencing remote ASes located a few AS hops away would be possible given that policy routing is shaped independently and differently by each AS [137]. Third, even if there was some manner in which this could be done, it would mean that a stub would need to frequently change its BGP announcements for potentially hundreds of ASes. If a significant percentage of the stub ASes begin to perform that kind of BGP tweaking, it is difficult to predict the global effect on BGP routing, which has already been demonstrated to suffer from convergence problems [116, 99].

Additionally, we showed in section 4.6 that the churn in the AS paths that carry the largest amounts of traffic is quite important. The uncertainty due to the traffic dynamics implies that controlling the interdomain traffic requires to influence more AS paths than what is believed based on an a priori knowledge of the important AS paths in the traffic. The literature typically argues that since popular destinations are pretty stable, traffic engineering should not be that difficult. As will be shown in Part III, the situation is more complex than that. Our study of the topological dynamics of the interdomain traffic indicates that to improve the certainty of influencing a large fraction of the traffic, more AS paths have to be used. To do this, one needs to influence more BGP routes, which in turn will potentially affect traffic that spans a larger fraction of the topology of the Internet. The global effect of BGP routing is a serious issue, especially with more and more multi-homed ASes.

Finally, tweaking BGP will impact on the intradomain traffic distribution of large ISPs [8]. Solutions that rely on intradomain routing to optimize the intradomain traffic of large ISPs [87] can be affected by the tweaking of BGP routing. The interactions between intradomain and interdomain traffic engineering [9, 46] must be studied before deploying interdomain traffic engineering tools.

4.8 Conclusion

In this chapter, we have studied the dynamics of the interdomain traffic on the AS-level topology, as seen by stub ASes. This was reached through the analysis of two interdomain traffic traces from two stub ASes so as to study the dynamics of the interdomain traffic on the AS-level topology.

This analysis revealed several properties of the interdomain traffic. First we showed that interdomain paths are stable from a routing viewpoint. Second, we showed that while a few AS paths were responsible for a large fraction of the interdomain traffic, capturing a large fraction of the total traffic requires several hundred AS paths. We then identified a transit core that aggregated large amounts

of traffic on the interdomain topology and that a small number of transit edges carry a large fraction of the interdomain traffic. Next, a study of the dynamics of AS paths needed to carry a given percentage of traffic demonstrated that the behavior of high traffic top AS paths is significant. However, it also showed that the stability of top AS paths over long time scales did not correlate to stability of these same paths on shorter time scales. In fact, we showed that the commonly expected stability of the AS paths did not actually closely coincide with observation. Lastly, the problem of capturing a given amount of traffic based on the previous behavior of specific AS paths was shown to require observational timescales on the order of days in order to limit short term variability. The only solution which limits this short term variability is to consider larger percentages of the total traffic on shorter timescales and a larger portion of the interdomain topology.

Further work should investigate similar aspects on transit ASes. We expect that due to more aggregation, transit ASes exhibit a behavior that is more favorable for interdomain traffic engineering. On the other hand, transit ASes might not have as strong incentives to engineer their interdomain traffic compared to stub ASes, as the cost of the interdomain traffic might not be such a big issue for transit ASes. Whether interdomain traffic engineering is possible for large transit ASes is an open issue [200].

Part III

Interdomain traffic optimization with BGP

In this third part, we wish to evaluate the feasibility of interdomain traffic engineering based on the tweaking of the BGP protocol. Chapter 5 is mainly introductory, in that it provides a first step into the evaluation of the problem of a multi-homed stub AS wishing to minimize the billing cost of its outgoing traffic. Chapter 5 also introduces the evolutionary meta-heuristics. Then, Chapter 6 takes the problem introduced in Chapter 5 and uses as objective the minimization of the number of BGP filters to be configured to attain an improvement in a single traffic objective. Chapter 7 studies the on-line optimization of a traffic balancing objective under constraint of minimizing the burden on BGP. Chapter 8 finally considers two traffic objectives defined on different timescales, also under the contraint of minimizing the number of changes in BGP.

Note that a part of Chapter 6 has been published in [197], Chapter 8 has been published as [192] and Chapter 7 has appeared as [196]. Material appearing in Chapter 8 is reprinted from [192] with kind permission from Imperial College Press.

Chapter 5

Interdomain traffic engineering with BGP: an Evolutionary Perspective

In this chapter, we propose the use of evolutionary optimization to perform traffic engineering and evaluate interdomain traffic engineering with BGP. We look at the the problem of optimizing a daily cost function defined on the interdomain traffic to understand the complexity of tweaking BGP to control the flow of the interdomain traffic.

This chapter is structured as follows. Section 5.1 introduces the problem of interdomain traffic engineering for stubs. Section 5.2 then details the nature of the single objective function to be optimized. Section 5.3 presents the different combinatorial views of the interdomain traffic engineering problem. Section 5.4 then introduces the multi-objective interdomain traffic engineering problem. Section 5.5 presents evolutionary algorithms. Section 5.6 describes our evolutionary algorithm for the single-objective problem. Section 5.8 presents the datasets we use to evaluate our algorithms. Section 5.7 shortly presents a few numerical results in the single-objective case. Section 5.9 explains how to extend the evolutionary algorithm used for the single-objective problem to the multi-objective case. Section 5.10 goes on to evaluate the evolutionary algorithm for the multi-objective problem. Finally, section 5.11 evaluates the simulation results.

5.1 The "simple" combinatorial optimization problem

To introduce the problem, let us start with an example illustrating the nature of the interdomain traffic engineering with BGP. Assume a stub AS having one link per provider. Our stub AS receives a full BGP routing table from each of its providers. To control the flow of its interdomain traffic, our AS can only rely on

the information found in these BGP routing tables.

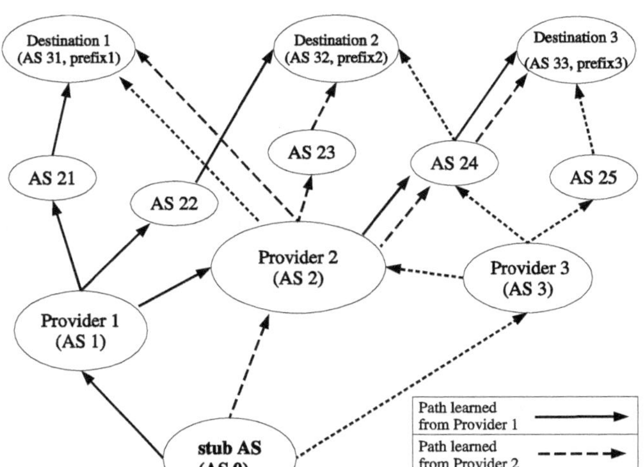

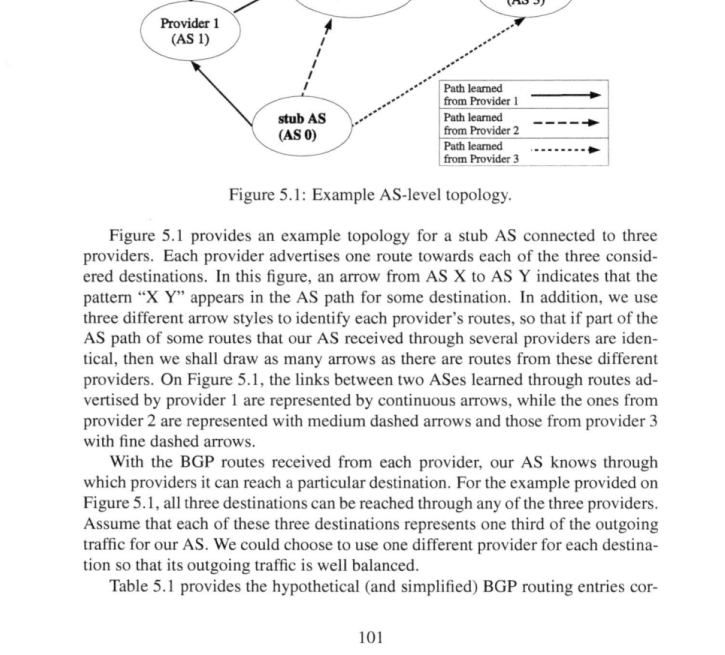

Figure 5.1: Example AS-level topology.

Figure 5.1 provides an example topology for a stub AS connected to three providers. Each provider advertises one route towards each of the three considered destinations. In this figure, an arrow from AS X to AS Y indicates that the pattern "X Y" appears in the AS path for some destination. In addition, we use three different arrow styles to identify each provider's routes, so that if part of the AS path of some routes that our AS received through several providers are identical, then we shall draw as many arrows as there are routes from these different providers. On Figure 5.1, the links between two ASes learned through routes advertised by provider 1 are represented by continuous arrows, while the ones from provider 2 are represented with medium dashed arrows and those from provider 3 with fine dashed arrows.

With the BGP routes received from each provider, our AS knows through which providers it can reach a particular destination. For the example provided on Figure 5.1, all three destinations can be reached through any of the three providers. Assume that each of these three destinations represents one third of the outgoing traffic for our AS. We could choose to use one different provider for each destination so that its outgoing traffic is well balanced.

Table 5.1 provides the hypothetical (and simplified) BGP routing entries cor-

responding to the routes for the topology given in Figure 5.1. Each destination AS

Table 5.1: Example BGP routing table.

Prefix	AS path	BGP next-hop	local-pref
prefix1	1-21-31	provider 1	100
prefix1	2-31	provider 2	100
prefix1	3-2-31	provider 3	100
prefix2	1-22-32	provider 1	100
prefix2	2-23-32	provider 2	100
prefix2	3-24-33	provider 3	100
prefix3	1-2-24-33	provider 1	100
prefix3	2-24-33	provider 2	100
prefix3	3-25-33	provider 3	100

shown on the topology of Figure 5.1 has a network prefix corresponding to the IP addresses that can be reached in this network (first column of Table 5.1). The second column of Table 5.1 provides the AS path that the IP packets sent towards these prefixes will follow. Next is the next BGP hop used to attain the destination. The next hop is the IP address of the BGP router of the peer AS with whom the BGP session is established, not a logical name like the one we used in Table 5.1 for the sake of clarity. Finally, the fourth column of Table 5.1 provides the value of the `local-pref` attribute of the routes.

Now we get back to the problem of our AS wishing to distribute evenly its outgoing traffic among the three providers. Upon receiving a BGP route, an AS may modify the value of some of the BGP attributes of this route through BGP filters. It is possible to apply a BGP filter that changes the value of the `local-pref` attribute of some route so that one route for a given destination received through one particular BGP peer be preferred over other routes for the same destination received through other BGP peers. The reason for choosing to rely on the `local-pref` attribute is that this attribute is local to the AS. It is not advertised outside the domain. Relying on other BGP attributes to achieve the same results is possible but then one must be careful that the attributes manipulation does not interact with intradomain routing. As can be seen on Table 5.1, the local AS sets to 100 the value of the `local-pref` attribute of all routes through input filters. In that case of all routes having the same (and lowest possible) preference, the first rule of the BGP decision process does not play any role in the choice of the best route.

If the stub AS shown on Figure 5.1 wants to achieve a good balance of its out-

going traffic, it can rely on the `local-pref` attribute to prefer some routes over others. Recall that we assume that each destination prefix receives one third of the traffic. Suppose that the restricted BGP decision process with only `local-pref` and the AS path length is considered. If the routes towards the three destinations have a default `local-pref` value of 100, then given the fact that the shortest AS path route would be chosen by BGP to reach a destination, the traffic towards destination 1 will be forwarded through provider 2 (AS path length of 2), the traffic towards destination 2 will be forwarded through providers 1, 2 or 3 (AS path length of 3), while the traffic towards destination 3 through providers 2 or 3 (AS path length of 3). This would mean that the routes for destinations 2 and 3 are non-deterministic (for this restricted BGP decision process) since there are several "best routes" among which BGP can choose. In practice, the BGP decision process ensures that only one route is used but this would not automatically lead to a good balance of the traffic. So to get back to our problem a solution is to put a higher value of the `local-pref` attribute in one of the routes learned from destinations 2 and 3 to ensure that only one best route remains and that the traffic balance be good among the three providers. For that purpose, our stub can attach a `local-pref` value of 110 to the route towards destination 2 learned from provider 1 and also attach a `local-pref` value of 110 to the route towards destination 3 learned from provider 3. This configuration would ensure that each provider gets one third of the outgoing traffic if the traffic is evenly distributed over the three prefixes.

5.2 Single-objective optimization

Let $x_i(t), i = 1,..,n$ be the amount of traffic to be sent towards ASi at time t, and let $C_i(y), i = 1,..,p$ be the cost of sending y bytes of traffic over link i. The objective is to find $min \sum_1^p C_i(l_i)$ under the constraints that $y_i \leq Cap_i$ and $\sum_1^p y_i = \sum_1^n x_j$ where Cap_i denotes the maximum amount of bytes that can be sent over link i. This maximum can be the limit of the physical link capacity or the limit imposed by policing. In other words, the objective is to distribute the traffic to be sent towards all ASes over the available links so that the total cost of sending it is minimum. Note that we do not pose the problem as a combinatorial optimization problem here because we are not aimed at finding the optimal solution to the problem, but to understand the problem of tweaking BGP to optimize a cost function.

The problem of choosing what fraction of the traffic should be sent over which link under minimal global cost is not difficult to solve, whenever the cost functions are sufficiently "well-behaved". For example, if the cost to send traffic towards a destination only depends on the provider and the amount sent to that provider,

not on the total amount sent for all traffic on a particular provider, then the cost function C_i can be expressed as weights c_{ij} so that the cost function is linear. In such a case, the problem is an instance of the assignment problem which can be solved through the Hungarian method in time complexity $O(n^3)$ where n denotes the number of destinations, see chapter 11 of [141] for a detailed presentation of weighted matching problems.

In order to have an idea of the pricing schemes currently used for billing providers, we asked the question of how providers are billed for interdomain capacity on the NANOG[1] mailing list (`nanog@merit.edu`). Here we provide a summary of the answers we received. First, most pricing schemes rely on the following procedure:

- collect samples of the traffic volume every t minutes (5 and 15 minutes are common);
- combine these t minutes samples into one combined sample;
- at the end of a billing cycle, compute the 95th[2] percentile of the combined samples, this number corresponds to the bandwidth L which will be used for the price.

The computation of the 95th percentile can be based on:

- 95^{th}(incoming)
- 95^{th}(incoming + outgoing)
- 95^{th}(max(incoming,outgoing))
- max(95^{th}(incoming),95^{th}(outgoing))

where `incoming` means the traffic entering the provider and `outgoing` the traffic leaving the provider. Now that we have introduced the percentile-based computation, let us describe the way billing is actually done:

1. percentile-based : x USD per y Mbps (n^{th} percentile) with a commitment of `c` Mbps. The price per Mbps can be different for the commitment and for the traffic above the commitment (also called "burstable").
2. average-based : same as point 1 but using an average (not 50th percentile) instead of a percentile.

[1]North American Network Operators' Group.

[2]The 95th percentile is the most commonly used in practice but any other percentile could be used as well.

3. volume-based : x USD per y bytes.

4. destination-based : x USD per Mbps for "local" traffic (national for instance) and y USD per Mbps for "non-local" traffic (international for instance).

5. max-based : flat rate based on the maximum available bandwidth, independent of how many bits are used.

It must be noted that the dynamics of the traffic can influence the value computed because for a given amount of bytes exchanged over some link, different values for the nth percentile can be found. We do not take into account such aspects in this chapter. Chapter 8 shall deal with more realistic traffic objectives. However, it must be noted that except for the volume-based billing scheme without any commitment and rate limit, all other common billing schemes are not linear. Solving this problem as an instance of a classical combinatorial problem is not forcibly the best way due to the intricate cost functions.

5.3 To group or to assign

The problem of interdomain traffic engineering as presented in section 5.1 can be seen as an instance of different combinatorial problems.

One way to view the previous problem is as an assignment problem. This problem consists in assigning the traffic t_i towards each destination i through one and only one provider j in order to minimize the total cost of the assignment $\sum_{j=1}^{P} C_j(\sum_{i=1}^{N} x_{ji}t_i)$, where C_j represents the total cost of sending traffic through provider j and x_{ji} represents through which provider j the traffic towards destination i is sent. Without constraints on how many changes to be made in the assignment and a linear cost function, this problem can be efficiently solved [141]. However, if one is interested in solving the assignment problem with constraints on the number of changes with respect to the default assignment found by BGP, then one deals with the generalized assignment problem which is known to be NP-hard [162].

The problem of interdomain traffic engineering can also be seen as an instance of a more general class of problems called "grouping" problems, also called partitioning or bin packing problems in the literature [74, 75]. These problems consist in grouping the items (the x_i of our problem) of a set U into mutually disjoint subsets U_i in such a manner that $\cup U_i = U$ and $U_i \cap U_j = \emptyset, i \neq j$. In addition, not all groupings are allowed, but hard constraints must also be met, like the sizes of the links $U_i \leq C_i$. A hard constraint is one that cannot be violated unless the solution

is unfeasible, in opposition to a soft constraint that is not mandatory. A cost function defined over the groupings must also be optimized. In our case, we want to minimize the total cost of the grouping which corresponds to the traffic sent over all links. These grouping problems are known to be NP-hard in the strong sense [93]. Henceforth, there is little hope to find an exact algorithm that will find an answer in polynomial time. This is why we rely in this chapter on an evolutionary algorithm [104, 96, 133] to find an approximation to the optimal solution.

In this chapter, we shall often refer to the problem of interdomain traffic engineering as a grouping problem. Viewing it as an assignment problem is as appropriate. It is only the way we developed our algorithm that explains that we see the problem in that way.

5.4 Multi-objective optimization

The assignment's view of interdomain traffic engineering does not consider the number of destinations that require tweaking their BGP routes. In practice, it might not be desirable to minimize the cost of the traffic if it requires changing the best route found by BGP for almost all destinations. Perhaps it is better to get the best improvement in the cost function for a given number of changes to be made to BGP routes. In addition, traffic demands change in time so that the optimal assignment also change to accomodate the new traffic pattern. If the objective of the algorithm is to find the set of BGP changes that provide the best improvement within a given number of BGP changes, then the strategy of re-optimizing each day might not be the optimal one. If one wishes to reduce as much as possible to number of BGP changes on the long run, limiting the daily improvement in the cost function each day might allow to spare many changes to be performed in BGP for consecutive days on the long-run.

So our new problem problem is now a multi-objective one, in that we try to find the minimum of the cost as well as the minimum of the BGP configurations necessary for this cost. These two objectives are conflicting since improving the cost function can only be done by applying changes to the BGP routes. In this chapter, we use "BGP configuration change" to mean that a given interdomain destination (be it a prefix or a destination AS) has to use another link than the one it had to use before the change.

There is no trivial way known in the literature to solve the conflicting objectives of multi-objective optimization problems [86]. There are however two main approaches. The first is to explicitly combine the two objectives into a single one, by a linear combination of the two cost functions. This solution consists in forming a new objective function F that is a weighted sum of the individual objectives

functions $f_i, i = 1, ..., N$

$$F = \sum_{i=1}^{N} \alpha_i f_i, \alpha_i \in \mathbb{R}^+. \tag{5.1}$$

This approach unfortunately forces one to set the relative importance of each criterion. If there exists a solution that succeeds to optimize each f_i independently this approach can be satisfactory. In general however, some of the f_i conflict with one another, requiring a degradation in terms of one objective in order to improve another. In some cases, trade-offs between the different objectives are meaningless. In such cases the f_i cannot made a scalar through weighting and the objectives are said to be non-commensurate.

The other approach relies on the concept of Pareto-optimality [142]: one now searches for solutions with respect to one objective, under the constraint that the solution cannot be worse in terms of the other objectives. This approach allows to find the region in the multi-objective plane that corresponds to values of the other objective that are less or equal (in the case of minimization) than a given value.

Pareto-optimality Pareto-optimality is named after the Italian sociologist and economist Vilfredo Pareto (1848-1923). Pareto optimality is a situation which exists when economic resources and output have been allocated in such a way that no-one can be made better off without sacrificing the well-being of at least one person. Pareto-optimality is thus related to the set of solutions whose components cannot be improved in terms of one objective without getting worse in at least one of the other components. More formally, a multi-objective search space is partially ordered in the sense that two solutions are related to each other in two possible ways : either one dominates or neither dominates. Consider the following multi-objective minimization problem:

$$\begin{array}{ll} Minimize & y = f(x) = (f_1(x), ..., f_n(x)) \\ where & x = (x_1, ..., x_m) \in X \\ & y = (y_1, ..., y_n) \in Y \end{array}$$

and where x is called the decision vector, y the objective vector, X the parameter space and Y the objective space.

In the context of this chapter, the decision vector $x = (x_1, ..., x_m)$ represents the allocation of each destination to a particular link where m is the number of destinations and each x_i is an integer representing the link over which the traffic of destination i is forwarded. The objective vector $y = (y_1, y_2)$ represents the pair of objectives to be minimized: the cost of the traffic and the number of "BGP configuration changes".

A decision vector $x_1 \in X$ is said to dominate another decision vector $x_2 \in X$ ($x_1 \prec x_2$), iff

$\forall i \in 1,..,n : f_i(x_1) \leq f_i(x_2) \quad \wedge$ $\exists j \in 1,..,n : f_j(x_1) < f_j(x_2)$. Domination is an important notion because it determines the result of the comparison of two decision vectors. A decision vector x is said Pareto-optimal *iff* x is non-dominated regarding X, i.e. $\nexists x' \in X : x' \prec x$

A Pareto-optimal decision vector cannot be improved in any objective without degrading at least one of the other objectives. These are global optimal points. In our context however, we are not interested in global optima but optimal points in some neighborhood of some of the objectives. More precisely, we aim at finding the Pareto-optimal points with respect to the cost of sending traffic over the available links in a neighborhood of the default allocation of the destination among the links found by the BGP routing protocol and having a distance of at most ϵ BGP configuration changes compared to this default BGP routing solution. Hence we do not search for globally Pareto-optimal decision vectors but locally Pareto-optimal decision vectors [63]:

Consider a set of decision vectors $X' \subseteq X$.

1. The set X' is denoted as a local Pareto-optimal set *iff*
$\forall x' \in X' : \nexists x \in X : x \prec x' \wedge ||x - x'|| < \epsilon \wedge ||f(x) - f(x')|| < \delta$
where $||.||$ denotes a distance metric, $\epsilon > 0$ and $\delta > 0$.
2. The set X' is called a global Pareto-optimal set *iff*
$\forall x' \in X' : \nexists x \in X : x \prec x'$.

Note that a global Pareto-optimal set does not necessarily contain all Pareto-optimal decision vectors. In the remainder of this book, we sometimes call non-dominated points Pareto-optimal while we have no explicit proof of their optimality due to large search spaces. To be rigorous, such points should be said non-dominated.

5.5 Evolutionary algorithms

In this section, we describe the evolutionary algorithm that searches for the Pareto-optimal solutions of the multi-objective minimization problem presented in section 5.4. One of the main reasons for relying on evolutionary algorithms when dealing with multi-objective optimization problems lies in that the search for solutions in terms of conflicting objectives is difficult for non-population based methods that search one solution after another in order to find an optimum [57]. In order to improve one objective, it is often required to degrade the performance in terms of another objective. It is therefore necessary to explore worse regions of the search space in terms of one or several objectives to find a subset of the Pareto-optimal front. This is why population-based methods have an advantage as they are able to better sample the Pareto-optimal front by exploring several re-

gions of the search space simultaneously. In addition, evolutionary algorithms are due to be less sensitive to the shape of the Pareto-optimal front than mathematical programming techniques. Furthermore, population-based search techniques can search the Pareto-optimal front in a single run.

The software used in this chapter is based on the GENESIS (version 5.0) system developed by J. Grefenstette [97] available at `http://www.aic.nrl.navy.mil/galist/src/#C`, which we modified for our particular needs.

Genetic algorithms [104, 96, 133] are stochastic search algorithms based on the mechanism of natural selection. They rely on a population of individuals who evolve with the successive iterations of the algorithm, and stronger individuals of the population are allowed to survive through selection mechanisms in a competing environment. The basic features of a typical genetic algorithm is as follows:

- population: a population of individuals stochastically searches the search space.
- generations: from one generation to the next, only the fittest individuals survive to ensure improvement with respect to the objective function.
- fitness: improvement in the search is measured by the "fitness" of any individual.
- selection: individuals of the current population are selected to survive to the next generation based on their fitness, inducing competition among individuals.
- crossover: pairs of individuals generate offspring that inherit from the "genetic" material of their parents.
- mutation: random changes in the individuals are performed to ensure a stochastic search.

The previous elements constitute the "core" of genetic algorithms. GAs are weak search methods, in the sense that they do not make assumptions about the problem at hand. They are however very powerful search techniques. Genetic algorithms are a paradigm, not instances of solutions. The issue with certain optimization problems is that they involve very large search spaces, that classical optimization techniques cannot handle efficiently. To properly leverage genetic algorithms, they must be transformed into an evolutionary algorithm [133, 177] which uses problem specific information allowing the algorithm to perform better than random search. Specific representation for the individuals together with genetic operators leverage the problem-specific knowledge in order to improve the search.

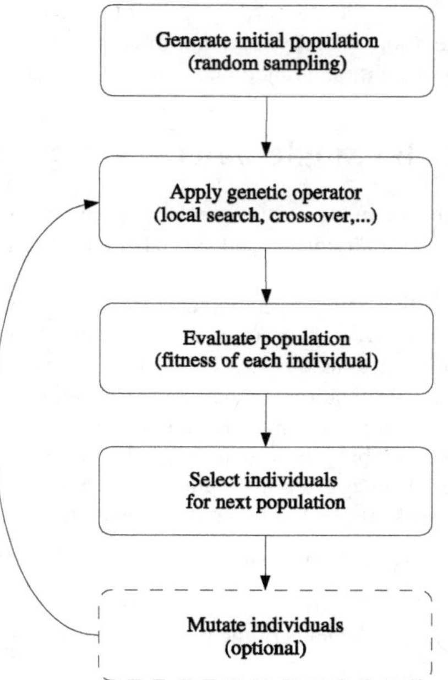

Figure 5.2: General working of an evolutionary algorithm.

Figure 5.2 illustrates the basic working of a typical evolutionary algorithm. First, an initial population is generated. This step is often carried through random sampling. Depending on the purpose of the algorithm as well as the feasibility of the sampling, random sampling is carried either in the objective space or the decision space. All individuals of the initial population might also be the same if the algorithm has to start with some default solution. Second, the genetic operators are applied on the individuals in search for improvement in the objective space. Third, all individuals of the population are evaluated and their fitness is computed. Fourth, the individuals are selected according to their fitness to constitute the population for the next iteration. Finally, optional mutations are performed on the individuals to provide an additional randomness into the search.

5.6 Algorithm for single-objective problem

Evolutionary algorithms are made of two important parts: encoding of the problem, i.e. data structures for individuals, and specialized operators, e.g. crossover, mutation, local search,...

We encode individuals as integer vectors: each value v_i of an element i of the vector v represents the link over which traffic for this destination is sent, $1 \leq v_i \leq l$ where l is the number of available links. This allows an explicit representation of the assignment of each destination to a particular upstream link.

The operators we use are particular versions of the classical crossover and mutation operators. We combine these two operators into a single one that tries to provide the intuitive advantage of both operators in the context of our problem. Each crossover phase works in the following way, illustrated in Figure 5.3:

1. select two parents (called mom and dad) to be used to generate two new offsprings;
2. copy the values of the elements of the parents into the offspring (offspring1 $\leftarrow$ mom, offspring2 $\leftarrow$ dad);
3. for each item i and both offsprings:
    a) compare each offspring with itself where the value of item i has been replaced by the value of item i of the other offspring, and take the individual among the original offspring and the modified one that has the smallest cost;
    b) if the cost of the minimum cost individual found in a makes the total of some link larger than its capacity, then generate a random integer value between 1 and l for the value of item i, i.e. `offspring[i] = randint(1,l);`

c) else put the value that minimizes the cost of the offspring found in a) in the element offspring[i].

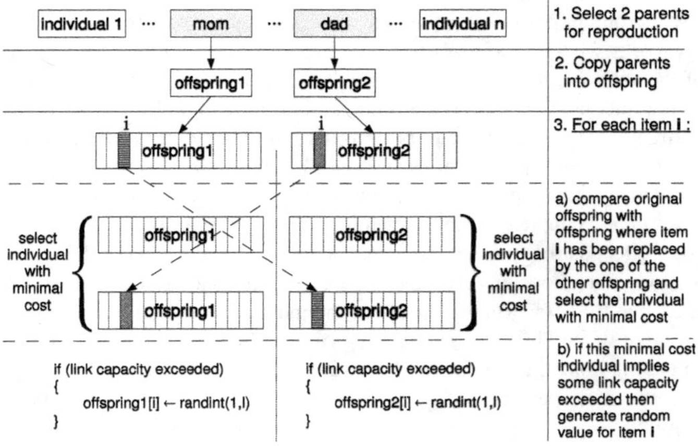

Figure 5.3: Description of single objective EA.

This procedure ensures that individuals improve by benefitting from the good groupings of the parents, and whenever the parents contain bad groupings (the capacity of some link is exceeded) a new random choice is made for the grouping of this item. The functions of both classical crossover (inheritance from the well performing groupings of the parents) and mutation (providing of new genetic material from random changes) are fulfilled. Whenever the parents allow to improve the offspring, then these changes are selected. When these changes make the solution unfeasible, then the item is mutated and its value is made random. The key of the performance of the algorithm lies in the search for improvement when evaluating each single element of the offspring.

Classical genetic crossover works on bitstrings, by injecting contiguous subparts of the parents in the offspring, on the grounds of the "building block hypothesis" [104, 96]. This hypothesis assumes that genetic algorithms perform adaptation by implicitly and efficiently implementing the recombination of building blocks, i.e. low-order schemata or substrings of the parents with above average fitness. This approach works well when optimizing numerical functions for instance, but not in the case of grouping or assignment problems [75]. The reason is that the semantics of grouping problems is to assign destinations into groups, hence the

object of interest is the group, not contiguous subsets of destinations. When performing classical crossover over our individual's representation, we would inject into the offspring a group of items that belong to different groups. The reason why an individual is better than another lies in the way items are attributed to groups, not because a group of items is attributed to some set of groups. The building bock hypothesis works when certain substrings of the data structure encode particular aspects of the solution, in the same way as DNA contains coding regions that are of vital importance and should be kept intact when transmitted from the parents to the children [14]. In our case, improving an individual can be performed by choosing a different group (link or provider) for some item. If one changes the grouping for many items that belong to many groups, the result on the fitness of the new individual is unclear, because changing many groupings at the same time can have as well a positive as a negative impact on the global fitness of the individual. This is why we chose our crossover to work on individual items, so that we can be sure what the effect on the fitness of the individual of changing the group of this particular item. This ensures that offspring are better than parents on every run, when paying no respect to the hard constraints.

The crossover we perform is "greedy", in that we do it on each item of the vector. Changing potentially each item means that we perform a local search but whose depth is relatively important. This strategy increases the probability that the offspring have a better fitness than the parents by trying to improve each offspring as much as possible. Our procedure does not preclude that the algorithm be stuck in a local optimum, but it tries to find the best grouping given the initial population. The larger the population, the more chances we have to find a better minimum (or the global optimum) when applying our crossover on the individuals. When the population is large compared to the number of groupings, it is likely that our crossover operator allows all groupings to be tested for each item. Note that we talked about "groupings" mostly because we allow several changes in the assignment of the destinations to some upstream provider. If one feels more comfortable with thinking about assignments, then this view is equivalent.

5.7 Example

Having described the working of the evolutionary algorithm for a single traffic demand, let us provide some simulations illustrating how well the algorithm performs. Assume a network has four upstream providers that bill the traffic differently:

- provider 1 uses flat rate billing: sending any quantity of traffic on that link costs 50 monetary units (irrespective of the quantity of traffic) and the maximal link capacity is 50 traffic units,

- provider 2 uses volume-based billing: sending traffic on that link costs 1.4 monetary units per traffic unit and the maximal link capacity is of 20 traffic units,
- provider 3 uses volume-based billing: sending traffic on this link costs 2 monetary units per traffic unit and the maximal link capacity is of 20 traffic units,
- provider 4 uses "burstable" billing: sending any quantity of traffic on that link below the commitment of 20 traffic units costs 20 monetary units, and any additional traffic exceeding the commitment of 20 traffic units costs 1.5 monetary units per traffic unit.

For a budget of 100 units, the optimal global cost should be around 113 with the following traffic allocation over the links: 50 traffic units for link 1, 20 traffic units for link 2, 10 traffic units for link 3 and 20 traffic units for link 4. The optimal solution that was the largest among all runs of the algorithm (30 different traffic demands were evaluated) found after 20 evaluations (19 generations) was 113.21. Note that we used a population of 50 individuals.

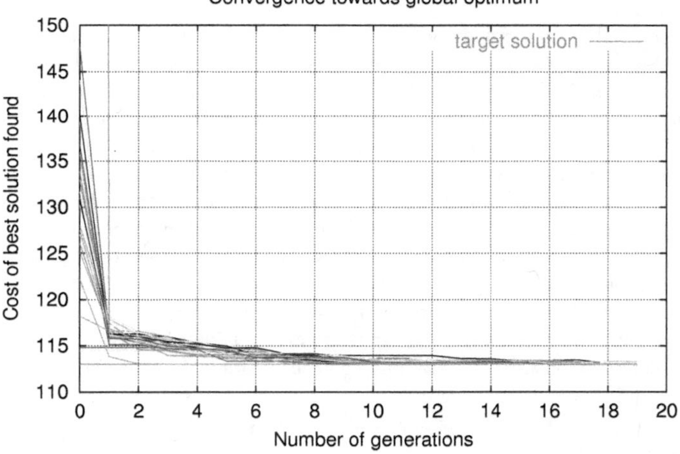

Figure 5.4: Convergence of the single-objective EA for 30 days of traffic demands.

Figure 5.4 shows the best result obtained with the evolutionary algorithm at each iteration (generation), i.e. the cost of the individual at that iteration having

the best solution. To simulate a realistic distribution of the traffic towards Internet destinations, we relied on the outgoing traffic from the Abilene network for each day of the month of September 2002 [1]. The Abilene network connects U.S. research and educational organizations. In this simulation run, we simulated a provider that had the Abilene traffic demands (amount of traffic towards each destination) each day and wanted to minimize its total cost of sending traffic each day. Except for the first day where the choice of the upstream link for each destination was random (uniform among the four links), the global optimum of each day was used to start the algorithm for the next day. The cost of the solution after the first evaluation can be above 1000 because we used a penalty function to prevent solutions to exceed the maximal capacity on any link. If any solution required to exceed the maximal capacity of any link, the evaluation function returned a cost of 1000 for this link. Note that the part of the "crossover" that attributes a random value for the value of an item has been seen to occur only during the first generation of the algorithm when having to adapt from one traffic demand to another. Once the largest destinations have been re-attributed onto least-loaded links, the algorithm spends time exclusively on minimizing the cost, not on trying to find feasible solutions. This shows that importance of the natural variability in the traffic demands, that can make the optimal solution of the previous day a bad solution for the next day.

Each curve of Figure 5.4 shows the results of the algorithm on a single day of the Abilene traffic demand. The results of Figure 5.4 for generation zero show that yesterday's best solution can be far from optimal on the next day if no re-optimization is performed. Not all days display this behavior, sometimes the optimal solution of the previous day is very close to the new optimum. As illustrated on Figure 5.4, the convergence towards the optimal solution is fast at the beginning. Further approaching the global optimum on the other hand appears more difficult, and more and more comparisons between different groupings are required to get closer to the optimal value. This convergence illustrates the nature of this combinatorial optimization problem, where it is not very difficult to get close from the optimum with our algorithm, but harder and harder to get closer to it. The running time of each day's optimization (20 evaluations) requires in the order of one minute on a PIII 1 GHz running Linux. Although there is no guarantee that the optimum will be found by the algorithm, evolutionary algorithms have the advantage of being typically able to scale to large problem sizes while finding acceptable solutions.

5.8 Abilene Netflow statistics

Having introduced the basics about the problem, we turn to more extensive evaluations. Again, we use in this chapter the data available from Abilene [1]. Abilene collects Netflow [53] statistics every day from its POPs and allow these statistics to be queried from the web at `http://www.itec.oar.net/abilene-netflow/`. We use four month of data, starting from June 1st 2002 to September 30th 2002. We rely on the "destination AS" reports (percentage without names) from September 2002. For the algorithm with time-varying demands, we use the whole "source-destination AS" traffic matrix because it provides total byte counts rather than percentages for each AS. Relying on byte counts instead of percentages is necessary since each day's traffic level matters for the computation of the expected long-term gain of the optimization.

To give an idea of the variability of the traffic demands from day to day, Figure 5.5 presents the amount of traffic in bytes for the largest destination ASes of Abilene. On Figure 5.5, we have ordered the destination ASes seen during the

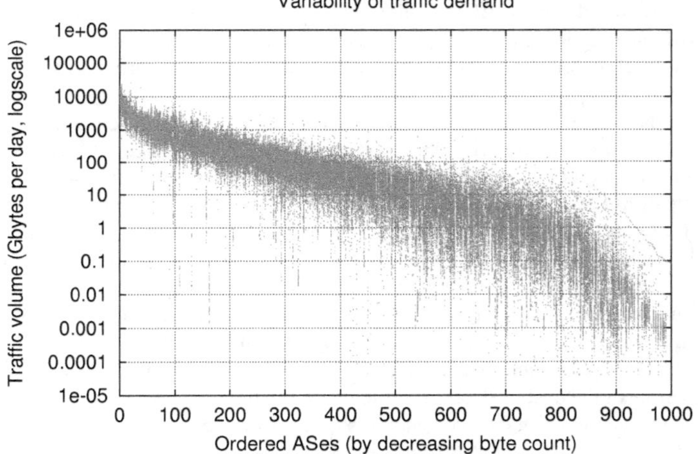

Figure 5.5: Traffic demands of Abilene destination ASes.

four months of the trace (x-axis) and we plot the amount of bytes (in logarithmic scale) sent to the one thousand ASes with the largest amount of traffic over the whole trace. We observe that the daily amount of traffic of the top destination ASes varies by a factor of 10 over the 4 months period. Less popular destination

ASes see even more variability in the traffic they receive from Abilene. This variability is consistent with our results of Chapter 4 where we showed that popular destinations of stub ASes also exhibit a significant variability in the traffic they exchange.

5.9 Minimizing BGP configuration changes

So far, we have only provided a sketch of our solution, we need to explain how to modify the evolutionary algorithm to adapt to changing traffic demands and minimizing the number of BGP configuration changes when re-optimizing the attribution of the destinations from one period to another. Because we have no way to explicitly provide a trade-off between the cost of sending traffic over links and the non-monetary cost of changing the BGP configuration in order to force the traffic of one interdomain destination to be forwarded over some upstream provider, we rely on Pareto-optimality introduced in section 5.4. Henceforth, we modify the evolutionary algorithm described in section 5.6 by adding the constraint of trying to improve the new objective while minimizing BGP configuration changes. Recall that we use the term "BGP configuration change" to mean changing the allocation of one interdomain destination to an access link other than the default one chosen by BGP, although we know that in practice the two things are not forcibly equivalent.

Due to the large size of the search space, we cannot realistically search for the Pareto-optimal front blindly. The knowledge of the number of BGP changes required for the single-objective problem should guide us towards a strategy to find a good trade-off between minimizing both the BGP changes and the cost of sending traffic. Our experience with the single-objective optimization showed us that many BGP changes are required for getting close to the global optimum (a few hundreds), while the pay-off for a large fraction of these BGP changes is small. This means that we cannot rely on the cost of sending traffic to find the Pareto-optimal front since it would over-sample the region of the front that requires many BGP changes. Nevertheless, we need to guide the evolutionary algorithm to improve in terms of the cost of sending traffic while minimizing the number of BGP changes. Our approach hence consists in starting from zero BGP changes (the solution found by BGP) and searching for the best solutions in terms of the cost of sending traffic. We allow a limited number of BGP configuration changes to be performed at each generation, so as to bias the search towards solutions that improve as much as possible the cost of traffic. The algorithm is very similar to single-objective algorithms except that at each generation, we only allow a given number of randomly chosen items to be changed.

The pseudo-code of the new evolutionary algorithm for a particular traffic de-

mand is the following:

1. Begin with the solution given by last iteration (previous BGP configuration).
2. For each generation and each individual of the population: select randomly n destinations among the destinations active during the forthcoming period and for each destination find the assignment that minimizes the cost of this solution.
3. If the number of generations is larger than the number of changes allowed by the provider: stop.

Point (2) requires some further description. We cannot for some given value of n compare all feasible solutions implying at most n BGP configuration changes. The reason is once more the size of the search space, that would represent about $Popsize \times l^n$ points at each generation. Given that we use $l = 4$, $Popsize = 100$ and $n = 10$, this would require $100 \times 4^{10} \sim 10^8$ evaluations per generation. Each evaluation requires as many additions as there are items in the vector (about 800) so even if the computer executes one operation per 10^{-9} second, one generation would require eighty seconds in theory. In practice, the time needed to process the population during one generation is more than that because we need to access to the individuals in memory, write to memory the best element of each generation to record the individuals of the Pareto-optimal front, without mentioning other overheads of the evolutionary algorithm.

Figure 5.6 presents the pseudo-code corresponding to point (2) above. For each individual and for each generation, we generate n times one item among the destinations that have traffic for the considered period. For this item, we determine the link that gives the smallest cost for this individual. This means that we perform n evaluations for each individual and each generation. The total number of evaluations for each generation is thus $Popsize \times l \times n$ instead of $Popsize \times l^n$, reducing much the number of evaluations. One may want to evaluate all the l^n solutions, in order to find the best solution requiring at most n BGP configuration changes. Assume that we have the best solution for the first n changes with respect to the default BGP solution. If n is not too large, then we have seen in the previous paragraph that it is possible to compute all the possible solutions. This would provide all points on the Pareto-optimal front for the first n values of the BGP configuration changes. However, this is impractical for large values of n: even for a rather small value of $n = 20$, about $4^{20} \sim 10^{12}$ evaluations are required. This would take more than 9 days under the optimistic assumption that each evaluation can be done in 10^{-7} second to find the first 20 points of the Pareto-optimal front. So assume we have the Pareto-optimal front up to 20 BGP configuration changes, the next generation has to start with a population made of the best individuals known up to that point. Given that we have to rely on a population of

```
foreach individual k {
   foreach item = 1 to n {
      // selecting one item randomly among the active destinations
      selected_item = active[randint(1,active_destinations)]
      // finding the link giving minimal cost for individual k
      best_link = min_cost(k,selected_item)
      // changing the link for item iff cost has decreased
      if (cost(k,selected_item,best_link) < current_cost[k]){
         // updating the value of this item
         k[selected_item] = best_link
         // updating the cost of the individual
         current_cost[k] = cost(k,selected_item,best_link))
      }
   }
   // computing the number of BGP configuration changes for this solution
   // compared to default BGP solution
   BGP_config[k] = required_BGP_config(k,selected_item,best_link)
}
```

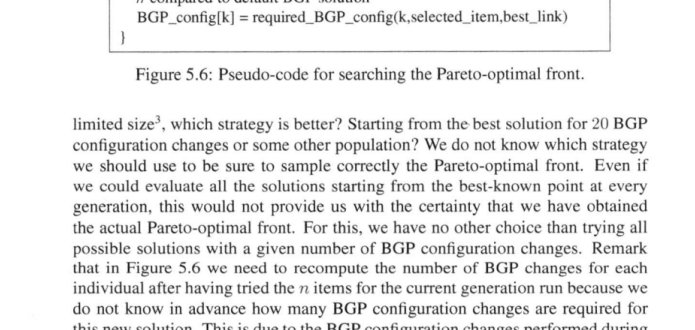

Figure 5.6: Pseudo-code for searching the Pareto-optimal front.

limited size[3], which strategy is better? Starting from the best solution for 20 BGP configuration changes or some other population? We do not know which strategy we should use to be sure to sample correctly the Pareto-optimal front. Even if we could evaluate all the solutions starting from the best-known point at every generation, this would not provide us with the certainty that we have obtained the actual Pareto-optimal front. For this, we have no other choice than trying all possible solutions with a given number of BGP configuration changes. Remark that in Figure 5.6 we need to recompute the number of BGP changes for each individual after having tried the n items for the current generation run because we do not know in advance how many BGP configuration changes are required for this new solution. This is due to the BGP configuration changes performed during the current generation can undo changes done in a previous generation, so both the number of BGP configuration changes and the cost of a solution can decrease from one generation to the next.

While this method does not provide the whole Pareto-optimal front, it has several nice properties. Firstly, we have a certainty about the maximum number of BGP changes the solutions found require. Each generation implies at most n BGP configuration changes so we can stop the algorithm when all individuals

[3]Increasing the population allows a better search at the cost of more memory consumption and larger overheads for the evolutionary algorithm.

of the current population require more than the higher bound specified in terms of BGP changes. Second, because we cannot search for all changes of upstream provider for all destinations, we allow the algorithm to select the fittest solutions at each generation. This allows the search to focus on the BGP changes that decrease the most the cost of sending traffic. Hence we try at each generation to decrease by the largest amount possible the cost of the solution, under the constraint that the number of additional BGP changes is smaller than n. This heuristic does not give a proof that we shall find the Pareto-optimal front but at least we try to get the best short-term improvement for a few number of BGP configuration changes.

Our approach is not the best one for searching the Pareto-optimal front in terms of both objectives. However, using a method that does not rely on the number of BGP changes to search for the multi-objective solutions has a high risk of not being able to find good trade-offs between the two conflicting objectives. We actually oversample the regions with the least number of BGP configuration changes because we know that we are not able to search the whole search space. The advantage of our solution is that it has a built-in upper bound on the number of BGP changes that some solution will require. In addition, we do not have to find a strategy for searching the "cost-BGP changes" plane while spending most of our time trying to decrease the cost while minimizing the number of BGP changes. To make a link between this section and section 5.4, X representing the set of feasible solutions x, what we are after here is the global Pareto-optimal set X' of non-dominated points x' that have the smallest cost of their groupings within a distance of ϵ ($\in \mathbb{N}$) BGP configuration changes with respect to the solution x^* found during the last period. If the function $cost(y)$ represents the cost of grouping y and $bgp(y)$ represents the smallest number of BGP configuration changes required to get from grouping x^* to grouping y, then
$\forall x' \in X' : \nexists x \in X : cost(x) < cost(x') \wedge bgp(x) < bgp(x')$.

The reader could have noticed that our evolutionary algorithm does not contain any crossover. We do not exchange information between parents and offspring, but each individual evolves (or dies) depending on its relative cost. There is a simple reason for that. The single-objective algorithm has shown that the effective part of crossover was the possibility of trying to improve each solution by changing as many items (destinations) as possible. The effective part of the algorithm is not the crossover. The choice among the two different groupings of the parents is not critical in the search. The "building block hypothesis" [104, 96] does not seem to hold for our problem. Allowing many items to be changed at each generation on the other hand ensures the largest improvements in the search. To search the Pareto-optimal front of a grouping problem, we have found that the most effective strategy is the one that allows as many changes of grouping for each individual. Restricting the number of BGP changes automatically provides a trade-off between the effectiveness of the search and the constraint on the number

of BGP configuration changes.

5.10 Predicting Future Demands

Up to now, we have made the assumption that the traffic demands were known beforehand. This is not true in practice. The scheme we have presented in section 5.6 works on the current traffic demand, assuming we already know it. However, how should we to predict the forthcoming traffic demand in practice?

Two things are unknown to the algorithm: the future traffic demand and the BGP routes. We do not know for which destinations we shall see traffic. Chapter 4 has shown that a large fraction of the AS paths were stable but a significant fraction was relatively short-lived. Chapter 4 has also confirmed the findings of [157] concerning the stability of the BGP routes for the largest fraction of the traffic. This stability unfortunately does not help us to predict over which upstream link we should send the traffic for a particular destination. It only tells us that relying on enough measurements should provide good results for the future, not how much data is required for best results. Furthermore, we believe that it is more desirable to know the total number of BGP configuration changes compared to the default BGP routing than to rely on the previous day's solution. Relying on the previous day's solution to build the new one might lead to the addition of too many configuration changes over time, so starting from the default BGP solution will be easier to manage. Another technical problem of relying on the last day's solution to find the current Pareto-optimal front is that we still do not know which solution of the front the provider wants to choose: how many BGP changes are allowed? So we should have to start from yesterday's Pareto-optimal front to find today's. This is not possible because today's front could be more costly, hence requiring the search to progress in many directions towards the new front. The last issue of this approach would be that we start from a potentially unfeasible region, we would then spend our time trying to find feasible solutions, instead of exploring the Pareto-optimal front.

So the main question we try to answer in this section is how long in the past should we go in order to be able to find the best prediction in terms of the upstream link we should use to send traffic towards a destination? To answer this question, we do not try to build models for the various unknowns of our problem. The reason is that various providers can have different traffic types according to their users' profile as well as different characteristics of their BGP peerings. So results obtained in one particular context will differ in another. Not knowing what are the important properties of the traffic for our particular problem, modeling does not seem reasonable. Modeling requires assumptions about the features of the modeled system, which we do not have here.

Left without a model, we rely on three months of Abilene data. For each day that has to be traffic engineered, we study the Pareto-optimal front found by our algorithm given that we rely on the Pareto-optimal front when the traffic of the last t days is used to predict the future traffic demand. What we do for each day is the following:

1. based on the BGP routing tables of the provider, determine through which upstream each destination's traffic should be sent if BGP was used to forward the traffic;
2. for each value of the time window t, compute the Pareto-optimal front for the traffic demand of the last t days;
3. compute the cost of the Pareto-optimal front found in the previous step for the forthcoming traffic demand.

We compute the Pareto-optimal fronts for the three months of the Abilene traffic data, one for each value of the time window. In practice, we do not have the opportunity to choose the right time window each day so that the "best" Pareto-optimal front will be found for the next day. Instead, a provider will have to choose a value of the time window and a maximum number of BGP changes to be made every day to get one point from the Pareto-optimal front, which will be used as the solution for traffic engineering. We expect that it is more desirable to have an algorithm whose parameters do not have to be tuned every day and be provided with an uncertain long-term planned gain, rather than having to choose a point on the Pareto-optimal front each day.

Figure 5.7 presents the global Pareto-optimal fronts for three months of the Abilene data [1]. We used the months of June, July, August and September 2002 to predict each day of July, August and September 2002. These traffic statistics were used together with four different BGP routing tables from four major tier-1 providers that were gathered on Oregon route-views [132] to simulate the BGP routes of potential upstream providers. We chose BGP tables from four major providers from October 3rd 2002: AT&T, Sprint, Level3 and Verio. Using a more recent BGP routing table than the traces was to ensure that all AS numbers of the traffic traces would be known. Each figure shows the Pareto-optimal fronts for each value of the time window. We show only values of 1, 5, 10, 15, 20, 25 and 30 days on these figures. Each point represents the total cost of sending traffic using this allocation of the destinations ASes over the upstream links (compared to the standard BGP routing) and its associated maximal number of BGP configuration changes required each day. A point (x,y) gives the expected gain $(1 - x)$ in terms of the cost of sending traffic compared to the classical BGP solution when at most y BGP configuration changes are applied each day.

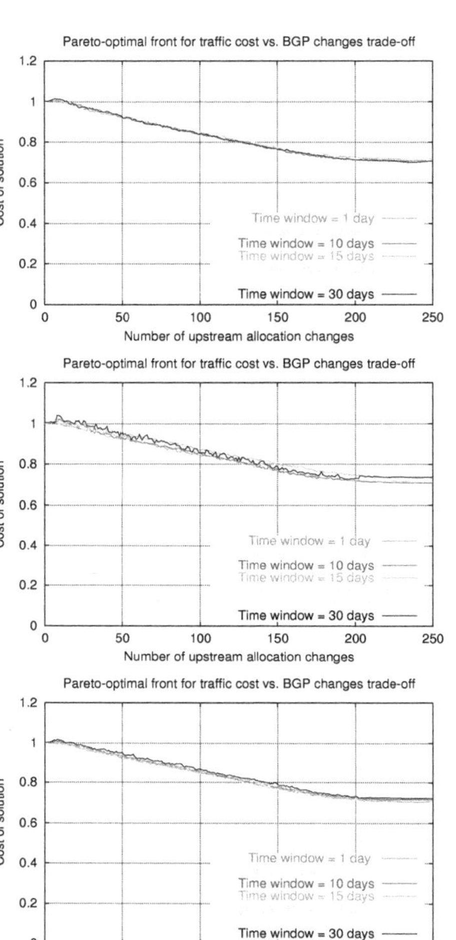

Figure 5.7: Long-term Pareto-optimal fronts for July (top), August (middle) and September (bottom) 2002.

Several points concerning the way we compute the global Pareto-optimal front need to be explained. First, for deciding which upstream to use for a given AS, we chose the upstream having the shortest AS path and a random choice of the upstream when several same length AS paths were found. This implies that there is not one initial cost for zero BGP configuration changes but many, the differences are however small compared to the difference between solutions on the Pareto-optimal front. Second, solutions on the Pareto-optimal front are monotonous in the case of one day, but not in the case of one month. This is because the sum of several monotonous functions is not monotonous anymore, generally speaking. Allowing several BGP configuration changes at each generation implies sampling the complete Pareto-optimal front for each day. One must be aware that there is a trade-off between the efficiency of the search and the sampling of the Pareto-optimal front. Constraining the search to improve the solutio with only one BGP configuration change at each generation dramatically slows down the improvement in terms of the cost.

Getting back to figure 5.7, the first thing one can notice is the absence of remarkable difference between the various Pareto-optimal fronts found for different values of the time window. This indicates that for this traffic demand and BGP tables, relying on more days in the past does not improve much the solution found on the long-term. This result is extremely valuable because it is an argument not to take into account a lot of data to predict the future. Although our results could be due to a particularly stable traffic demand of Abilene, we doubt that relying on a large amount of traffic statistics from the past will help finding less costly solutions on average. As both popularity and amount of traffic exchanged with destinations change over time, using old data might not be useful in our context. The second thing we observe is that the best BGP solutions (0 upstream allocation change) can be better than relying on a few BGP configuration changes (up to 20). This means that if a provider does not want to rely on more than twenty BGP configuration changes, it is not worth trying to optimize as traffic demand variability will invalidate previously good solutions. On the other hand, if a provider is ready to rely on many BGP configuration changes, then the expected gain can be significant, depending on the relative cost of its upstream links. The gain shown on figure 5.7 only reflects our choice of the relative cost of the four upstream providers we simulated for the Abilene traffic demand. This gain cannot be generalized to other situations. In our case, the cheapest upstream provider had a cost 4.5 times smaller than the most expensive one. Differences of interdomain bandwidth prices of a factor two or three are not uncommon in the current Internet (e.g. `http://www.telegeography.com`).

The computation of the Pareto-optimal fronts for each day's traffic demand (more than 850 destination ASes per day) took about 20 minutes on a PIII 1 GHz running Linux.

5.11 Evaluation

In this section we evaluate the advantages and drawbacks of the approach we used in this chapter to tackle the interdomain traffic engineering problem.

First, the cost objective can be replaced by any other objective that can be evaluated on each interdomain destination: quality of the routes (delay, bandwidth, hop count distance,...), load balancing, optimizing simultaneously the requirements of different traffic types,. . . There are many ways to easily adapt our scheme for different particular needs. For optimal load balancing for instance, the objective would be to minimize the maximum of the traffic sent on any link (or the load of any link). For that, instead of minimizing the cost of one solution, one would try to minimize the maximum traffic sent over any link (or the load). This would automatically distribute the traffic (or load) on all the available links. The main advantage of our technique is its generality with respect to the main objective. We do not claim that the gain is worth the effort but in some cases where the relative cost among the upstream providers can be large we are confident that the long-term gain can be significant without implying a high burden on BGP. The main advantage of our solution is that it can be very easily adapted to other traffic objectives. In addition, since we do not require a particular way to tweak the BGP routes[4], our solution can work even if other types of BGP routes attributes manipulations are performed. We do not require that the solution we start with be the default traffic distribution of BGP. Any default solution will do. In practice, one could imagine running this algorithm on a central route server [32, 42] which whould insert the right `local-pref` values into the BGP routes and redistribute them to all BGP edge routers dynamically.

Second, it must be noted that the main target of this chapter are multi-homed non-transit providers. These providers could benefit from such a scheme without too much problems regarding the behavior of their BGP routing. For transit providers on the other hand, the benefits of our approach are less clear. Large transit providers do not necessarily have such large cost constraints on their interdomain links, so minimizing the cost of the traffic may not apply or be less important than intradomain issues [87]. Some benefits could be achieved with transit providers if some cost function represents some real interest, for instance if a large transit provider wants to find a better selection among its peers for each source or destination AS, evolutionary algorithms could find a better and more robust solution than the one classically found by IP protocols.

Finally, on the combinatorial optimization side, we presented in this chapter an evolutionary algorithm capable of solving grouping problems efficiently. Our

[4]We rely on `local-pref` attribute of the BGP routes in our simulations but in practice other attributes can be used [155].

single-objective optimization algorithm described in section 5.6 demonstrated that this class of NP-hard problems could be solved in a greedy way, by combining crossover and mutations into a single "genetic operator". Although this operator is not suited to perform interdomain traffic engineering with BGP if the number of BGP filters is to be minimized, it exhibited interesting convergence properties towards the global optimum that make its efficiency likely on other grouping problems.

5.12 Conclusion

In this chapter, we provided a first evaluation of the feasibility of tweaking BGP for interdomain traffic engineering. We explained in section 5.3 the combinatorial nature of the interdomain traffic engineering problem. Then, we described our first evolutionary algorithm in section 5.6. We showed the main elements for the success of solving the single-objective problem (minimizing cost only). We then introduced in section 5.9 another evolutionary algorithm that tried to rely on as few BGP changes as possible when re-optimizing the traffic objective for the successive days. We carried out simulations on real traffic demands for both algorithms and evaluated their effectiveness. Finally, our study of the long-term gain on three months of traffic demands from Abilene under changing traffic demands showed that there was no gain at using more than the previous day to optimize the next day's traffic demand. This long-term study also showed that, on the traffic demands we used, a large number of interdomain sources or destinations could have to be influenced in order to get close to the global optimal value of the cost objective. We will consider in the next chapters how to reduce the number of necessary configuration changes.

Chapter 6

Interdomain traffic engineering with minimal BGP configurations

6.1 Introduction

In this chapter, we first explain in section 6.2 the limitations of the previous chapter. Section 6.3 then presents the evolutionary algorithm aimed at solving a new interdomain traffic engineering problem, where we aim to minimize the number of BGP filters to be applied on the BGP routes to perform traffic engineering. Section 6.4 presents simulations that study the ability of limiting the number of BGP configurations when engineering the interdomain traffic of an AS.

6.2 The multi-objective combinatorial optimization problem

In the previous chapter, we have dealt with what we called the "simple" problem. The reason why we called this problem "simple" is not because it is easy to solve, but because it is purely combinatorial. It does not make any explicit mention of the number of BGP route changes that are required on the BGP routers: potentially as many as the number of destination prefixes for which we need to override the choice made by BGP in order to distribute the traffic as we would like among the available providers. Our aim however would be to utilize as few BGP changes as possible in order to achieve the "best" (according to the objective function) distribution of the traffic. This problem is thus a multi-objective combinatorial optimization problem, since we want to optimize the objective function while minimizing the number of BGP changes at the same time. In fact, we are not interested in the pure combinatorial problem since the optimal solution

to the assignment problem could involve too many BGP changes. On the other hand, we do not know the cost of a BGP change, this objective is not measurable. Hence our two objectives, cost of traffic and number of BGP changes, are *non-commensurate*.

Given that we want to find the optimal distribution of the interdomain traffic for any number n of BGP changes to be applied, our objective is to find them among the best n changes for the m possible destinations and the p available providers, but starting from the default BGP configuration. So the actual problem we have to solve is not an unrestricted search by trying to swap the destinations among the providers but we need to know how to tweak BGP and find the successive changes that allow to reduce the objective function starting from the default solution distribution found by BGP.

Before we think about the solution to the problem, we should have a rough idea of the size of the search space that we are going to work in. Assuming there are about one thousand ASes with which the local AS exchanges a significant amount of traffic each day, among the fifteen thousand ASes that composed the Internet in 2003, and given that we have a worst case of p providers having each one route towards each destination, finding the best 10 BGP filters to apply to the routes would require to search for the best 10 BGP filters among 1000 times p possible ones. For the case $p = 2$ and the first 10 BGP filters, this amounts to $\frac{2000!}{10!1990!}$ different possibilities. Of course, the number of filters that can actually be applied is smaller since in the case of two providers there will be always one default BGP route towards any destination and only one filter can be applied to change the provider used by the BGP default routing. Still, the number of possible filters is $\frac{1000!}{10!990!}$, a number with 23 decimal digits! Since we do not know how many BGP filters will be required to find an "acceptable" solution to the multi-objective problem, we certainly need to reduce the size of the search space.

To reduce this number of BGP changes, we need to aggregate the interdomain destinations into coarser objects. The AS-level topology obtained from the BGP routing tables constitutes a sound basis for that, by aggregating the destinations into what we call "AS sink trees". So let us define some preliminary concepts that will be useful in the remainder of this chapter. A "terminal AS" (destination) is *an AS that appears as the last AS in the AS path of a BGP route.* A "sub-ASx path" of an AS path is *the suffix of that AS path whose first AS is the first occurrence of ASx.* Note that we cannot remove the "first occurrence" part of the previous definition because of AS path prepending [175]. Even though loops do not occur in the paths of a path vector protocol like BGP, repetitions of the same AS are allowed. The "ASx sink tree" is an instance of an "AS sink tree", defined by *the set of all ASes that appear in all sub-ASx paths in a given BGP table*. A practical, but important reason to rely on these definitions is that all existing BGP routers can be configured to attach `local-pref` values to learned routes whose AS

path matches such a regular expression [100]. It must be noted that the AS-level topology as seen by a stub AS is not a tree nor a forest, but an acyclic directed graph. So there are chances that an AS might appear in several "AS sink trees". In the case that several BGP filters defined on "AS sink trees" where some common ASes belong to them, then we do not try to solve this conflict of which BGP route will be chosen, but we let the BGP decision process solve them.

The rationale for introducing the concept of AS sink tree is the hope of a gain in the BGP changes required, given the size of the search space. In the simulations proposed in section 6.4 with a little less than one thousand terminal ASes, we have about more than $11,000$ AS sink trees while when taking into account only the AS sink trees for ASes located two AS hops away from the local AS (which we call "level 2" AS sink trees), this number reduces to about 150 for two providers. This means that considering "level 2" AS sink trees would result in a number of possible filters of $\frac{150!}{10!140!}$ for two providers, a number with 15 decimal digits instead of 23. This is still a large number but the gain is significant, especially because some AS sink trees represent larger traffic aggregates. We cannot perform an exhaustive search nor a random search in such a large space. An exhaustive search would take too long while a random one would not guarantee that we find a good solution. Biasing the search by trying first the most important destinations in terms of their traffic provides good solutions for a few BGP filters only and does not provide any advantage for a few tens of filters. Using an integer program to solve this problem is quite difficult, mainly because our problem depends on the behavior of BGP when a filter is applied. An integer program would require that we be able to explicit the behavior of the BGP decision process which is by no means simple. Our algorithm simulates a subset of the real BGP decision process and takes as input real BGP routing tables. We do not rely on any theoretical assumption concerning the way BGP works, which we would have to do with an integer program. The "evolutionary" approach provides a simple way to search for good BGP filters in a way that is close semantically to the problem we are trying to solve. It must be noted that we cannot afford to try to find, in addition to the best AS sink trees, the optimal value of the `local-pref` attribute for each AS sink tree. Doing this would make the size of the search space increase and make the search even more difficult. Although BGP encodes the `local-pref` attribute as 32 bits integer, we only utilize two different `local-pref` values in this chapter, a default low value for all routes (e.g. 100) and a higher value (e.g. 110) for the routes on which a BGP filter is applied.

Relying on the AS path attribute to aggregate the traffic is only one particular way to define BGP filters. The objective of BGP filters is to be able to select traffic aggregates in order to control the amount of traffic that is exchanged with each provider. Other BGP attributes might be used for this purpose, like MED or communities [155].

6.3 The evolutionary algorithm

Given that our problem is largely combinatorial, we rely on an evolutionary algorithm that will perform a stochastic search using a population of solutions to find among the BGP filters the most interesting ones. By most interesting, we mean those filters that improve the global cost of the traffic as much as possible.

We encode an individual as an integer vector v: each value v_i of the vector v represents the provider to be used to reach a particular destination. An individual is equivalent to the output of the BGP decision process.

Our algorithm works as follows. We initialize the search by running the BGP decision process once and all individuals are then set according to the results of the BGP decision process. At each iteration and for each individual, we choose randomly (and uniformly) an AS sink tree and compute through which provider we should sent the traffic for all the destinations of the AS sink tree. For each provider, we apply a `local-pref` value higher than the default one to all the routes of this provider for these destinations and run the BGP decision process to know the new distribution of the traffic among the providers. If this change leads to an improvement of the objective function, we actually change the `local-pref` value of all the routes of this AS sink tree and this new individual replaces the old one. The main issue is that we cannot know in advance how the distribution of the traffic among providers will change due to both the BGP decision process and the overlap of the destinations among different AS sink trees. In theory, it should be possible to compute the actual amount of traffic that overlaps among two AS sink trees, but given the complexity of the BGP decision process it is not trivial to assess whether this overlap in the AS sink trees will help in determining the exact effect on the traffic distribution. If some destinations have the same value of `local-pref` due to different BGP filters, it will be up to the tie-breaking rules of the BGP decision process to decide about the final traffic distribution. Among these individuals, not all lead to an improvement in the objective function, we do not remove the individuals which did not lead to an improvement during the current iteration because they could permit improvements during the next iterations. It is up to the selection procedure of the evolutionary algorithm to decide which individuals must survive.

Figure 6.1 provides a pseudo-code version of the search procedure used in our evolutionary algorithm. The most important parts are lines 9 to 13. In line 9, we search in the AS-level topology of provider *provider* all the destinations that are in the chosen AS sink tree. The AS-level topology built from the BGP routes is know to be relatively flat: most destinations are within 3 or 4 AS hops away from the local ISP [195]. So we use a recursive breadth-first search to find the destinations within a given AS sink tree. In line 11 we apply the BGP filter by putting a `local-pref` value of 110 to all routes of this AS sink tree and in line

```
foreach individual k = 1 to Popsize {
// try uniformly a random AS sink tree
sink_tree ← randint(1,num_sink_trees)
// try all providers for this AS sink tree
foreach provider = 1 to p {
    // check whether this provider is connected to this AS
    if (connected(provider,sink_tree) == TRUE){
        // compute the destinations for this AS sink tree and provider
        destinations ← dest(provider,sink_tree)
        // change local-pref for the routes of the destinations
        solution ← change_local-pref(k,provider,destinations)
        // run the BGP decision process
        distribution = BGP_decision_process(solution)
        // compute objective function value
        new_cost[k] ← compute_obj_value(distribution)
        // check whether we improved in the objective function
        if (new_cost[k] < old_cost[k]){
            // update the individual k with improved solution
            k ← solution
            old_cost[k] ← new_cost[k]
        }
    }
}
}
```

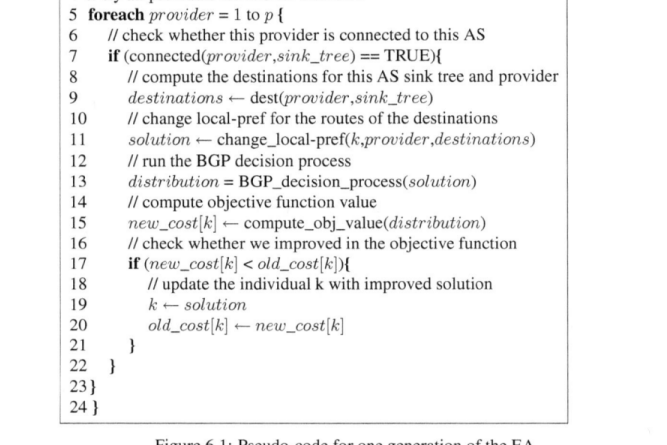

Figure 6.1: Pseudo-code for one generation of the EA.

13 we run the BGP decision process but only on the BGP routes having as last hop in their AS path an AS number with whom traffic has been exchanged, reducing the number of routes on which the BGP decision process has to be applied.

Two crucial parts of the algorithm are not shown in the pseudo-code above: the selection process and determining the population size. First, the selection process is key to sample efficiently the search space. We use the classical stochastic universal sampling [22] to perform the selection of the individuals that survive from one generation to the next. The selection process is critical to focus the search on interesting individuals (from the viewpoint of the objective function) while still retaining a sufficiently large diversity of solutions in the population at each generation. So it must select the individuals according to their relative fitness without biasing the selection too much towards highly fit individuals.

The second important part is the size of the population. In order to ensure that all AS sink trees are tried at least once at every generation, the size of the population must be larger than the number of available AS sink trees that the current aggregation level contains. If the population size is smaller than this number of available AS sink trees, then it is possible that we do not find the best BGP filters during the early generations of the algorithm. So a large enough population size is important to get the best out of the early generations of the algorithm. As the number of BGP filters applied increases, the fraction of the search space sampled becomes small. In that case, increasing the population size does not provide any interesting pay-off, hence the population size should be chosen to efficiently sample only the first few BGP filters.

6.4 Simulations

The purpose of this section is to study the behavior of our algorithm to understand the impact of different parameters, like the number of providers, the objective function and the AS sink tree aggregation used during the search.

From the discussion of section 6.2, we have seen that increasing the number of providers dramatically increases the size of the search space. The type of objective function can also make the life harder for our algorithm, for instance if the improvements in the objective are a complex non-linear function of the amount of traffic that is switched among providers, then the algorithm might take more generations to improve the objective. Finally, we would also like to determine the "ideal" AS sink tree aggregation to perform interdomain traffic engineering. Allowing all possible AS sink trees to be tried (any AS in the AS-level topology) is due to provide the largest flexibility for the search since the algorithm could find the minimal number of BGP filters (i.e. AS sink trees) to obtain a given value of the objective function. However, allowing more AS sink trees means to increase

the size of the search space, hence slowing down the search.

In our simulations we use two different objective functions. The first objective function is the maximum amount of the traffic exchanged on any of the interdomain links. This objective function tries to balance the traffic sent to the upstream providers. The second objective function is a volume-based billing cost defined on the traffic sent to the upstream providers.

The scenario of our simulations consists of a stub provider connected to two to four large tier-1 providers. The traffic demand of this would-be stub provider was gathered on the web from the Abilene Netflow statistics page [1]. We have used one aggregated demand of one month of daily traffic demands, from August 22nd to September 21st 2002. This traffic demand concerned 977 different destination ASes. This dataset has already been described in the previous chapter in section 5.8. We gathered four BGP routing tables from Oregon Route views [132] from four major tier-1 providers from October 3 2002: AT&T (AS7018), Sprint (AS1239), Level3 (AS3356) and Verio (AS2914).

6.4.1 Traffic balance

Figure 6.2 presents the results of the optimization with the traffic balance objective with two, three and four providers. For the case with two providers, we confront AT&T and Sprint. For three providers, we confront AT&T, Sprint and Level3. The x-axis of the graphs of Figure 6.2 shows the number of generations required to reach a given value of the objective function. The y-axis gives the value of the objective function *max(traffic(i))*, $i = 1, .., 4$ where *traffic(i)* denotes the amount of traffic exchanged with provider i. The total traffic of the one month traffic demand corresponds to about 640 units, one unit being one billion of bytes, not considering $1/1000$ packet sampling performed by Netflow. So the unit corresponds roughly to one terabyte of traffic.

Figure 6.2 illustrates the effect of two parameters: the number of providers and the AS sink tree granularity. First let us deal with the number of providers. The default solution found by the BGP decision process appears for generation zero. As the number of providers increases (middle and bottom graphs), BGP does not provide a good traffic balance among all the available providers when the default values of the route attributes are used, with a distance between the optimum and the BGP solution does not decrease when the number of providers increases. Increasing the number of providers a stub would have thus does not ensure that the traffic balance among these providers shall be good. This is a consequence of the way BGP chooses its best route among several available. The BGP decision process does not directly include a notion of the monetary cost of the traffic, nor of the distribution of the traffic among the available exit points. The local-pref attribute is typically used to express how preferred neighboring

ASes are. This preference may reflect the actual monetary cost. For taking into account the traffic distribution on the other hand, no BGP attribute explicitly allows to express how much traffic is carried over a given neighbor or exit point. No mechanism in BGP relates the choice of the best route with the traffic load inside the network. Typically, IGP weights are used to match the traffic demand with the network topology, as IGP weights are used to compute the shortest paths inside the network. However, given that the IGP cost towards the exit point is only used as a tie-breaking rule in the BGP decision process by the routers, tuning the IGP weights as to achieve a particular distribution of the traffic across the internal network is complex [87, 46].

Figure 6.2 shows that a larger number of generations (hence of BGP changes) is required for the algorithm to converge towards the optimum: 20 generations for two providers, a little less than 150 generations for three providers and more than 150 generations for four providers. Since the number of generations provides an upper bound on the number of BGP changes, we see that a good traffic balance for a small number of BGP changes can be obtained with a few BGP changes only in the case of two providers. This larger number of BGP changes required to improve the traffic balance has two causes. The first is the inability of BGP to balance the traffic because of the tie-breaking rules of the BGP decision process that systematically favor one of the upstream providers. The second cause is the distribution of the traffic for the interdomain destinations. We know from Chapter 4.5 that the cumulative percentage of the traffic for interdomain destinations requires more and more destinations to capture increasing percentages of the total traffic. The first few BGP changes are able to move a significant amount of traffic from the default upstream provider found by BGP to the one providing the largest gain in the objective function. Additional BGP changes on the other hand work on smaller amounts of traffic, so that the improvements in the objective function become smaller and smaller. This behavior does not depend on the optimality of the search technique, but only on the distribution of the traffic for the interdomain destinations. Given that the distribution of the traffic among the popular destinations seems to be an invariant of network traffic and not a property of the considered network, we do not expect that any optimization technique will be able to be significantly better than what is shown on Figure 6.2.

The other parameter studied in Figure 6.2 is the AS sink tree granularity. Each graph of Figure 6.2 compares three different ways of aggregating traffic using AS sink trees. The first way consists of allowing all ASes seen in the AS paths for all traffic destinations to be an AS sink tree. This means that we allow AS sink trees to be anywhere in the AS-level topology, from direct neighbors to destination ASes. This corresponds to the curves labeled "all" on the graphs of Figure 6.2. There are in total respectively 1075, 1112, 1118 and 1125 different AS sink trees for the simulations with one, two, three and four providers. The total number of

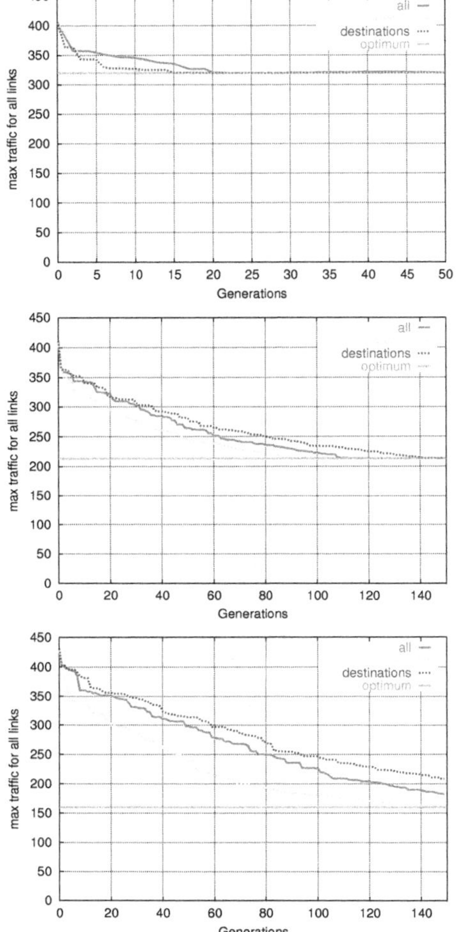

Figure 6.2: Simulations for traffic balance: two providers (top) , three providers (middle) and four providers (bottom).

AS sink trees does not change much for different numbers of providers, as each provider as more or less the same view of the AS-level Internet. The second type of AS sink trees corresponds to destination ASes only. There are 977 AS sink trees (for any set of providers) corresponding to the curve labeled "destinations". The last traffic granularity we study is based on our knowledge of the AS-level topology. The coarsest aggregation possible for AS sink trees is the one of the upstream providers themselves. Aggregating traffic on a per upstream provier basis unfortunately does not provide a sufficiently fine-grained granularity to allow optimal load balancing. Because most of the traffic destinations for interdomain traffic are located two to four AS hops away from a stub ISP [76, 195], aggregating traffic destinations one AS hop behind the upstream providers is due to provide a significant aggregation of the destinations on the AS sink trees, while still providing a fine enough traffic granularity. Figure 6.2 confirms this intuition by showing the decrease of the objective function for a given number of BGP changes to be applied. There are respectively 80, 155, 185 and 225 AS sink trees at a distance of two AS hops (curve labeled "level 2"). This search space is the smallest among the three AS sink tree aggregation levels.

When comparing the curves of the three types of AS sink trees on Figure 6.2, we observe that the AS sink trees located at two AS hops provide the smallest value of the objective function for few BGP changes. This confirms our intuition that these AS sink trees constitute a good balance between the number of AS sink trees and their traffic granularity. The important thing to bear in mind when reading this figure is that the main difference between the three curves comes from the traffic aggregation for a limited number of BGP changes. For a relatively small number of BGP changes, it is the amount of traffic that can be moved by BGP changes that matters most. We see on Figure 6.2 that the "level 2" aggregation works better than the other two most of the time. We also see that the "all" aggregation works better than the "destinations" one. The reason is that the "level 2" aggregation is the coarsest, then follows the "all" aggregation, then the "destinations". Both the "all" and "destinations" aggregation levels suffer from a too large search space compared to the "level 2" aggregation. The "all" and "destinations" aggregation levels would need a larger population size in order to properly sample the search space and find better solutions for a few BGP changes. This would however take too much time when the number of BGP changes grows.

Relying on large traffic aggregates in the heuristic provides good results only for a limited number of BGP changes. Biasing the search towards large traffic changes is effective when the traffic is not well balanced among the providers, as is usually the case when the default BGP solution is used. Once the largest traffic moves have been performed by the first few BGP filters on the other hand, the objective of the search is to find how the small traffic destinations can be combined to further reduce the imbalance between the upstream providers. When one

upstream provider receives substantially more traffic than the others, the algorithm has only to find the largest traffic aggregate and move it to the least loaded among the other upstream providers. When the differences in the amount of traffic received by each upstream provider become relatively small, there are many ways to move several small traffic aggregates to reduce the traffic disbalance. The combinatorial nature of the problem then plays an important role and the size of the search space impedes on the effectiveness of the heuristic. One must bear in mind that including domain-specific information that works well on average in the search heuristic will bias sooner or later the search, there is no free lunch in optimization [212].

6.4.2 Cost of traffic distribution

In this section we make the task of our algorithm a bit more complex, by providing a different cost function for the traffic exchanged through each provider. Each traffic unit exchanged through a given provider has a fixed cost (volume-based billing). Such a cost function is more complex than the traffic balance objective because improvements depend not only on the amount of traffic that is switched from some provider to another but also on the relative cost of the traffic for the providers.

Figure 6.3 presents the simulation results with two (top), three (middle) and four (bottom) providers. Each graph of Figure 6.3 shows the best solutions found by the algorithm at each generation, expressed as a percentage of the cost of the solution found by the default BGP routing. For each graph we provide simulations for the three traffic aggregation levels and two different cost settings of the providers. The cost of one unit of traffic for each provider is provided on the legend of the graphs.

The results of the two providers simulations (top of Figure 6.3) illustrate the important points about the convergence of the algorithm. First, the larger the difference in the costs of the two providers, the larger the gain for a limited number of BGP changes. For instance, with a cost of two for the first provider and one for the second provider (denoted by (2,1) on the labels of the top graph of Figure 6.3), the total gain after 140 generations is about 23 percent for all AS sink trees while about 17 percent for the "level 2" aggregation. For the (4,1) cost setting, the gains are a little under 40 percent for the "all" AS sink trees aggregation and a little under 30 percent for the "level 2" aggregation.

The second thing we notice concerns the quality of the solutions found by the three AS sink tree aggregation levels. The three aggregation level exhibit a similar behavior as in the traffic balance objective, with a better solutions found by the "level 2" aggregation for a small number of filters while for a large number of BGP changes the "destinations" aggregation provides the best results. For a

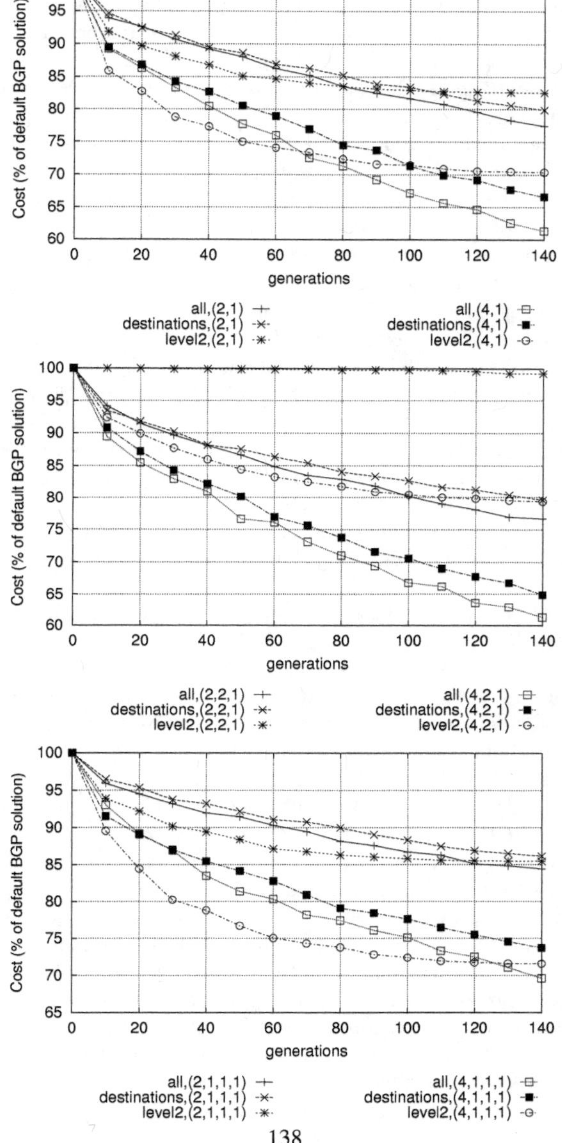

Figure 6.3: Simulation for cost function: 2 providers (top), 3 providers (middle) and 4 providers (bottom).

small number of BGP changes, the most important factor is the amount of traffic that can be moved in one BGP change. For a large number of BGP changes on the other hand, the "level 2" aggregation is stuck in its search because it works with too large traffic aggregates. The "all" aggregation has a larger search space than the "destinations" aggregation, hence its improvements are slower than the "destinations" aggregation. The "level 2" aggregation works with larger traffic aggregates and in a smaller search space, hence it is quite efficient for a limited number of BGP changes. The other two aggregation levels work in larger search spaces so are slower but they work on the finest traffic aggregation possible, giving them an advantage for a sufficiently large number of BGP changes.

As shown on the bottom graphs of Figure 6.3, the simulations with four providers give gains similar to the two providers case. When one provider has a cost significantly larger than all others, the algorithm can find improvements as in the two providers case. When the cost setting is more complex however, moving the traffic on a "level 2" AS sink tree has a non-trivial impact on the improvement in the cost function because many destinations are involved. When one provider has a larger cost than all others, switching a "level 2" AS sink tree towards a smaller cost provider is certain to reduce the total cost of the operation since some traffic goes from this most expensive provider to less costly ones while the other destinations whose traffic change between equal cost providers do not increase the cost. When the cost setting is more complex, moving some traffic to a less costly provider may also move some other traffic from a still less costly provider to a more costly one, making the net result of the change difficult to predict. Relying on traffic aggregation for traffic engineering makes the technique scalable, but also more complex to control.

An illustration of the main disadvantage of the "level 2" AS sink trees appears in the simulation with three providers and the (2,2,1) cost setting. This simulation is shown to be stuck in a local optimum because of the difficult cost setting that prevents the algorithm to improve the search by moving a large traffic aggregate among providers. Only when the relative costs among providers are different can one be sure that large traffic aggregates will allow improvement in the cost function with a single BGP change. If we had allowed the algorithm to apply several BGP changes at each generation, the algorithm would not have been stuck for the "level 2" AS sink trees. In practice, it suffices that the largest traffic aggregates be reachable through upstream providers having the same cost to slow down the improvements of the search. Note that the other simulation with three providers and the "level 2" AS sink trees performs also significantly worse than the other two AS sink tree aggregations, confirming that the particular settings with three providers we used are responsible for the bad results of the "level 2" AS sink trees. Tuning traffic to optimize an objective requires care with respect to the granularity of the traffic aggregates used in the optimization.

6.5 Conclusion

We presented in this chapter an evolutionary algorithm that minimizes an objective function by finding the successive BGP changes to be applied to the BGP routes. We evaluated this algorithm on two objective functions and a real interdomain traffic demand and real BGP routing tables. We studied the behavior of our algorithm for three parameters: the number of providers, the AS sink tree aggregation for the interdomain traffic and the type of objective function.

Our simulations showed that the default way BGP performs traffic balancing between providers is poor, and it does not improve as the number of providers increases due to the tie-breaking rules in the BGP decision process. We showed that, notwithstanding the combinatorial complexity of the problem, improving the traffic balance compared to the default BGP solution is a tractable problem in practice. However, when the number of upstream providers increases, more and more BGP changes are required to properly balance the traffic even when traffic aggregation is used.

For a cost objective, our simulations showed that the gain of the optimization depends a lot on the relative costs among the upstream providers. We showed that reducing the cost of the interdomain traffic by a significant percentage can be achieved with a limited number of BGP changes.

Chapter 7

On-line interdomain traffic engineering with BGP

The traffic engineering performed in the previous chapters was not taking into account short-term traffic dynamics. Daily traffic demands were used as a predictor for the next traffic demand, and the optimization of the traffic was made on the previous traffic demand, not a prediction on the future. Over timescales of days, traffic prediction appeared not useful, see Section 5.10.

The goal of the traffic engineering technique presented in this chapter is to track the optimal distribution of the interdomain traffic on timescales of minutes under the constraint of minimizing the number of BGP configuration changes. The two considered objectives are:

1. minimize the number of BGP route changes,
2. optimize an objective function defined on the outbound traffic sent to the BGP neighbors.

The first objective concerns BGP routing. Given that there are many remote networks with which a stub AS exchanges traffic on timescales of minutes to days [155], engineering the interdomain traffic should try not to influence the traffic of a too large fraction of these remote networks. A reasonable objective of any BGP-based traffic engineering technique should be to minimize as much as possible the burden on BGP. Tweaking the value of local BGP attributes should have minimal impact on the BGP advertisements sent to BGP peers of a stub AS, so we consider it as the preferred way of implementing our traffic engineering solution.

The traffic objectives to be optimized by stub ASes might differ depending on their size and their main business. In this chapter, we use two representative objectives. The first is to equally balance the total traffic over the available providers, specified as $min(max(tr_i))$, $i = 1, ..., n$, n being the number

of available providers and tr_i denotes the amount of traffic sent to provider i. Because traffic balancing might be a simplistic objective, we also evaluate the optimization of a second objective of balancing the cost-weighted traffic, specified as $min \sum_{i=1}^{n} c_i \times \frac{tr_i}{\sum_{i=1}^{n} tr_i}$ where c_i denotes the cost of one unit of traffic sent to provider i. The minimum for the second traffic objective happens when $c_i \times tr_i = c_j \times tr_j, \forall i \neq j$. We thus call this second objective "cost-weighted" traffic balancing since at the minimum the values of the traffic of each provider weighted by its cost are equal among the providers. The second traffic objective models the volume-based billing performed by transit ISPs. Note that the content of this chapter has also been published as [196].

7.0.1 Stub ASs and multi-homing

As usual in this book, the ASes we focus on in this chapter are stub ASes that constitute the majority of the ASes in the Internet [176]. Among these stub ASes, we focus on ASes connected to the Internet through several different providers. Because stub ASes do not offer transit service, they do not advertise via eBGP the routes they learn from their peers. In the case of a stub, there should be no interactions between the local BGP route tweaking performed on the eBGP routes learned from the peers and the rest of the Internet.

Figure 7.1 plots the distribution of the number of providers stub ASes have according the the BGP tables gathered from several vantage points on December 12 2003 [176]. On Figure 7.1, we only consider the ASes from the gathered BGP routing tables that were not considered as provider for any of their peering according to the heuristic proposed in [176]. According to the data of [176], $14,287$ ASes were considered as stub ASes. Among these stubs ASes, 5671 (40 %) are single-homed and 6748 (47 %) dual-homed. These multi-connected ASes need to distribute their interdomain traffic among several interdomain links. Almost all ASes have to pay large ISPs to get a global connectivity on the Internet. The cost of the traffic on these links can become important. These ASes would find interest in redistributing some traffic from more expensive providers towards less expensive ones or simply better balancing their traffic among the various available access links. Furthermore, the number of providers per stub is not expected to grow much in the future since adding providers has been shown to be of limited value [12].

The remainder of this chapter is structured as follows. In section 7.1 we discuss the difficulty of on-line traffic engineering with BGP. In section 7.2 we provide simulations to evaluate the feasibility of outbound on-line traffic engineering with BGP. Section 7.3 discusses extensions to our interdomain traffic engineering solution.

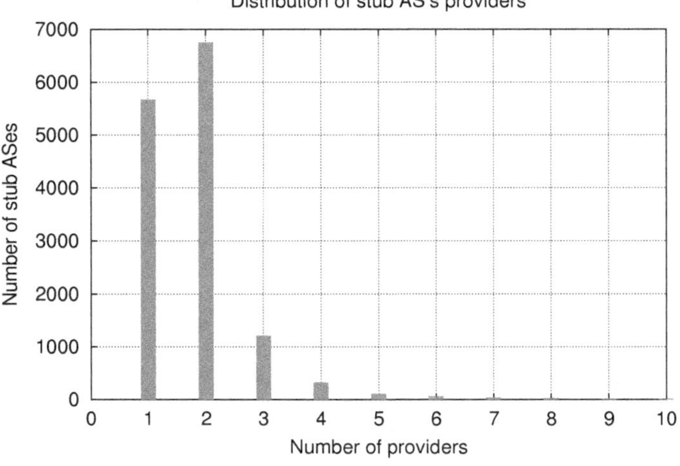

Figure 7.1: Number of providers for stub ASes.

7.1 Difficulty of on-line TE

Most optimization problems dealt with in the literature are static [93, 141]. Dynamic optimization on the other hand mainly addresses continuous problems [37]. For discrete and combinatorial problems in dynamic environments, evolutionary algorithms and swarm optimization techniques are widely used [35, 44]. Unfortunately, no optimization technique is known to be able to deal with all dynamic problems. The specificities of each problem make some techniques more suited than others.

The problem we tackle in this chapter consists in controlling the traffic and its dynamics by tweaking the BGP routes. This problem is discrete and the control of the traffic is indirect, by tweaking the BGP routes. Furthermore, the dynamics of network traffic is significant [145]. Tracking the traffic increases the difficulty of the problem.

In this chapter, we rely on an evolutionary algorithm (EA) to perform on-line traffic engineering. The choice of evolutionary algorithms lies in their ability to solve complex problems and particularly multi-objective optimization problems [57]. Because our problem is a dynamic multiple-objective one, a population-based heuristic makes the search easier as explained in the next sections.

The principles of our scheme are as follows. For each short-term time interval,

a current BGP configuration drives the distribution of the outbound interdomain traffic. By BGP configuration, we mean through which provider the traffic towards some BGP prefix is sent. A BGP configuration can be equivalently expressed as the attributes of the BGP routes of the local AS. During each time interval, the algorithm must find a non-dominated front of solutions (with respect to the objectives defined earlier in this chapter) that will be used to choose the BGP configuration to be applied for the next short-term time interval. There are three main issues to be tackled during the design of the algorithm: 1) tracking the traffic distribution for the next time interval, 2) computing the Pareto front and 3) choosing the next BGP configuration among the solutions from the non-dominated front.

Let us deal with the third issue right now. As will be shown later on, the choice of the solution to use during the next time interval among those on the non-dominated front is an issue only whenever large trade-offs exist between the number of BGP route changes and the value of the traffic objective. It will be shown in the next sections that even if these two objectives are conflicting, allowing additional BGP route changes provides a decreasing marginal gain in the traffic objective, similarly to results found in the previous chapters. Henceforth, a very limited number of BGP route changes are required in practice and network operators will only have to choose the upper bound on the number of BGP advertisements allowed.

7.1.1 Tracking the traffic

Network traffic is known to be bursty (see [145] and references therein). Correctly predicting the amount of traffic that will be sent through any provider will be critical to be able to find a good BGP configuration. For the algorithm to work in most practical situations, not having to rely on a particular model for the data is desirable, since different stubs may have very different user profiles, popular applications and thus different traffic types [211, 72].

[164, 150] have studied the predictability of network traffic. [164] studied the multi-step ahead predictability of network traffic relying on the ARMA and MMPP models, to evaluate the possibility of multi-step ahead prediction for traffic control. [164] showed that smoothing and traffic aggregation help the prediction. [150] on the other hand studied the multiscale predictability of network traffic, for only one-step ahead prediction. The benefits of smoothing the time series for prediction from [164] were confirmed in [150]. However, [150] also found by studying many different traffic traces that the predictability of network traffic highly depends on the considered trace. The performance in the prediction was evaluated based on the ratio of mean squared error to variance in [150, 164]. The results of [150] indicate that the performance of different predictors does not improve much with their increasing complexity. Relying on the last value of the time

series or a simple moving average is as efficient as more complex AR, ARIMA, and ARFIMA models [34]. Since smoothing might provide some benefits while complex predictors do not, we compare in this chapter the performance of adaptive exponential smoothing with a simple predictor: the previous value.

Several flavours of adaptive exponential moving averages have been compared in [39], for bandwidth prediction. [39] showed that Internet traffic could be predicted with a particular adaptive EMA algorithm: the low pass EMA (LpEMA). The basic idea is to modify the classical EMA formula

$$e_i = (1 - \alpha) \times e_{i-1} + \alpha \times tr_i \tag{7.1}$$

where e_i is the estimate at time i, α the weight of the moving average and tr_i the traffic at time i. Instead of a fixed weight α, the weight is made adaptive and computed as follows:

$$\alpha_i = \alpha_{max} \times \frac{1}{1 + \frac{|m_i|}{m_{norm}}} \tag{7.2}$$

where α_{max} is the maximum weight, m_i is the gradient of the traffic at time i ($\frac{tr_i - tr_{i-1}}{t_i - t_{i-1}}$) and m_{norm} is a normalizing gradient. In this chapter, we computed m_{norm} as the mean gradient over a time window of one hour (six 10 minutes bins). We chose to rely on a time granularity of 10 minutes in this chapter.

Let us now have a closer look at the behavior of the LpEMA predictor. First, we compute m_{norm} as the mean gradient (in absolute value) over these last 6 time intervals. The adaptive EMA weight, α_i, is computed according to Formula 7.2. The name Low pass EMA is due to the behavior of α_i that reduces the impact of large changes in the traffic by smoothing out the time-series, having as reference the mean gradient as an indication of the mean variability of the time-series. If on the other hand relatively limited changes in the signal occur during the current time interval, then the value of α_i comes closer to α_{max} and the EMA better tracks the short-term changes in the time-series. The rationale behind this adaptivity is that large changes in the traffic should not be followed unless they correspond to longer-term trends. The greater adaptation of the EMA whenever the shifts in the traffic correspond more closely to the average behavior of the signal (mean gradient) are meant to follow these short-term variations that better match the dynamics of the traffic. The main reason for the smoothing behavior of the LpEMA concerns the fact that Internet traffic is bursty. Tracking large shifts in the traffic is useless for prediction purposes. Unless these changes are longer-term trends, adapting the prediction based on traffic shifts will merely destabilize the prediction while not help in predicting such random traffic surges.

To assess the benefit of the LpEMA predictor to track the traffic of each provider, Figure 7.2 provides both the mean and the standard deviation of the prediction error for different values of α_{max}. On Figure 7.2, the prediction error has

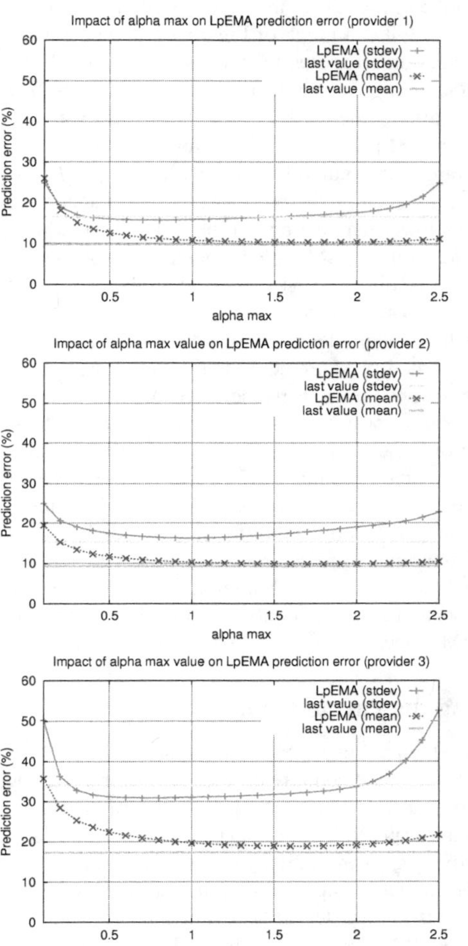

Figure 7.2: Effect of α_{max} value on prediction error: provider 1 (top), provider 2 (middle) and provider 3 (bottom).

been computed for each time interval as the relative error between the predictor and the true value of the traffic. The traffic traces used in this section are described later on, in Section 7.2.1. Each graph of Figure 7.2 plots the mean and standard deviation of this prediction error over the 6 days of the trace. Both the mean and the standard deviation of the error for the LpEMA predictor exhibit a first decreasing part but increase after some value of α_{max}. The "last value" predictor has a lower mean error than the LpEMA one (no matter the value of α_{max}) for the three providers. On average, the prediction error made by LpEMA is always worse than using the last value as a predictor. However, the standard deviation of the LpEMA error can be lower than the one of the last value predictor, see provider 1 and 3. The behavior of the LpEMA predictor can be explained in the following way. Increasing the value of α_{max} first allows the predictor to adapt to the natural variability of the traffic. Then, as the value of α_{max} increases beyond some value, both the mean and the standard deviation of the error increase. This is due to the greater adaptability of the LpEMA prediction for increasing values of α_{max}.

Figure 7.2 suggests two things. First there is no benefit to rely on the LpEMA predictor since its mean error is always larger than the one of the last value. Second, the ideal value of α_{max} seems to depends on the particular provider, hence it is also due to depend on the particular traffic trace. This section suggest that complex and adaptive techniques do not provide any benefit in terms of the accuracy in the traffic prediction, at least on the traffic traces used. The last value predictor already constitutes an accurate predictor, whose performance is comparable to an adaptive exponential moving average, both in its average and standard deviation. The simulations of section 7.2.2 will confirm that the last value predictor performs better than the LpEMA one when used with the traffic engineering scheme.

7.1.2 Searching the Pareto front

Once the method used to predict the traffic distribution during the next time interval has been chosen, we need to choose how to search for the changes in the BGP routes that will optimize the traffic objective while limiting the number of BGP route changes as much as possible. For each time interval, the algorithm must find the BGP configuration that is more likely to optimize the traffic objective, given that we assume the predictor to provide a reliable estimate of the traffic distribution. The objective is to find, during each time interval, a non-dominated front representing the trade-off between the number of BGP route changes to be applied and the value of the traffic objective. The main issue with these two objectives is their conflicting nature: the more BGP route changes allowed, the better the expected value of the traffic objective. Changing BGP routes modifies the amount of traffic sent over each provider at each time interval. Increasing the number of BGP route changes will not always improve the traffic objective due to a chang-

ing traffic pattern. If some traffic from a previously congested provider has been moved to another (previously uncongested) provider, it could be necessary to undo the concerned traffic move to improve the traffic objective if congestion has now appeared on the previously uncongested provider.

We have several options concerning how the search can take place. We can start at each time interval with the last BGP configuration and search for the optimal one and sample the Pareto-optimal front for the next time interval. Another option is to start with the default BGP configuration (no BGP route change) and find the Pareto-optimal front from scratch for each time interval. Finally, we can also start with the last Pareto-optimal front in search for the new Pareto-optimal front.

The first option has the drawback of requiring to potentially undo BGP route changes that are not optimal anymore due to changes in the traffic patterns. By having to start with a BGP configuration that already contains BGP route changes, there is a risk of having to remove some BGP route changes that were good for the last interval but need to be removed for the current time interval. To illustrate this issue, Figure 7.3 shows the two Pareto-optimal fronts for two consecutive time intervals, $n-1$ and n. The x-axis of Figure 7.3 represents the number of BGP route changes and the y-axis gives the value of the traffic objective of the best possible solution having a given number of BGP route changes. Figure 7.3 shows the two Pareto-optimal fronts for the two time intervals, containing both the default BGP solution (0 BGP route change) and the BGP configuration chosen at each time interval. Suppose that we have to start the search at the BGP configuration at time $n-1$, containing a few BGP route changes. To join the Pareto-optimal front for time interval n, we need to remove some BGP route changes from the BGP configuration at time n. On Figure 7.3, we show the image at time n of the BGP configuration at time $n-1$. Because the traffic pattern changes, a good solution during the last time interval might have a relatively large value of the traffic objective during the next time interval. This issue however is theoretical. We will show in the next sections that this first option is the one that is the most efficient in terms of the number of required BGP route changes.

Figure 7.3 shows that the search path between the BGP configuration at time $n-1$ and the Pareto-optimal front at time n requires to remove BGP route changes. The effect of removing these BGP route changes is to improve the traffic objective. In practice, removing previous BGP route changes might also worsen the traffic objective. Finding the path between the last BGP configuration and the current Pareto-optimal front takes computational time that will not be spent on searching for the Pareto-optimal front of time interval n. Additionally, the search would have to allow any intermediate solution to add or remove BGP route changes, implying a larger search space compared to a search that only adds BGP route changes.

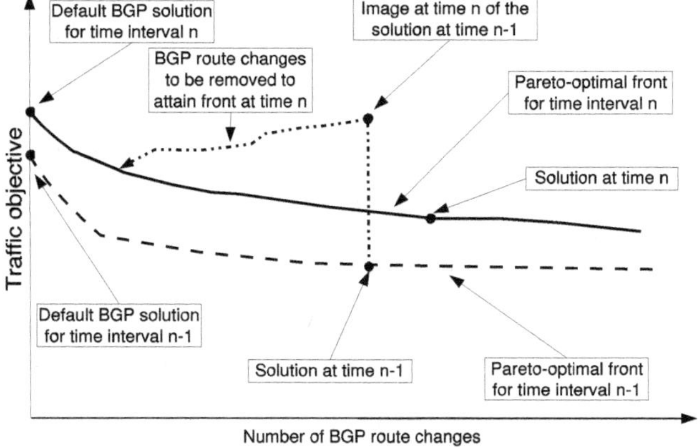

Figure 7.3: Search of the Pareto front from one time interval to another.

In the case of the second option, we start from the default BGP solution without BGP route change, and build the Pareto-optimal front from scratch for each time interval. This solution has the advantage of limiting the size of the search space because the algorithm can iterate by only adding successive BGP route changes to find the Pareto-optimal front. The drawback of this method is that one does not leverage the knowledge of potentially good solutions found during previous time intervals.

The last option consists in building the next Pareto-optimal front by starting from the last one. This solution has the same potential drawback as the first option, in that the size of the search space is very large because the search must allow moves backward and forward with respect to the previous BGP route changes. An important constraint is to work with a non-dominated set of solutions during the search of the Pareto-optimal front. This requires that the search effort be distributed over the different points of the front. Some of the points of the old Pareto-optimal front however will not easily join the new Pareto-optimal front, so that valuable computational time may be wasted.

We experimented the three options to determine which search heuristic algorithm to choose. Our simulations indicated that allowing to undo BGP route changes and starting with a Pareto-front when searching for the next front had problems to sample the Pareto front. In the remainder of the chapter, we do not allow to undo BGP route changes and always start with a single solution at each time interval in search of the Pareto front.

7.1.3 Search algorithm

The previous section discussed the choice of the search method. In this section, we describe in more details the working of the search algorithm. Figure 7.4 provides the pseudo-code describing the main operations during one particular time interval. In line 2 of Figure 7.4, we initialize the current population at the default BGP solution (0 BGP route change). Line 4 of Figure 7.4 contains the main loop that iterates over the population and adds BGP route changes one at a time. The loop on line 4 of Figure 7.4 iterates from the first BGP route change until *max_iter* BGP route changes, the maximum number of BGP route changes allowed. Within an iteration, the algorithm performs a local search by adding an additional BGP route change to the individuals of the population *pop* (line 6 of Figure 7.4). Individuals who were added a BGP route change that improve their traffic objective value are put in *improved_pop*. Then, this population of improved individuals is used to update the Pareto-optimal front of the current time interval (line 8 of Figure 7.4). At most one individual is added to the Pareto front at each iteration since each successive generation deals with a population that contains *iter* BGP route changes. Finally, selection is performed in line 10 of Figure 7.4 where individuals from *improved_pop* are chosen to make the population for the next iteration (line 12 of Figure 7.4).

```
1 // Population ← default BGP solution
2 init_pop(pop)
3 // Each iteration adds a BGP route change
4 foreach iter = 1 to max_iter {
5  // Trying to add a BGP route change
6  improved_pop = try_add_filter(pop)
7  // Updating current front
8  update_front(improved_pop)
9  // Selecting individuals
10 new_pop = select_population(improved_pop)
11 // Replacing old population by new one
12 pop = new_pop
13 }
```

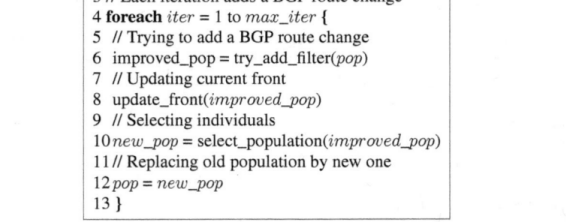

Figure 7.4: Search algorithm for one time interval.

In the pseudo-code of Figure 7.4, there is no explicit mention to the minimization of the BGP route changes. Actually, the number of iterations *iter* is a higher bound on the number of BGP route changes since the algorithm might be unable to find an improved solution at some iteration. Making the objective in terms of

the number of BGP route changes a dimension of the search provides an explicit control on the number of BGP route changes that are required by the solutions found by the algorithm.

The performance of the search algorithm is sufficient to be used as an on-line traffic engineering scheme. The current prototype Perl version of the algorithm takes about one second per iteration on a P4 2.4 GHz. A C version would probably be at least an order of magnitude faster but does not seem necessary for our purposes.

7.2 Performance evaluation

This section provides results about the performance of the on-line interdomain traffic engineering scheme introduced in the previous sections. Section 7.2.1 first presents the typical context in which the scheme is to be used. Section 7.2.2 then discusses the impact of the traffic variability on the performance of the scheme. Section 7.2.3 proposes a modification of the basic on-line traffic engineering scheme to reduce the number of required iBGP advertisements. Section 7.2.4 then goes on to show that relying on the MED attribute of the BGP routes performs almost as well as relying on the `local-pref-on-line` attribute. Finally, section 7.2.5 evaluates the performance of the scheme for the second traffic objective introduced in this chapter.

7.2.1 Scenario

The scenario of our simulations consists of a stub AS connected to three would-be providers, as illustrated by Figure 7.5. The local stub has one physical link with each provider over which a BGP session is established. In addition, the traffic engineering scheme centralizes the BGP routes announced by the BGP router of each provider and collects traffic statistics at the granularity of the BGP routes. The collection of traffic statistics could be performed in various ways (see [118] and references therein). The on-line traffic engineering algorithm runs on the machine labeled "optimization box". BGP routers `R01`, `R02` and `R03` each have a BGP session established with a different provider. The "optimization box" has eBGP multihop sessions established with the border BGP routers of each provider. It also receives the BGP routes from the providers and runs the algorithm to find how to tweak the BGP routes. Finally, the "optimization box" modifies the `local-pref` attribute of the routes in its BGP routing table and advertises these modified routes to the local routers `R01`, `R02` and `R03`. In practice, the "optimization box" might be a modified route reflector implementing the functions required for the on-line traffic engineering [32].

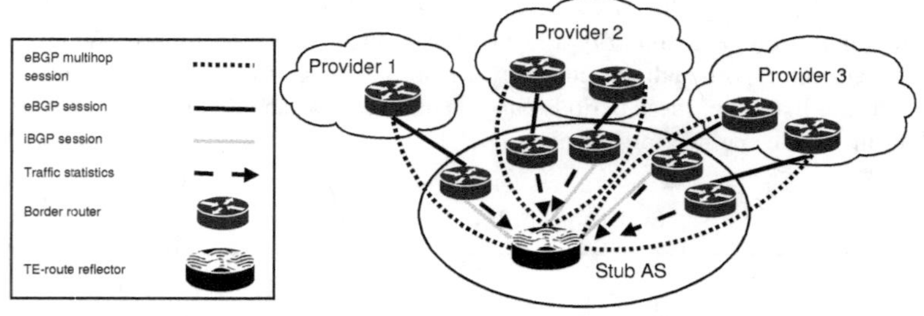

Figure 7.5: Simulation scenario.

The iBGP messages exchanged between the "optimization box" and each of the border BGP routers during each time interval are composed of two different kinds of updates. The first type of iBGP updates are sent by the "optimization box" to each of the border BGP routers to undo the route tweaking performed during the previous time interval. The second kind of iBGP updates are those concerning the BGP routes that are preferred for the next time interval. In the remainder of this chapter, we always speak in terms of the total number of iBGP messages exchanged between the "optimization box" and each border BGP router. This means that all figures that show the value of some traffic objective as a function of the number of iBGP updates in this section contain an even number of iBGP updates since for each new BGP route to be changed by the algorithm, a previous BGP update must be undone.

The first traffic trace used in our simulations is a six days trace of all the outbound traffic from the University catholique de Louvain (UCL), which we call *stub1* in the remainder of this chapter, starting from March 19nd 2003 00:00 CET. During this 6 days period, UCL sent 1.7 terabytes of traffic or an average of 27 Mbps. This 6 days long trace was cut into time intervals of 10 minutes. For each 10 minutes interval, we recorded the amount of bytes sent to each BGP prefix.

The second traffic trace is an 18 days long trace of all the outbound traffic from BELNET, the Belgian research network, which we call *stub2* in the remainder of this chapter, a gigapop stub with 5 Gbps of capacity with its providers, starting from October 30th 2003 12:00 CET. The trace was collected using the NetFlow sampling available on the access routers of *stub2* with a $1/4000$ packet sampling rate. Figure 7.6 shows the evolution of the normalized outbound traffic of *stub2* during the 18 days of the trace. One unit on the y-axis of Figure 7.6 corresponds to the average traffic over the whole duration of the trace. Since packet sampling was used, we multiplied the byte count per second of the sampled trace by 4000 to obtain the expected total traffic. The purpose of the trace from *stub2* is to evaluate

the benefit of on-line traffic engineering for larger stub ASes. Several traffic disruptions appear on the graph of Figure 7.6, due to maintenance operations on the Netflow collector. Although no actual traffic disruption occurred on the routers of *stub2* during that time, these periods can be considered as shutdown periods for the on-line traffic engineering technique.

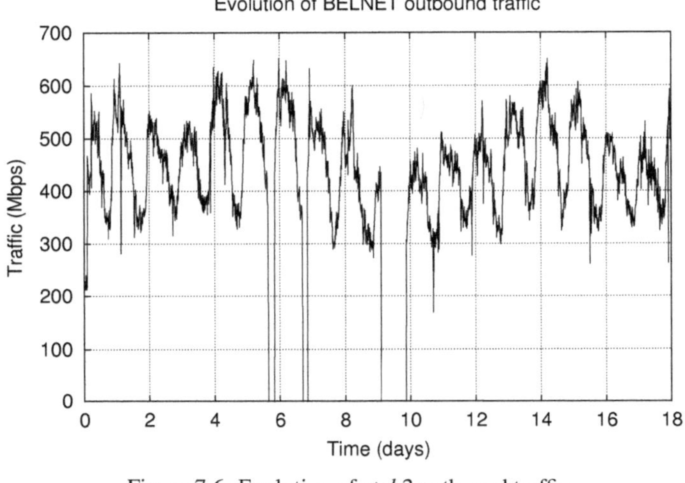

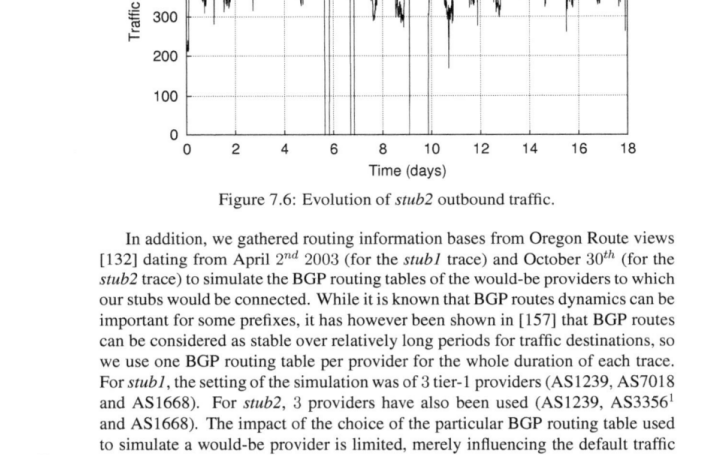

Figure 7.6: Evolution of *stub2* outbound traffic.

In addition, we gathered routing information bases from Oregon Route views [132] dating from April 2nd 2003 (for the *stub1* trace) and October 30th (for the *stub2* trace) to simulate the BGP routing tables of the would-be providers to which our stubs would be connected. While it is known that BGP routes dynamics can be important for some prefixes, it has however been shown in [157] that BGP routes can be considered as stable over relatively long periods for traffic destinations, so we use one BGP routing table per provider for the whole duration of each trace. For *stub1*, the setting of the simulation was of 3 tier-1 providers (AS1239, AS7018 and AS1668). For *stub2*, 3 providers have also been used (AS1239, AS3356[1] and AS1668). The impact of the choice of the particular BGP routing table used to simulate a would-be provider is limited, merely influencing the default traffic imbalance. Chapter 6 showed that for the optimization of a traffic objective, the number of available providers complexifies the search. Studying the impact of the

[1]We could not rely on AS7018 for these time periods because it was not present in Route views at the time.

choice of the particular provider on the default traffic distribution found by BGP is irrelevant in the context of this chapter since its purpose is to demonstrate the feasibility of on-line traffic engineering with BGP, not to quantify the expected gain of on-line traffic engineering. The latter depends on the traffic pattern as well as the cost function to be optimized. The performance evaluation has been carried out for both stub ASes but we present each result for one of them only.

7.2.2 Impact of traffic uncertainty

In this section, we study the effect of the impact of the traffic predictors on the quality of the achieved traffic balance. We also compare the results of the algorithm with and without the uncertainty of the traffic demands to understand to what extent the traffic dynamics prevents the algorithm to approach the optimal traffic balance.

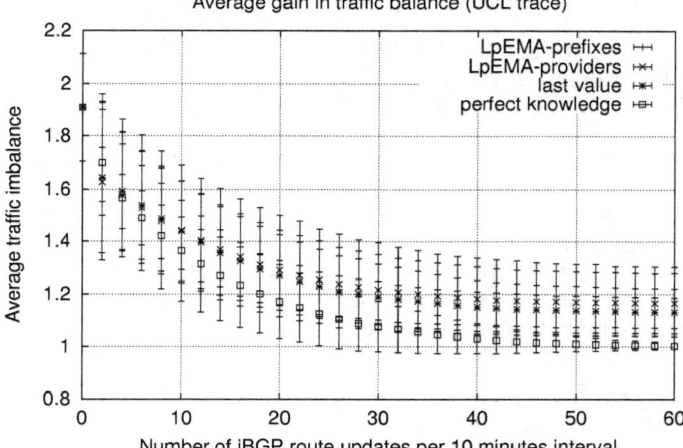

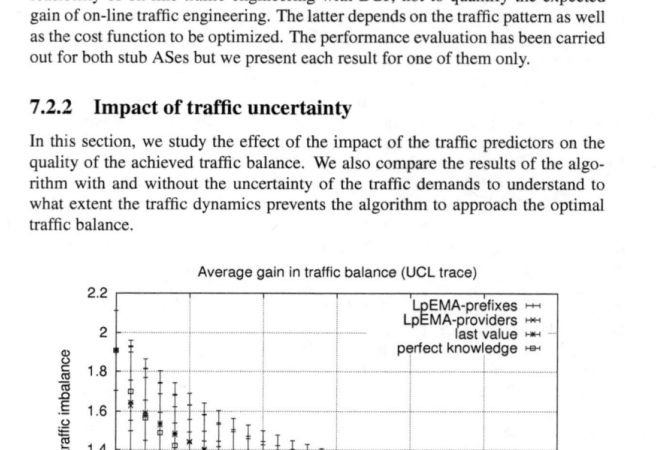

Figure 7.7: Average gain in traffic balance as a function of the number of iBGP updates.

Figure 7.7 plots the average traffic imbalance as a function of the number of iBGP updates allowed per 10 minutes interval. We call "imbalance ratio" the maximum amount of traffic sent through any of the three providers divided by the ideal traffic balance (total traffic divided by the number of providers). To compute the average, we took the whole non-dominated front for each time interval, and

averaged the imbalance ratio for the 6 days of the trace. Because all values of the number of iBGP updates are not present in the non-dominated fronts for each time interval, whenever some value for a given number of the iBGP updates was not present we used the value of the imbalance ratio for the solution having the larger number of iBGP updates smaller than the current number of iBGP updates. For example, if some solution for 60 iBGP updates did not exist, then we used the solution with the larger number of iBGP updates smaller than 60.

The four curves of Figure 7.7 start with the same value of the average traffic imbalance of about 1.9. The default route choice by the BGP decision process chose to send on average 1.9 times more traffic on one of the three providers than the average traffic over the three providers. With 10 iBGP updates, the average traffic imbalance is below 1.5 for all traffic tracking schemes. With 20, 30, 40, 50 and 60 iBGP updates, the average traffic imbalance is respectively below 1.33, 1.25, 1.21, 1.20 and 1.19. So for all practical purposes, relying on more than 40 iBGP updates provides a very limited improvement for the traffic balancing objective. The difference in the average traffic imbalance between the "last value" predictor and the prefix-based LpEMA is of about 0.05. This section confirms the results of Section 7.1.1 showing that the LpEMA predictor was not better than the last value. Even under perfect knowledge of the future traffic demand, the improvement provided by more than 40 iBGP updates is of at most 0.03. Relying on many iBGP updates to try to improve the traffic objective will not provide a significant gain. Note that some curves on Figure 7.7 have error bars that go below an imbalance of 1 because of the asymmetric variance around the mean. The error bars show the standard deviation around the mean. In reality imbalance cannot take a value smaller than 1.

The number of iBGP updates required for the optimization is small compared to the level of the "BGP noise". By BGP noise, we mean the iBGP updates that are routinely exchanged between BGP routers. A recent study at a large ISP [8] reports a BGP noise of more than one hundred iBGP updates per minute.

7.2.3 Tabu prefixes

In the previous section, we saw that relying on complex predictors was not worth for practical purposes and that a limited number of iBGP updates are enough to approach a good traffic balance. An issue with the on-line traffic engineering algorithm is that some BGP route might be tweaked back and forth during consecutive time intervals. This means that traffic flows would have to switch from one provider to another. In addition, if the on-line optimization algorithm decides to change the provider used to carry the traffic towards some prefix, then it would be desirable to stick to this choice during a sufficient amount of time to limit the effect of the on-line optimization on the instability of the topological distribution

of the traffic.

For that purpose, we added a "tabu list" that maintains the set of BGP prefixes for which the on-line traffic engineering modified the route attributes during the last x time intervals. The traffic for the BGP prefixes present in the "tabu list" cannot be moved to another provider during x time intervals, starting from the time interval during which this BGP route change was applied. The tabu list contains the BGP routes changed by the algorithm during the last x time intervals, hence BGP prefixes and the associated preferred provider.

To evaluate the practical interest of the "tabu list" method, we need to address the following questions:

1. How many new BGP route changes should be accepted per time interval?

2. For how long should a BGP route change be present in the tabu list?

Figure 7.8 plots the average traffic imbalance for several values of the number of the iBGP updates per time interval (10, 20, 30 and 40) and the lifetime of the tabu list entries (0, 1, 3, 6, 9, and 12 time intervals). On the x-axis of Figure 7.8, the total number of BGP routes that have been tweaked by the on-line traffic engineering scheme for any time interval are shown. This total number of BGP routes that have been tweaked is not the same as the number of iBGP updates. With the tabu list method, we have no iBGP messages for the entries of the tabu list that stay the same between two consecutive time intervals. With the tabu list method, there are iBGP updates for the BGP routes whose attributes are reset to the default value (entries of the tabu list that expire) as well as iBGP updates for the BGP routes that are tweaked by the on-line traffic engineering scheme (entries that enter the tabu list).

Let m represent the number of time intervals during which an entry of the tabu list remains and n represent the number of new tabu list entries accepted per time interval. Each time interval, there are thus $2n$ iBGP updates, n for the new tabu list entries (newly tweaked routes) and n for the old tabu list entries that expire. An entry of the tabu list expires when it has been present in the tabu list for more than m time intervals. During each time interval, there are $n \times m$ tabu list entries that have not yet expired, as well as n new ones selected to enter the tabu list, a total of $n \times (m + 1)$ BGP routes that have been tweaked.

Figure 7.8 presents the effect of the parameters of the tabu list method on the performance of the scheme. The important variable that guides the value of the traffic objective on Figure 7.8 is the total number of BGP routes that have been tweaked by the traffic engineering scheme, not the number of iBGP updates per time interval. The number of BGP routes that have been changed by the traffic engineering scheme corresponds to the number of entries in the tabu list. On the traffic trace of $stub1$, there is ample choice for the number of iBGP updates

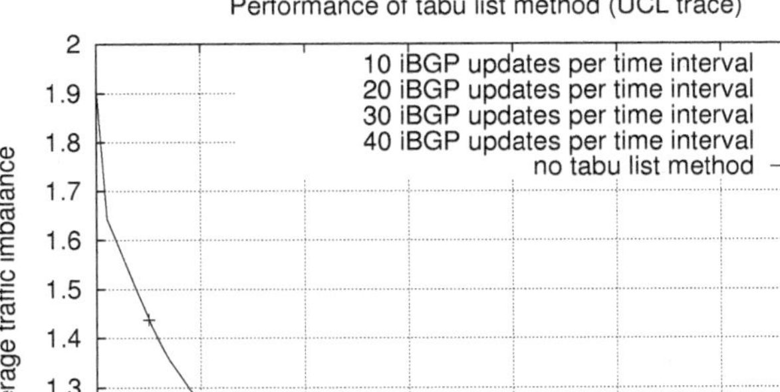

Figure 7.8: Performance of tabu list method with last value predictor.

allowed as well as the number of tabu list entries present during each time interval. The network manager has thus the choice to choose the acceptable burden on BGP as well as how close he wants to get from the optimal traffic balance achievable, simply by selecting the right size of the tabu list. The two questions above are thus related and only the total number of modifications with respect to the default BGP routes seems to be relevant for what concerns the value of the traffic objective achieved. On Figure 7.8, we also reproduce the results obtained without relying on a tabu list ("last value" curve of Figure 7.7), for up to 60 BGP route changes. The tabu list results are seen to match the "no tabu list" ones, with the points for the tabu list results closely following the "no tabu list" curve. This confirms that the value of the traffic objective seems to depend largely on the total number of tabu list entries present during each time interval, not on how many BGP changes are added during each time interval.

The benefits of the tabu list method are obvious when comparisons are made with respect to the number of iBGP updates. Figure 7.9 gives for the *stub2* trace the imbalance ratio as a function of the number of iBGP updates per time interval with the tabu list method. The comparison made on Figure 7.9 is intentionally unfair towards the "no tabu list" method because the tabu list method limits the number of iBGP updates per time interval. The continuous curve on Figure 7.9 labeled "no tabu list" gives the improvement in traffic balance when the optimiza-

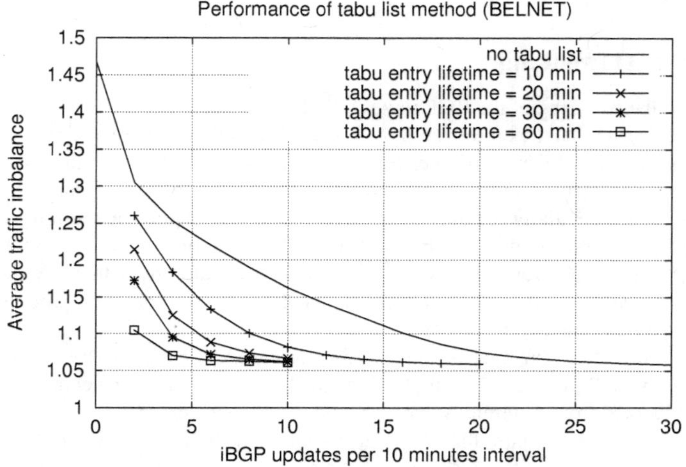

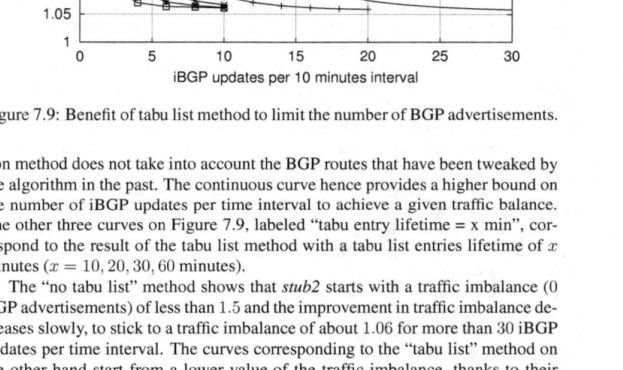

Figure 7.9: Benefit of tabu list method to limit the number of BGP advertisements.

tion method does not take into account the BGP routes that have been tweaked by the algorithm in the past. The continuous curve hence provides a higher bound on the number of iBGP updates per time interval to achieve a given traffic balance. The other three curves on Figure 7.9, labeled "tabu entry lifetime = x min", correspond to the result of the tabu list method with a tabu list entries lifetime of x minutes ($x = 10, 20, 30, 60$ minutes).

The "no tabu list" method shows that *stub2* starts with a traffic imbalance (0 BGP advertisements) of less than 1.5 and the improvement in traffic imbalance decreases slowly, to stick to a traffic imbalance of about 1.06 for more than 30 iBGP updates per time interval. The curves corresponding to the "tabu list" method on the other hand start from a lower value of the traffic imbalance, thanks to their larger number of total BGP routes tweaked per time interval (entries of the tabu list). With only 10 iBGP updates per time interval, the "tabu list" method is able to attain a traffic imbalance a little above 1.06 for a tabu list entries lifetime of 30 and 60 minutes.

The tabu list method performs well compared to the method that allows all the BGP routes to change for each time interval. The tabu list method has the advantage of allowing to reduce the burden on the number of iBGP updates to be sent during each time interval, without impacting on the quality of the traffic balancing. Maintaining a list of tabu BGP routes is thus extremely interesting

from a practical viewpoint.

7.2.4 MED tweaking

Up to now, the tweaking of the BGP routes was done by using the `local-pref` attribute, ensuring that the optimization will override all other BGP tweaking performed by the AS. This way of tweaking the BGP routes could make the BGP decision process choose a route with a longer AS path in order to optimize the traffic balance, potentially leading to a worse end-to-end quality of the routes. If the on-line optimization is required not to override the other types of traffic engineering (e.g. MED or IGP cost), then ideally one would prefer that the optimization tweaks a BGP attribute that is evaluated by the BGP decision process just before the last rule.

In the case of non-transit ASes, the optimization can rely on the MED attribute. According to the BGP Internet draft [156] the MED attribute cannot be compared for routes learned from different ASes[2], the route with the lowest MED being best. Allowing to always compare the MED attribute among BGP routes is however common [100]. Although there is no reason a priori to compare the MED attribute of routes learned from different neighboring ASes, stub ASes could rely on the MED attribute or another attribute having a meaning local to the AS to perform traffic engineering. Only limited modifications to BGP would be needed for this.

In this section, we allow the BGP instance running on the "optimization box" of the local AS to set the MED attribute of the routes received from the different neighboring ASes to prefer one route over another. The working of the optimization technique is similar to the one presented in section 7.2.3, except that the preferred route for the optimization algorithm gets a lower MED compared to the other routes towards a given prefix instead of having a higher value of the `local-pref` attribute. The main difference with the algorithm of section 7.2.3 is that only routes having the same `local-pref` value and AS path length will have their MED value compared. The optimization technique has thus a more limited choice in the prefixes that can be used to balance the traffic, compared to the `local-pref` way.

Figure 7.10 presents the simulation results of the tabu list on-line optimization with the MED attribute for *stub*1. Figure 7.10 shows the average traffic imbalance as a function of the number of iBGP updates per 10 minutes interval. Figure 7.10 is the counterpart of Figure 7.9. Figure 7.10 shows that the performance of the tabu list on-line optimization with MED is quite comparable to the one

---

[2]Unless the always-compare-med deterministic-med options are used, see Cisco's website at `http://www.cisco.com/en/US/tech/tk365/technologies_tech_note09186a0080094925.shtml` for more details.

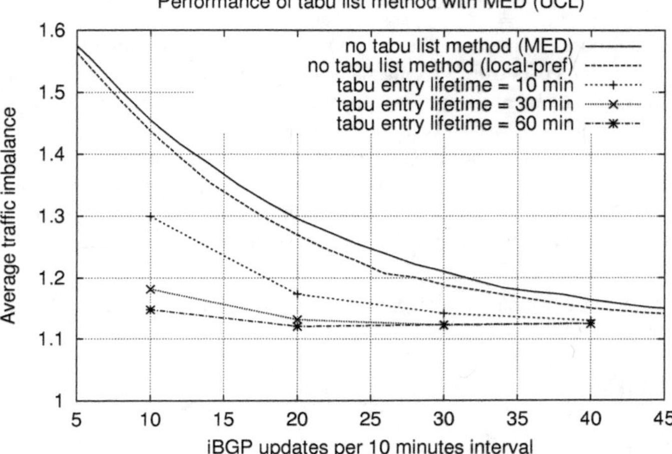

Figure 7.10: On-line optimization with MED.

with the `local-pref` attribute, only slightly worse. The two continuous curves on Figure 7.10 give the simulation results for the "no tabu list" method, with `local-pref` and MED. The dotted lines of Figure 7.10 on the other hand provide the simulation results for the tabu list method by tweaking MED. Increasing the size of the tabu list (by increasing the entries lifetime) provides a significant reduction in the number of iBGP updates to be announced.

7.2.5 Cost-weighted traffic balancing

The previous sections have provided simulations with a relatively simple traffic objective. In this section, we provide simulation results with the cost-weighted traffic balance objective introduced earlier in this chapter for the traffic of *stub2*. This objective is related to the economical cost of the interdomain traffic, hence it is important in practice. The cost-weighted traffic balancing objective is defined as

$$Minimize \quad \sum_{i=1}^{n} c_i \cdot \frac{tr_i}{\sum_{i=1}^{n} tr_i}$$

$$subject\ to \quad \sum_{i=1}^{n} \frac{tr_i}{\sum_{i=1}^{n} tr_i} = 1$$

where $c_i > 0 \ \forall i$, denotes the cost of one unit of traffic sent to provider i. We used the following costs: $c_1 = 1.5, c_2 = 1.25$ and $c_3 = 1$. These costs yield

the following traffic distribution at the minimum: $tr_1 = \frac{10}{37} \times \sum_{i=1}^{n} tr_i$, $tr_2 = \frac{12}{37} \times \sum_{i=1}^{n} tr_i$, and $tr_3 = \frac{15}{37} \times \sum_{i=1}^{n} tr_i$.

We left the costs c_i fixed in the following simulations but nothing in our scheme prevents the costs from varying with time or depending on other parameters. For instance, a larger cost can be attributed to the traffic sent to a provider during the busy hours or the cost could depend on the relative load of the access links. The sole constraint on the objective function is that the impact on the objective function of a change in the best BGP route for some destination prefix must be known for the search algorithm to be able to improve it. No constraint on the look of the objective function (convexity, piecewise linearity,...) is made.

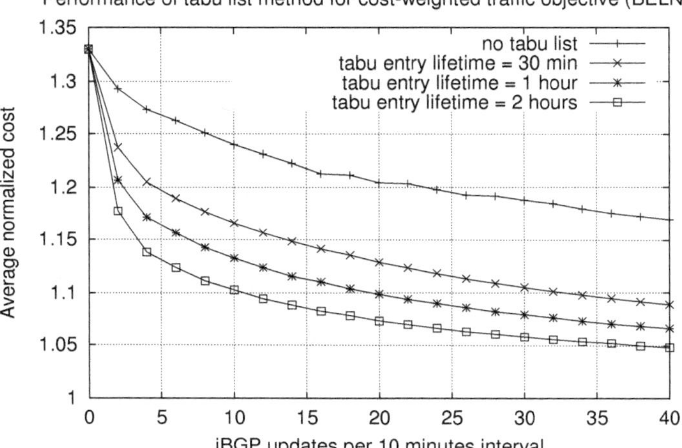

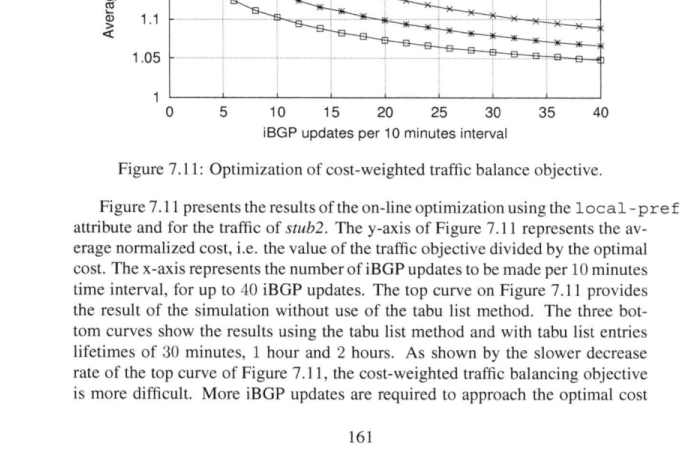

Figure 7.11: Optimization of cost-weighted traffic balance objective.

Figure 7.11 presents the results of the on-line optimization using the `local-pref` attribute and for the traffic of *stub2*. The y-axis of Figure 7.11 represents the average normalized cost, i.e. the value of the traffic objective divided by the optimal cost. The x-axis represents the number of iBGP updates to be made per 10 minutes time interval, for up to 40 iBGP updates. The top curve on Figure 7.11 provides the result of the simulation without use of the tabu list method. The three bottom curves show the results using the tabu list method and with tabu list entries lifetimes of 30 minutes, 1 hour and 2 hours. As shown by the slower decrease rate of the top curve of Figure 7.11, the cost-weighted traffic balancing objective is more difficult. More iBGP updates are required to approach the optimal cost

compared to the traffic balancing objective of the previous sections. The initial value of the cost-weighted traffic objective found by BGP is 1.33. Without the tabu list method, 20 iBGP updates per 10 minutes interval yield an average normalized cost of 1.2, 100 iBGP advertisements yields an average normalized cost of 1.1 (not shown on Figure 7.11), and 180 iBGP updates an average normalized cost of 1.06 (also not shown on Figure 7.11).

As expected, the tabu list method allows to drastically reduce the number of iBGP updates required to achieve a given gain in terms of the traffic objective. The larger the lifetime of the tabu list entries, the smaller the number of BGP advertisements required to reach a given average normalized cost.

7.3 Beyond outbound traffic engineering for stub ASes

In this chapter, we have proposed the utilization of an optimization box inside stub ASes to control the flow of their outgoing interdomain traffic. As presented in figure 7.5, this optimization box would collect the traffic statistics from the border routers and maintain multi-hop eBGP sessions with all the providers of the stub. In practice, the optimization box could be co-located with a Route Reflector [26] provided that it could receive all eBGP routes advertised by the providers of the stub. This could be achieved by using the BGP extension proposed in [204].

Besides controlling their outgoing traffic, stubs also need to control their incoming traffic. Several BGP tweaking techniques have been proposed to achieve this control [155, 77, 11] but they all suffer from the same drawback. To control its incoming traffic, a stub must influence the decision process of many remote ASes. Remote ASes, because of their routing policies or outbound traffic engineering, may inclucence the flow of the incoming traffic of our stub AS. The approach that we propose cannot be used to control the flow of the incoming traffic. We studied the feasibility of prepending for a stub AS in [152], and observed that in general it is very difficult to control the incoming traffic. BGP communities allow a better control than path prepending, but they must be supported by all the upstream providers to work. Inbound traffic control requires an explicit cooperation between distant ASes as proposed in [10].

Although the focus of this chapter is on stub ASes, transit ASes may also benefit from such on-line traffic engineering with BGP. Those providers have different sizes and different traffic engineering requirements. Most of those transit ASes are small national or regional ASes that are connected to larger providers. Those transit ASes will also need to optimize the traffic on their provider links. Finally, the tier-1 ISPs at the core of the Internet do not have any provider, but they often have multiple peering links with other tier-1 ISPs that need to be engineered. On-line traffic engineering might pose problems for transit ASes having a large

number of downstream customer ASes since changing only a few BGP routes will require to announce the modified BGP routes to every downstream customer. If this number of customers is large, then even a few BGP advertisements every few minutes might constitute a burden. For transit ASes with a limited number of customers on the other hand, announcing a small number of BGP routes every few minutes will not hassle the BGP routers of the customers.

The main issue with transit ASes is that an iBGP change in the AS would cause a modification of the eBGP routes advertised by the transit AS to its BGP neighbors [77]. This would potentially increase the instability of BGP in the Internet. Furthermore, if the traffic engineering technique changes the best route towards some destination prefix, then downstream ASes with respect of the traffic direction could also change their best route to reach this particular destination prefix. Suppose that on Figure 7.12 transit AS 6 relies on a traffic engineering scheme to balance its outbound traffic. Assume also that the best BGP route chosen by transit AS 6 to reach destination prefix `X/Y` is the one through AS 2. Transit AS 6 thus advertises to AS 7 a BGP route to reach prefix `X/Y` with an AS path `6-2-1` of length 3. AS 7 receives another BGP route to reach prefix `X/Y` through transit provider 5, with an AS path `5-2-1` of length 3. If the traffic engineering scheme that transit AS 6 implements chooses to use the BGP route through AS 4 and AS 3 to reach prefix `X/Y`, then transit AS 6 will advertise to AS 7 a BGP route for prefix `X/Y` with an AS path `6-4-3-1` of length 4. If AS 7 does not tweak its BGP routes with the `local-pref` attribute, the BGP decision process of AS 7 will choose as its best BGP route to reach prefix `X/Y` the route learned from transit AS 5 that has a shorter AS path. Hence the effect of the tweaking of the BGP routes by transit AS 6 to balance its outgoing traffic will be to deflect part of the traffic it carried having as destination prefix `X/Y`.

A possible solution to avoid this problem would be to ensure that an iBGP change inside the AS does not cause an eBGP change. This could be achieved by using AS Sets [156] and BGP aggregation rules [156]. In BGP, aggregation is used to aggregate several prefixes together to reduce the size of the BGP routing tables. For example, assume that in figure 7.12 AS 2 advertises 10.0.0.0/8 to AS 6 and AS 4 advertises 11.0.0.0/8. In this case, AS6 could advertise prefix 10.0.0.0/7 to AS 7 with an AS path of 6-{2,4}. If AS6 engineers its outgoing traffic between its providers AS 2 and AS 4, then it could advertise prefix X/Y with an AS path of 6-{2,4,3}-1 to indicate that it can use different routes to reach X/Y. Since this advertisement covers both available paths, an iBGP change inside AS 6 would not need to be propagated to AS 7. By using this aggregation, AS 7 receives a less precise AS Path than without the aggregation. This might be a small price to pay compared to the benefit of reducing the number of eBGP messages. Other techniques can be used to obtain the exact AS-path to reach a given destination [127, 126, 136]. In practice, a small to medium transit AS

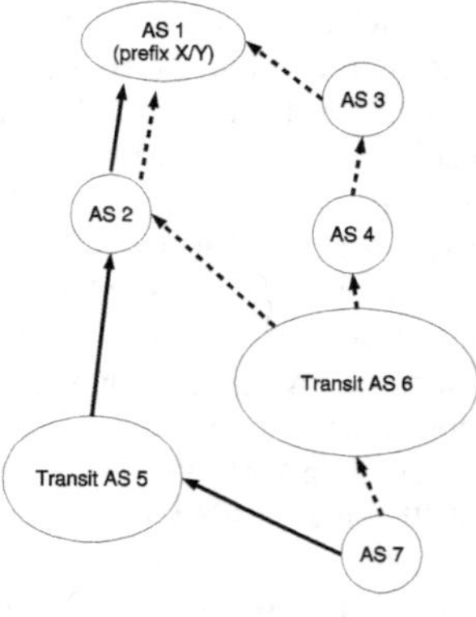

Figure 7.12: Interdomain traffic engineering with BGP for transit ASes

would probably only use an on-line traffic engineering technique to control the traffic sent via its providers and not the traffic sent to its peers. The proposed aggregation should only be used for routes learned via providers. This can easily be achieved by marking those routes with a special BGP community value [165].

Another concern with traffic engineering techniques is that local objective functions might impact other aspects of the end-to-end paths followed by IP packets like path length [167, 178, 172] or path performance [106]. For instance, an AS optimizing the traffic balance or some cost function over its inbound or outbound traffic without respect to AS path length might increase the length of the AS-level or IP-level path to reach some destinations. Hence care must be taken when optimizing particular objectives to prevent some traffic engineering function from impeding on other important aspects of the flow of the interdomain traffic.

7.4 Conclusion

In today's Internet, for both cost and performance reasons, Internet Service Providers often need to engineer their interdomain traffic. This interdomain traffic engineering is often done by manually tweaking the configurations of the BGP routers on an error-prone trial-and-error basis.

In this chapter, we have proposed a systematic approach to solve this important operational problem in stub ASes. Our approach allows the network operator to define objective functions on the interdomain traffic. Those objective functions are used by an optimization box placed inside the AS that controls the traffic.

For this, it relies on the BGP routes and the traffic statistics received from the border routers of the AS. Based on the traffic statistics from the last period, the optimization box predicts the traffic for the next period. Then, it uses an efficient evolutionary algorithm to select the required iBGP changes to optimize the traffic during the next period. Those iBGP changes are then distributed inside the AS.

We have evaluated the performance of our evolutionary algorithm through simulations with real traffic traces from two stub ASes. Our simulations with two different objective functions show that the outgoing traffic can be efficiently engineered on a 10 minutes timescale by advertising not more than a few iBGP messages per minute. This is lower than the BGP noise in the Internet. Moreover, our prototype implementation written in Perl can be used in real-time since it requires only a few seconds of CPU on a P4 2.4 GHz to select the required iBGP changes for each 10 minutes period.

Finally, although our technique was designed for stub ASes, we have discussed how a different type of AS path aggregation could be used in small to medium transit ASes to allow them to engineer the traffic sent to their providers or peers without sending eBGP advertisements.

Chapter 8

Multi-objectives interdomain traffic engineering

In this chapter, we extend the problem of interdomain traffic engineering with BGP for stub ASes with as few BGP filters as possible. We study the optimization of two simultaneous objective functions, one defined on the short-term traffic dynamics and another on the daily traffic demands. We propose a multi-objective evolutionary algorithm aimed at sampling the objective space and illustrate its working with realistic traffic billing procedures. Note that part of the simulation results and figures of this Chapter are taken from [192], with the kind permission of Imperial College Press.

In section 8.1, we describe the problem we deal with in this chapter. Section 8.2 presents our multi-objective evolutionary algorithm on which we rely to study the problem of multiple-objectives interdomain traffic engineering by tweaking BGP. Section 8.3 then presents the scenario of the simulations used to evaluate the problem described in section 8.1. Finally, section 8.4 presents the results of our simulations.

8.1 Problem statement

The problem this chapter tries to answer is the one of studying the trade-offs between the following three objectives :

1. the number of BGP filters,
2. the cost or balance of the traffic over the available providers on long timescales,
3. optimize the traffic balance over the available providers on short timescales.

The first objective concerns BGP routing. Given that there are many remote ASes with which an AS exchanges traffic on timescales of hours to days (see Chapter 4.5), one cannot require that engineering the interdomain traffic needs hundreds of BGP filters or even more. An objective of an interdomain traffic engineering technique should be to minimize as much as possible the burden on BGP. Tweaking the value of local BGP attributes should have no impact on the BGP advertisements sent to BGP peers for stub ASes. This does not preclude however the possibility of misconfigurations is the traffic engineering technique requires that router configuration be changed [121].

The second objective is related to long-term interdomain traffic engineering. With the rapid changes in traffic patterns and the significant cost of Internet connectivity, multi-homed ASes will become more sensible to the usage of their upstream providers in the future [12]. For this, BGP could be used to achieve long-term traffic engineering to optimally use their upstream capacity.

Finally, the third objective is more or less related to Quality of Service and network performance. From the description of "route optimization" techniques, it appears that their objective is to find the best performing (e.g. in available bandwidth or delay) path for each destination, based on both active and passive measurements [15]. While we have assessed the feasibility of optimizing the daily traffic demand of stub ASes in Chapters 5 and 6, as well as of shorter-term traffic engineering in Chapter 7, it is unclear how well long-term and short-term traffic engineering work together.

An important concern for interdomain traffic engineering is to know whether long-term and short-term objectives are conflicting, and how much BGP filters would be required to achieve a "good" traffic engineering, namely minimizing the traffic objectives while not putting too much a burden on BGP. We would thus like to find out whether the three objectives are conflicting, neutral or harmonious [149]. This knowledge would tell us what we can realistically expect from traffic engineering when dealing with several objectives of a different nature like the three presented in this section. Note that we shall recurrently use the term "traffic objectives" for the second and third objectives related to the interdomain traffic.

8.2 BGP tweaking

Given that we do not wish to prefer some objective compared the other two objectives, we would prefer a search technique that will not bias the search towards subset of the three-dimensional search space. We know however that the sole parameter over which we have control in the setting of our problem is the BGP rout-

ing. The only thing we can do with BGP is to manipulate the BGP routes[1]. Due to the way the BGP decision process chooses the best route towards a destination, we have ample choice concerning how to tweak the BGP routes. Nonetheless, we decided in this chapter to tweak the value of the `local-pref` attribute. For this, we rely on what we call a "BGP route change". A BGP route change is a pair *<prefix,provider>* indicating that the traffic having the prefix *prefix* as destination will be forwarded through provider *provider* among the providers which advertised a BGP route towards this prefix. What we call a filter hence concerns a single BGP prefix. The actual effect of a BGP route change on the BGP routes is to force the value of the `local-pref` attribute of the BGP route towards prefix *prefix* which was learned via provider *provider* to be set with a value of 110, hence a higher preference. All other routes for prefix *prefix* (learned from other providers) have their `local-pref` attribute (re)set to the same value which will be strictly smaller than 110.

Our previous work in trying to find good BGP route changes ([188, 197] and Chapter 6 showed us that the tie-breaking rules of the BGP decision process were the crucial part of tweaking BGP for interdomain traffic engineering. Most routes received by a non-transit AS from its upstream providers have a similar AS path length. The rules of the BGP decision process (see section 1.4) that are situated after the AS path length will decide about the best BGP route. If a non-transit AS does not propagate the IGP metric into the BGP routes, the seventh rule of the decision process that will choose the best BGP route, namely the smallest `router-id` of the `BGP next-hop` used to attain the destination, or the lifetime of the particular route depending on the particular router vendor's BGP implementation. This means that the "tie-breaking" rules of the BGP decision process, especially the cost towards the closest exit point [183], will often bias the choice of the best route towards a given neighbor.

The previous discussion concerning the BGP decision process is not just a BGP technical detail. This aspect is of utmost importance because it implies that the complete search space permitted by BGP is vast since several rules of the BGP decision process could be modified by a BGP filter. However, our choice in this chapter is to try to understand the relationship between short-term and long-term interdomain traffic engineering and the BGP filters. Here our focus is on tweaking BGP in the most controllable way. In that respect, the `local-pref` attribute provides us with the certainty that setting a higher value of this attribute for the route of a given provider compared to its value for other provider's routes will force traffic towards this destination to get through this provider. We might also use MED as in Chapter 7 to prevent the traffic engineering to override the

[1]This is actually not true, we could also change the way the BGP decision process works, but this is outside the scope of this thesis.

routing policies implemented by the AS.

8.2.1 The search space

Having explained the main problem-related issues of our problem, this section is dedicated to the problem of finding the Pareto-optimal points for our three objectives. Given that the search is limited by the granularity of a BGP filter, we must now decide how we shall search the space of our three objectives. Several methods can be envisioned.

The first one consists in enumerating all possibilities. This solution is unmanageable for the simple reason that the size of the search space is too large. Assuming there are about one thousand ASes with whom the local AS exchanges a significant amount of traffic each day among the $25,000$ ASes that compose the Internet, and assuming that there are only two providers having each one route towards each destination of the one thousand networks, the number of possible BGP configurations for as few as 10 BGP route changes amounts to $\frac{1000!}{10!990!}$ possible combinations. This number has 23 decimal digits. We thus need to rely on another method to sample the Pareto-optimal front. Note that the number of possibilities is due to our restriction that several route changes cannot be tried for a given prefix in the course of the search, and thus the order in which the BGP route changes are applied is unimportant since they are independent of one another.

A second method would be to constrain the values of the two traffic objectives not to increase while applying the BGP route changes, for applying a route change should allow to improve in at least one of the two traffic objectives. This approach seems at first appropriate in that it will prevent the search to take place in regions whose BGP route changes seem at first uninteresting. Unfortunately, we cannot guarantee that it will not be necessary to worsen the value of some objective through some BGP route change to further improve in the search and prevent the search to get stuck in a local optimum, in particular if the objectives are conflicting.

Thirdly, we could also work on the traffic objectives by trying local improvements of one of the two objectives and finding out how this relates to the BGP route changes. The issue there concerns the discreteness of both traffic objectives. We cannot search for all solutions that are within some ϵ since this would require that we already know the effect of a BGP route change on the traffic objectives, which will not be true except for simplistic traffic objectives. One must however realize that although we can predict the effect of changing the value of the `local-pref` attribute of some route on the way traffic will be distributed over the different providers over the long-term, the extent of the change in the traffic objectives will depend on the particular objective functions. These objective functions not being forcibly linear and depending on the traffic dynamics on the

short-term, basing the search on the traffic objectives could be even more difficult than the initial problem we aim to solve.

The previous discussion focused on the problem-specific constraints of our search. By now, it should be more or less apparent that the search of the Pareto-optimal front cannot be achieved through a blind search nor a local search in the traffic objectives. For such kind of Pareto-optimal front sampling, we need again an evolutionary algorithms, capable to find many points on the Pareto-optimal front in a single run [86, 57, 64]. Additionally, relying on the "evolutionary" paradigm allows to leverage the mechanisms of population-based search and selection among individuals. The advantages of evolutionary search can be measured by the number of recent studies that rely on such techniques in the literature. A list of references on evolutionary multi-objective optimization maintained by Carlos A. Coello Coello can be found at `http://www.lania.mx/~ccoello/EMOO/EMOObib.html`.

8.2.2 Data structure

The first issue one must tackle when developing an evolutionary algorithm is to think about the data structure used in the search. This aspect is important for two reasons. First, the data structure that represents an individual should allow easy manipulations during the search process. Second, the choice of the data structure must be concerned with the large amount of time an evolutionary algorithm often spends at evaluating individuals. So the structure must be close to the problem at hand, but also efficient for evaluating the fitness of the individual.

The aim of the individual's representation is to allow fast fitness evaluation, while at the same time containing information with respect to the three objective functions so that we do not need to recompute the individual's objective values each time we wish to manipulate it. The presence of the value of the two traffic objective functions of this individual seems valuable to efficiently compute new objective values for additional BGP route changes. The main question however concerns how to encode the BGP route changes defined in section 8.2.

We can choose to directly encode which provider is preferred for each prefix. This would mean that the size of an individual would be in the order of the number of prefixes with whom our local domain exchanges traffic. We know however that there can be up to one thousand prefixes. A less costly choice in terms of the size of the encoding consists of putting in the individual only the prefixes for which the preference is changed compared to the default BGP solution, namely the BGP route changes. The representation of an individual we use is shown on Figure 8.1.

For what concerns the efficiency of the search, it is crucial that one does not have to recompute the complete traffic distribution among the providers when there are many prefixes with whom traffic is exchanged, especially since the num-

Long-term objective value	Short-term objective value	Number of BGP filters	BGP filter 1		BGP filter n

Figure 8.1: Representation of an individual.

ber of BGP route changes is due to be smaller than the number of such prefixes. This is only possible when the BGP route changes have an effect on independent traffic aggregates, the prefixes in our case. If we had to try to further minimize the number of BGP route changes as we did in Chapter 6, then we would have to work on traffic aggregates that leverage the way traffic is distributed over the AS-level topology. Such BGP route changes however would not be independent because several route changes might share a common prefix in the topology. In that case the value of the `local-pref` attribute of the various routes for the corresponding prefixes is not enough to decide about the best route. The full BGP decision process must be run. Putting complexity in the objectives' evaluation is not a good idea since this operation is performed frequently.

8.2.3 Search procedure

Depending on the relationships between the two traffic objectives which might be conflicting, harmonious or neutral, the search on the Pareto-front will have to differ [149]. Recall that we do not know beforehand the relationship between the traffic objectives, and our aim is to find based on our search of which type this relationship actually is. We must enforce that our search method is not biased towards any of the two traffic objectives, in order to sample the search space in the best possible manner.

Our search works as follows. At the first generation, we start with a population of individuals initialized at the default solution found by BGP. At generation zero, all individuals have the same values of the two traffic objectives and contain no BGP route change. At each generation, we use a random local search aimed at improving the current population by applying one additional BGP route change.

Each individual of the population is non-dominated with respect to the other members of the population for what concerns the traffic objectives. Recall that a solution is said non-dominated with respect to a set of solutions if no member of the set has better objective values for all objectives simultaneously. In addition, the current population is always made of individuals having the same number of BGP route changes. At each generation, we parse the whole population and for each individual we try to apply one additional BGP route change. Whenever a BGP route change provides improvement with respect to at least one of the traffic objectives, we accept this improved individual and put it in the set of accepted

individuals. We iterate this procedure until we find a target number of improved individuals, or stop when we have performed a target number of tries (the variable `iter`). Figure 8.2 provides a pseudo-code description of the search procedure. Note that the pseudo-code given at Figure 8.2 describes only one generation, the purpose of variable `iter` is not to count the generations, but to ensure that the search will not loop indefinitely during the current generation.

```
accepted = 0
iter = 0
while ((accepted < MAXPOP) AND (iter == MAXITER)){
 foreach individual k {
    // Trying a random BGP route change
    BGP_route_change.prefix = rand_int_uniform(1,MAXPOP)
    BGP_route_change.provider = rand_int_uniform(1,NUM_PROVIDERS)
    // If effect of route change is improvement accept it
    if (improved(k,filter)){
       accept(k,route_change)
       // Update counter for accepted improved individuals
       accepted++
    } // end if
} // end foreach individual
// Update iteration counter
iter++
} // end while
```

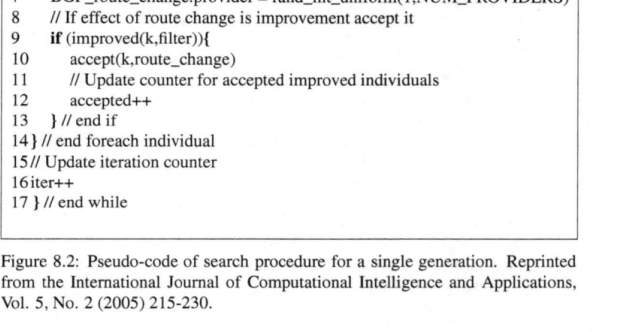

Figure 8.2: Pseudo-code of search procedure for a single generation. Reprinted from the International Journal of Computational Intelligence and Applications, Vol. 5, No. 2 (2005) 215-230.

8.2.4 Pareto-optimal front*

The previous section described the procedure to search for BGP route changes that improve the individuals of the previous population with respect to any of the two traffic objectives. These improved individuals however are not non-dominated. Some of them can be dominated as we did not check for non-domination when accepting an improved individual. Improvement was sufficient to accept an individual. The next step is to check for non-domination [173] on this population of improved individuals to obtain a non-dominated front. For that purpose, we rely on the fast non-domination check procedure introduced in [65]. This procedure

has time complexity $0(MN^2)$ where M is the number of objectives and N is the size of the population. We do not describe this procedure in details but refer to [65] for the original idea and to [64] for a detailed explanation. We only mention the main points here. Let P denote the set of non-dominated individuals found so far at the current generation. P is initialized with any of the individuals among the accepted ones. Then we try to add individuals from the set of accepted ones one at a time in the following way:

- temporarily add individual k to P
- compare k with all other individuals p of P:
  - if k dominates any individual p, delete p from P
  - else if k is dominated by other members of P remove k from P

This procedure ensures that only non-dominated individuals are left in P. The number of domination checks is in the order of $0(N^2)$ while for each domination check M comparisons are necessary (one for each objective). The maximum time complexity is thus $0(MN^2)$.

Having found the non-dominated front for a given number of BGP route changes, we are left with determining the individuals of the next population. Actually, the number of non-dominated individuals from the set of improved ones is due to be smaller than the size of the population we use during the search process (MAX-POP). To make the population for the next generation, we have to decide how many individuals in the next population each non-dominated solution will produce. Because non-dominated individuals are not comparable between one another, we must choose a criterion that will produce MAXPOP individuals from the set of non-dominated ones. On the one hand, we would like to include at least every non-dominated individual in the population. On the other hand, depending on the way the accepted solutions are spread over the non-dominated front, we must sample differently different regions of the non-dominated front for a given number of BGP route changes. This notion of sampling the non-dominated front is close to an idea of distance between neighboring individuals in the objective space. Maintaining diversity on the non-dominated front requires that individuals whose neighbors are farther apart be preferred over non-dominated individuals whose neighbors are close. The rationale behind this is that less crowded regions should require more individuals to be correctly explored than regions having more non-dominated individuals. The computation of the crowding distance for each individual is done according to [64] pp. 248. First the non-dominated individuals are sorted according to each objective. Then the individuals having the smallest and largest value for any objective are given a crowding distance d^m of ∞

to ensure that they will be selected in the population. For each objective m, the crowding distance of any individual i, $1 \leq i \leq (|P| - 2)$, is given by

$$d_i^m = \left| \frac{f_{i+1}^m - f_{i-1}^m}{f_{max}^m - f_{min}^m} \right| \tag{8.1}$$

where f_i^m denotes the value of individual i for objective m, f_{max}^m (respectively f_{min}^m) denotes the maximum (respectively minimum) of the objective value m among individuals of the set P. The global crowding distance for all objectives is the sum of the crowding distance for all objectives. For our two objectives, this crowding distance represents half the perimeter of the box in which individual i is enclosed by its direct neighbors in the objective space.

To achieve our goal of sampling the regions of the non-dominated front according to their crowding as well as to generate a population of MAXPOP individuals from the non-dominated front, we perform a tournament selection based on the crowding distance of each individual. Figure 8.3 provides a pseudo-code describing the selection process. Until we have a population size of MAXPOP,

```
1 pop = 0
2 while (pop < MAXPOP){
3  // randomly choosing two non-dominated individuals in P
4  first = rand_int_uniform(1,|P|)
5  second = rand_int_uniform(1,|P|)
6  // largest crowding distance individual wins
7  if (crowd_dist(first) >= crowd_dist(second)){
8     population[pop] = first
9  }
10 else {
11    population[pop] = second
12 }
13 // incrementing population counter
14 pop++
15 } // end while
```

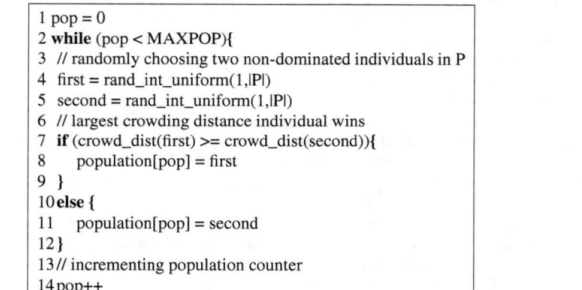

Figure 8.3: Pseudo-code for crowding distance based selection.

we iterate the stochastic tournament (line two of Figure 8.3). A tournament selects uniformly two individuals in the set P (lines four and five of Figure 8.3) and the winner of the tournament is the individual with the largest crowding distance (lines seven to twelve of Figure 8.3).

8.2.5 Practical issues*

While the previous sections provided the main features of our search procedure, several practical issues need to be mentioned. First, in order to sample correctly the search space, a sufficiently large population must be used. By "sufficiently large", we mean a few times larger than the number of possible BGP prefixes. The reason is that when the non-dominated front grows larger, the non-dominated individuals having the largest crowding distance must get the opportunity to have a large number of BGP route changes tried during the next iteration. So the population size must scale up with the size of the non-dominated front in order to be able to progress in the search. On the other hand, the larger the population size, the more CPU time it takes at each generation. With a population limited to $10,000$ individuals, it took up to one day on a P4 2.4 GHz with 1 GBytes of memory to perform one simulation (100 generations). Reducing the population size is dangerous since the sampling of the search space will suffer. The other issue concerns the number of non-dominated individuals found. The larger the non-dominated front, the more time it takes to sort its elements. Although the crowding distance procedure is quite fast $O(MN^2)$ with M being the number of objectives and N being the number of non-dominated individuals, with a non-dominated front made of less than one hundred individuals and two objectives, it is still scalable but when it goes higher than 1000 individuals the crowding procedure becomes a bottleneck. So both the size of the population and the non-dominated front need to be limited for practical purposes.

We used a population size of $10,000$ and limited the non-dominated front size to 1000 individuals. In addition, we do not accept individuals as soon as they improve the individual from which they come, but we perform the non-domination check just after line 9 of Figure 8.2 so that the number of accepted individuals does not grow too large. This prevents individuals that would be discarded during a later non-domination check phase to be accepted in the population of improved individuals. The issue with this way of accepting individuals is that we slow down the search in order to prevent the size of the population of improved individuals to grow too large. If one allows dominated individuals in the population of improved individuals, this will fasten the acceptance of improved individuals, hence a gain in CPU time. On the other hand, not doing this non-domination check at acceptance time permits uninteresting improved individuals to be accepted while they will be discarded later anyway. So in practice some tuning is required to decide if the domination check should be performed during the acceptance of improved individuals or later.

8.3 Simulation scenario

The scenario of our simulations is the same as the one of section 7.2.1. The traffic trace used in our simulations is a one day trace of all the outbound traffic from the Université catholique de Louvain starting from April 2 2003 17:20 CET. During this one day period, 215 GBytes of traffic were observed with an average bit rate of 19.9 Mbps. This one day long trace was cut into time intervals of ten minutes. For each 10 minutes interval, we have one file giving the amount of bytes for every BGP prefix towards which the university sent some traffic.

In addition, we gathered a complete routing information base from Oregon Route views [132] dating from April 2 2003 to simulate the BGP routing tables of different providers to which our stub would be connected. While it is known that BGP routes dynamics can be important for some prefixes, it has however been shown in [157] and Chapter 4 that BGP routes can be considered as stable over relatively long periods, so we use one BGP routing table per provider for the whole day.

8.4 Simulations

Having provided the necessary background to understand the working of our search algorithm, we now proceed to analyze the behavior of the search for real datasets. Our objective is not to validate the algorithm or to prove that our search technique works. The aim of this section is to provide a case study concerning the search of "good" BGP route changes to understand the relationships between the three considered objectives: the number of BGP route changes, the long-term traffic objective and the short-term traffic objective. It should be seen only as an illustrative case study on our particular traffic demand and several traffic objectives. Choices of different traffic demands, objective functions and providers might lead to different results. No conclusion should be drawn from these simulations concerning the quantitative improvements achieved in terms of any of the objectives.

8.4.1 The Broad Picture

Let us first understand how our algorithm performs when we simply sample the objective space without trying to optimize one particular objective, namely when running the basic algorithm described in section 8.2.3. In this section, we use a scenario where our stub AS is connected to four different large tier-1 providers. We relied on the BGP tables of AT&T (AS7018), UUNet (AS701), Verio (AS2914) and Sprint (AS1239). The long-term cost function corresponds to volume-based

billing with costs per traffic unit of 0.5, 1, 1.5 and 2 for UUNet, AT&T, Verio and Sprint respectively. The amount of traffic that a "traffic unit" represents is not important, only the relative cost is. For the short-term objective, we computed for each 10 minutes period the maximum amount of traffic sent through any of the 4 providers, and summed these maxima over the whole day (traffic balancing objective). While this scenario is probably not representative of a real stub AS, since being connected to 4 different large tier-1's is unlikely in practice [176], it can be considered as a difficult setting for our algorithm due to similar AS path lenths for the BGP routes between tier-1 providers for many destinations. The cost functions used in these first simulations need not be the most realistic since our focus here is about understanding the behavior of the search algorithm on a first scenario. More realistic simulations will be carried in the next sections. Note that throughout this chapter we do not consider a prefix if it has less than one megabyte of traffic during a 10 minutes interval in order to limit the number of prefixes that have to be taken into account. This would be done in practice due to Netflow sampling and because removing such small traffic prefixes is marginal with respect to the total traffic. By doing this, we still consider about 90% of the total daily traffic while reduced the number of different prefixes seen over the day from more than $80,000$ to less than 3000.

We plot on Figure 8.4 the non-dominated front found for 100 generations of the algorithm in the previously described scenario. Recall that the successive generations correspond to successive values of the number of BGP route changes. The point corresponding to the default BGP routing has 0 BGP route changes and a value of the two traffic objectives equal to 1. We normalized the objective values of all points with respect to the value found by the default BGP routing (with 0 BGP route change) in order for the graphs to be independent of the actual value of the objective functions that depend on the amount of traffic considered, the length of the time interval, and the particular costs defined for the objective functions. Because our focus here is on the look of the non-dominated front and not the actual numerical values of the objective functions, we decided not to show any actual value of the traffic objectives but only normalized ones.

Globally, we see two regions on Figure 8.4. The first region concerns point for the first few BGP route changes. These points start at the top right of Figure 8.4 (default BGP solution) and converge to the non-dominated front which constitutes the second region of the graph. The second region of the non-dominated front indicates that the two traffic objectives are conflicting for a number of BGP route changes larger than 20. The conflicting nature of the objectives can be seen by a relatively linear (slightly convex) trade-off between the two traffic objectives, for a given number of BGP route changes. Finding a solution providing a smaller cost on the long-term for a given number of BGP route changes requires to worsen the short-term objective value. In the same way, finding a solution providing a

smaller value of the short-term objective function for a given number of BGP route changes requires that one worsens the value of the long-term objective function. But this is valid only for a sufficiently large number of BGP route changes.

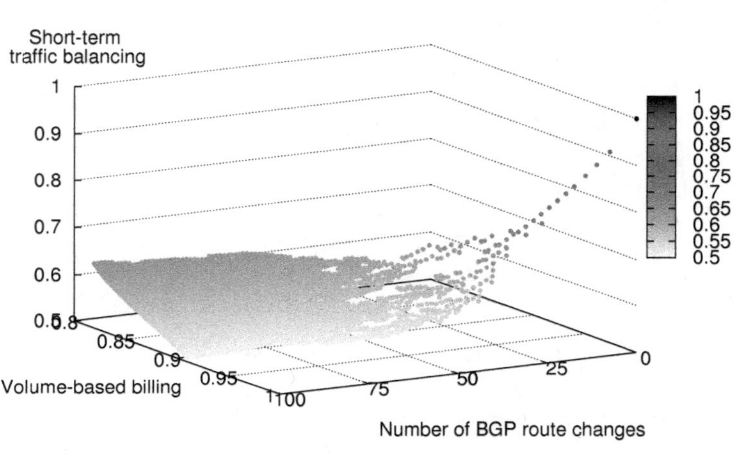

Figure 8.4: Solutions of the search without preference for any of the two traffic objectives. Reprinted from the International Journal of Computational Intelligence and Applications, Vol. 5, No. 2 (2005) 215-230.

The 3 dimensional plot of Figure 8.4 does not show precisely the value of each objective for a given number of BGP route changes, so Figure 8.5 provides the projection of the points of the non-dominated front on each traffic objective. On Figure 8.5, we can see that for the first 20 BGP route changes, the non-dominated front for a given number of BGP route changes is limited to one or two points. This means that all other solutions found for this number of BGP route changes were dominated by the points shown on Figure 8.4. In addition, Figure 8.5 showing the evolution of the values of each traffic objective as a function of the number of BGP route changes for the non-dominated front tells us that the first few BGP route changes allow a significant gain in terms of both traffic objectives at the same time. There are two explanations for this behavior.

First, BGP is a reachability protocol which was not designed to perform traffic engineering. The main feature of BGP is policy routing, which is aimed at

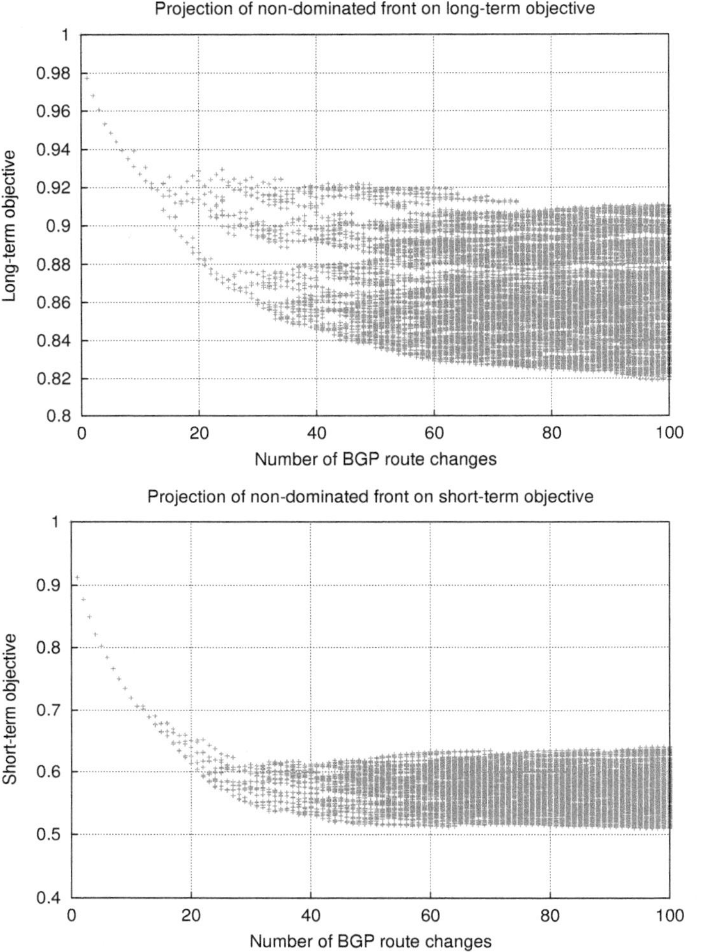

Figure 8.5: Projection of non-dominated front on each traffic objective (no preference among traffic objectives).

enforcing local policies in terms of routing. This is in contrast to our traffic engineering objectives, which do not care about routing policies but the distribution of the traffic on the available providers of the local AS. Being able to improve quite importantly the two traffic objectives means that this part of the search mainly counteracts the "inability" of the default BGP choice to optimize objective functions defined on the traffic. We insist that this "inability" of BGP to optimize traffic objectives is rather normal, in that BGP was not designed to do that. Relying on BGP route changes to perform interdomain traffic engineering really is about "tweaking" BGP. Getting deeper into the working of the BGP decision process, we observed in Chapter 6 that the tie-breaking rules of the BGP decision process were extremely important to explain the difficulty for BGP to optimize a cost function defined on the traffic. The reason is that many prefixes have the same `local-pref` and AS path for the different providers of many stubs. [31] has compared the BGP routes of 20 tier-1 providers based on the route-views BGP tables [132], and found that about 60 percent of the routes had the same AS path length when comparing any two tier-1's. Thus it is likely that the tie-breaking rules of the BGP decision process are an important factor when trying to optimize a traffic objective by tweaking BGP, although it will depend on the different types of providers each stub is connected to.

Second, the size of the population used in our search also impacts the quality of the solutions found during the first few generations of the search. We used a population of $10,000$ individuals for 2832 prefixes having more than 1 MB of traffic during at least one 10 minutes time interval for the considered day. From this population of $10,000$ individuals, we allowed the number of solutions improving the individuals of the current population to be arbitrary large. Each time a new individual improved the existing solution in the population from which it was improved, we checked for non-domination in the set of already improved and non-dominated individuals. If the newly found individual was non-dominated, we included it among the non-dominated front to prevent the same individual to be included more than once. Note that we are not certain that the non-dominated points we find for these small numbers of BGP route changes are globally optimal with respect to the three objectives. The size of the search space, even for a few BGP route changes, increases too fast for that. For instance, with 4 providers and 2832 prefixes that can be reached through any of the 4 providers, we have a worst case of $\frac{(2832\times3)!}{10!((2832\times3)-10)!}$ different possible solutions for ten BGP route changes that could be applied, a number with 32 digits. Even for a limited number of BGP route changes, the size of the search space grows quite fast. For the few first BGP route changes, it is possible to enumerate all solutions, but this cannot be done for up to 20 BGP route changes.

8.4.2 Biasing the search towards one traffic objective

In the previous section, the algorithm was used to sample the non-dominated front for successive numbers of BGP route changes. This provided a broad view of the relationship between the two considered traffic objectives. If one is interested in obtaining better solutions in terms of one traffic objective only, then a small change has to be made to the algorithm. When trying to sample the Pareto-optimal front, we aim to select solutions for the next generation that sample as broadly as possible the non-dominated front found at the current generation. When trying to find good solutions for a given traffic objective, we want to preferably select individuals based not on a crowding distance but on the value of the traffic objective we are interested in. So the only change to be made to the algorithm described in section 8.2.1 is within the selection procedure, where the individual that wins the stochastic tournament is the one having the best objective value for the traffic objective on which we focus.

Figure 8.6 provides the non-dominated front found with the selection process favoring the long-term traffic objective (top) and the short-term traffic objective (bottom). It is apparent on Figure 8.6 that when preferring one particular traffic objective, the sampling of the non-dominated front is poor compared to the one found when a crowding distance is used (Figure 8.4). If we compare Figure 8.6 with Figure 8.4, we can distinguish two different fronts: one that samples the better values of the short-term objective and another that samples better values of the long-term billing. We can also see those two fronts on the top of Figure 8.5. When there is no bias against any of the two traffic objectives, the selection tries to sample non-dominated solutions, whether or not they would have a poor value in one of the traffic objectives. This re-assures us that the non-biased search is working properly. For what concerns the results of the search biased towards a specific traffic objective, we expected the observed behavior since selective pressure at each iteration favors the preferred objective only, hence obtaining a representative non-dominated front for the two objectives will not happen whenever the two objectives are conflicting.

Figures 8.7 and 8.8 provide the corresponding projections of the non-dominated fronts onto each traffic objective. The solutions found in terms of the particular traffic objective towards which the search was biased are not particularly better than those found with the crowding distance based selection (Figure 8.5). Only for large numbers of BGP route changes do these biased search techniques find better solutions in terms of the preferred objective. Even when we prefer one particular traffic objective, the aim of the search is still to understand the relationship between the two objectives so our search does not guarantee that we find the best solutions in terms of the preferred objective. If one wants to obtain the best route changes in terms of one particular objective, then a single-objective optimization

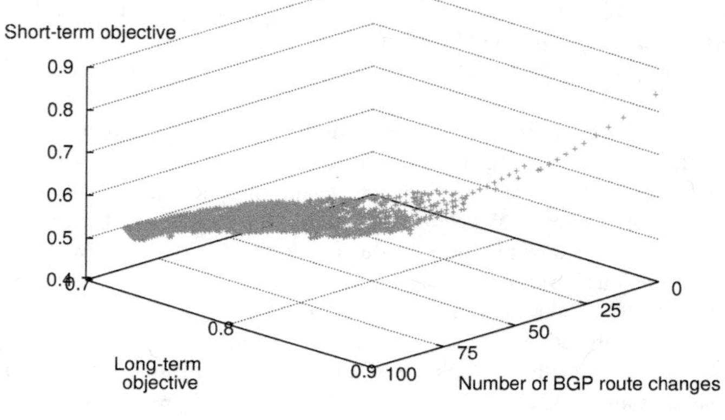

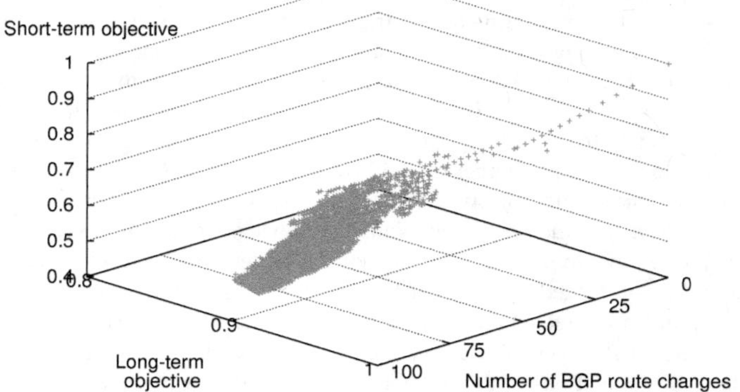

Figure 8.6: Solutions of the search for long-term preferred objective (top) short-term preferred objective (bottom).

technique should be used.

When one compares the best solution in terms of the each traffic objective found with and without bias, the biased search finds slightly better solutions, but at the price of driving the search towards BGP route changes having a bad value of the other traffic objective (compare top of Figure 8.7 with top of Figure 8.5 and bottom part of Figure 8.8 with bottom of Figure 8.5).

The non-dominated front found when biasing the search towards one particular traffic objective indicates that the conflicting nature between the two traffic objectives used is not an artifact of the sampling of the non-dominated front we performed through the crowding distance based selection of section 8.4.1. The two objectives are really conflicting since when sampling the "best" solutions in terms of the short-term traffic objective, we find that for a given number of BGP route changes, solutions having a better value of the short-term objective have a worse value of the long-term objective.

When we compare the points found by the algorithms with and without bias towards any traffic objective, we also find that the best 15 BGP route changes are found by the three methods. This constitutes further indication that the first non-dominated points found for the few first BGP route changes are not locally sub-optimal. For additional BGP route changes on the other hand, the search space is so large large search space that it is likely that any search method would have a hard time finding the optimal solution, while the expected gain is probably quite small. As said previously, the more BGP route changes to be applied the smaller the expected gain in terms of any traffic objective, because of the distribution of the traffic among the prefixes. Only relying on some additional traffic aggregation by leveraging the knowledge of the AS-level topology might provide larger traffic shifts with a single BGP route change. Chapter 6 found that that the gain of leveraging the AS-level topology to reduce the number of BGP route changes works for a limited number of BGP route changes, and to approach the optimal value of one traffic objective will require a significant number of BGP route changes, no matter the search technique. Drawbacks of taking into account the AS-level topology are the increased size of the search space and the overlap among the traffic influenced by different BGP route changes. We do not believe that global optimization techniques will provide significantly better results than our heuristic. It might however be an interesting further work to carry out.

Based on the simulations of this section, we feel confident that the crowding distance based method works well with conflicting traffic objectives since it finds good BGP route changes and it seems to sample well the non-dominated front. This is not forcibly valid for other traffic objectives. We investigate this aspect in the next section.

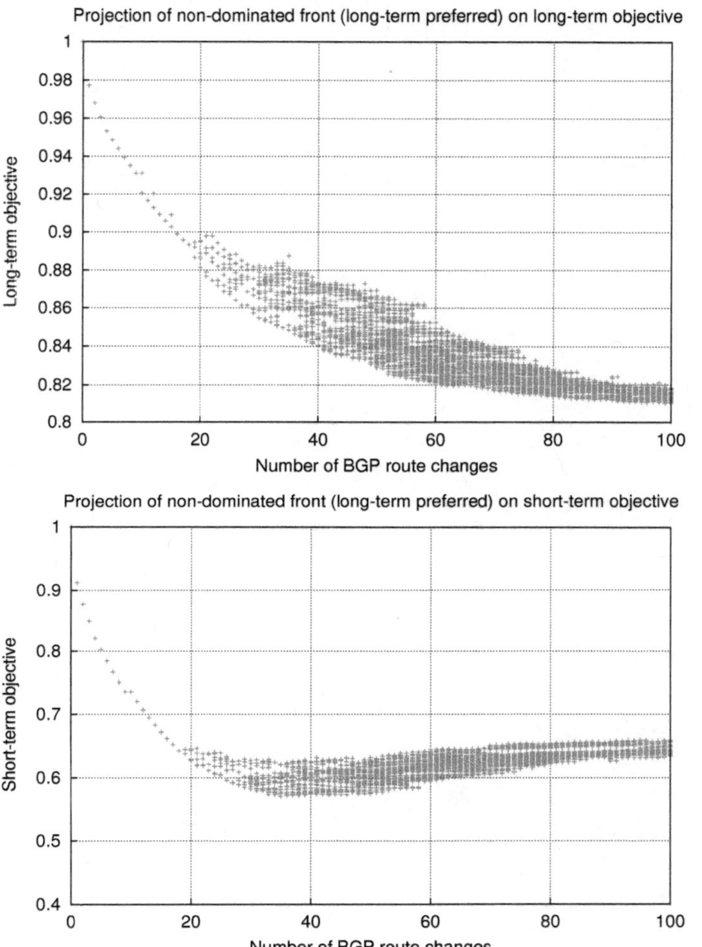

Figure 8.7: Projection of non-dominated front on each traffic objective for long-term preferred objective.

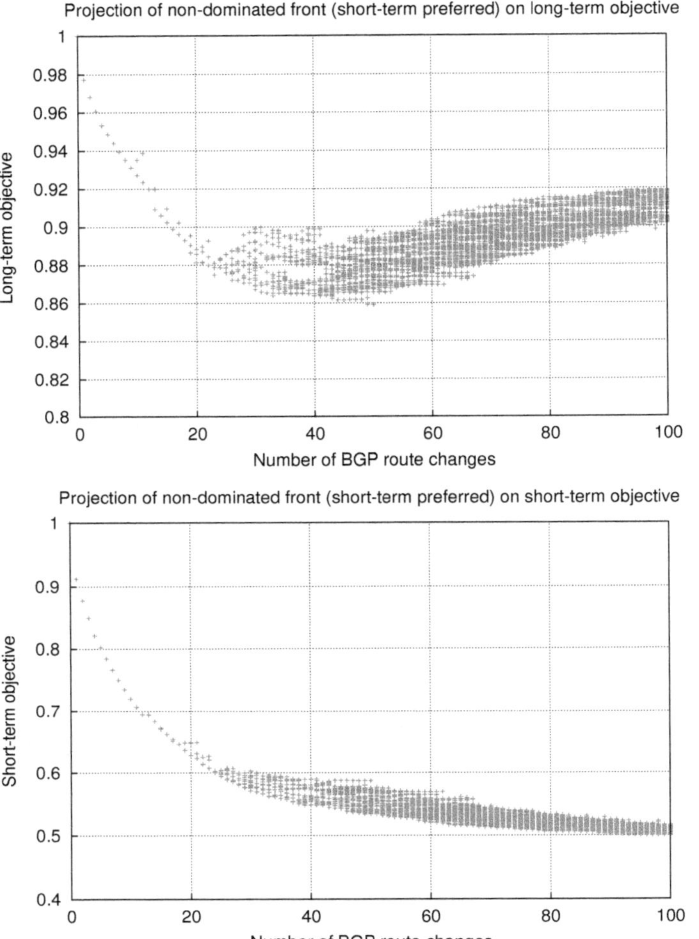

Figure 8.8: Projection of non-dominated front on each traffic objective for short-term preferred objective.

8.4.3 Percentile-based billing

Up to now, we have used one short-term and one long-term objective function which are not those that are most likely to be used in practice. In this section, we compute a few non-dominated front for more realistic traffic objective functions to confirm the behavior of our algorithm. Note that in practice the short- and long-term traffic objectives need not be conflicting, so the shape of the non-dominated fronts could be different than those found in the previous section.

Volume-based billing is a very simple long-term cost function. In practice, billing is often done according to a 95th percentile of the average bandwidth (in Mbps during each time interval) computed over equal time intervals of a few minutes. We use 10 minutes time intervals in this section. We also use different bandwidth prices for the commitment and the traffic above the commitment, and different prices for the different providers. For the short-term traffic objective, we still use the load balancing objective that tries to minimize the sum over each short-term time interval of the maxima of the traffic sent through any provider. In addition, we change the different providers available by reducing their number to 3 and choosing three different types of providers : a tier-1 (AS1239), a tier-2 (AS1668) and a tier-3 (AS11608) provider. The reason for this choice is to limit the effect of the tie-breaking roles of the BGP decision process when relying on tier-1s only. The classification as tier-1, tier-2 and tier-3 for these ASes comes from [176].

There was a total traffic for the considered day of about 215 gigabytes, a little less than 20 Mbps on average. We use a value of 7.5 Mbps for the commitment for each provider so that we can see during the search how many BGP route changes are needed in order for the 95th percentile to be under the commitment. If the search is to find BGP route changes that allow the 95th percentile to be under the commitment, then the long-term traffic objective cannot be improved, so only the short-term objective can improve from that point. Any traffic above 7.5 Mbps on average over the whole day gets billed 50 percent more than the price below the commitment, for any provider. The traffic below the commitment gets billed using a flat rate, each provider bills the commitment whether or not the traffic is well above it or equal to it.

The non-dominated front for the long-term percentile-based traffic objective is provided on Figure 8.9. We see on Figure 8.9 that there is no smooth non-dominated front even for a large number of BGP route changes, in contrast to the results of the long-term cost objective of section 8.4. The reason is the stochastic nature of the percentile-based objective that largely depends on the short-term dynamics of the traffic. The value of the 95th percentile depends on the distribution of the values of the traffic for each provider and each short-term time interval. Changing the provider used to carry the traffic for some prefix has a non-trivial

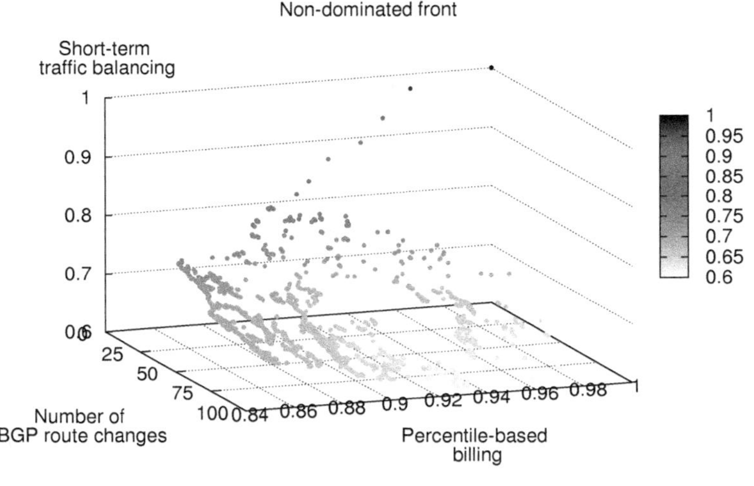

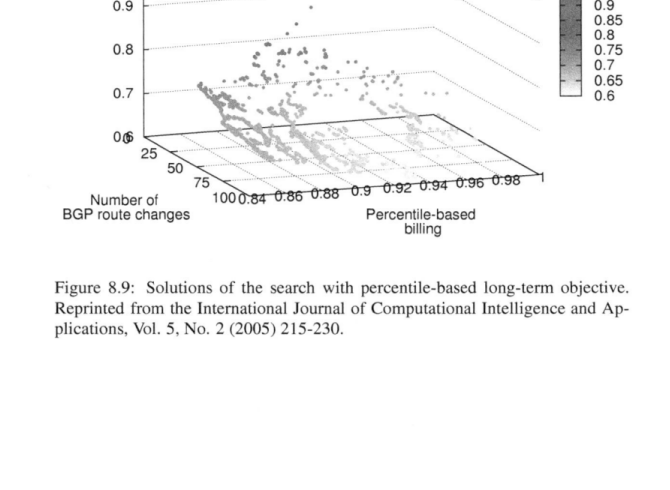

Figure 8.9: Solutions of the search with percentile-based long-term objective. Reprinted from the International Journal of Computational Intelligence and Applications, Vol. 5, No. 2 (2005) 215-230.

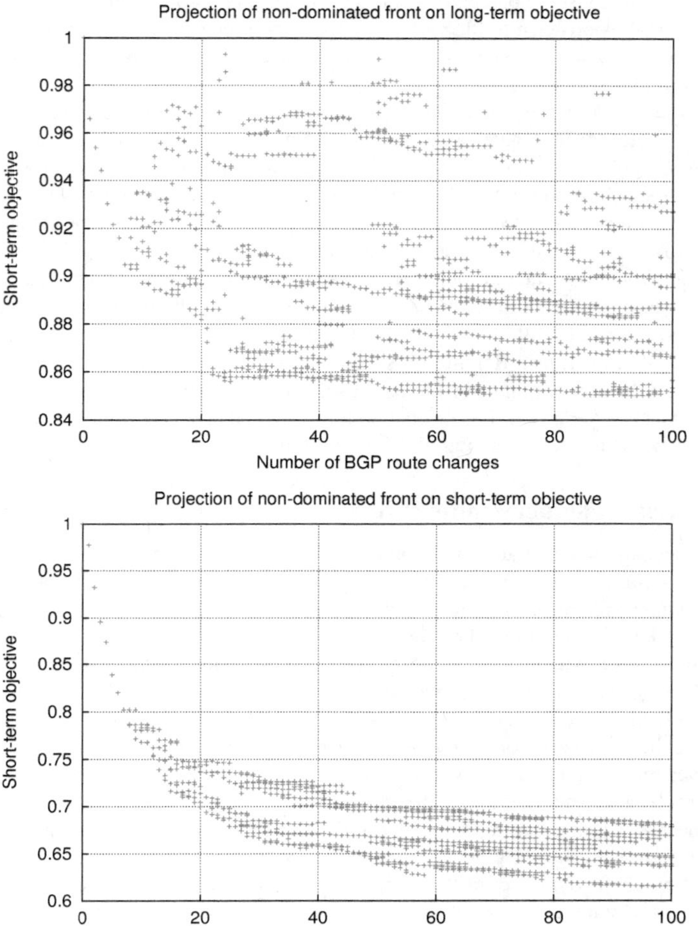

Figure 8.10: Projection of non-dominated front on each traffic objective for percentile-based long-term traffic objective.

effect on the value of the percentile. A cost function as used in section 8.4 is insensitive to the short-term traffic variability for some prefix, while the percentile is. A percentile-based cost function is hence a difficult long-term traffic objective to sample and optimize.

Figure 8.10 shows the projection of the non-dominated front for the percentile-based long-term traffic objective on the two traffic objectives. The projection on the long-term objective (top of Figure 8.10) illustrates quite well the difficulty of sampling evenly the non-dominated front for the percentile-based objective. The stochastic nature of the percentile-based objective prevents our search to be able to find improved individuals at successive generations. Correctly sampling the non-dominated front with such an objective is likely to be an interesting challenge. For what concerns the short-term load-balancing objective, the projection of the non-dominated front (bottom of Figure 8.10) has a look more similar to those of the previous simulations. Relying on several sub-populations each focusing on a subpart of the non-dominated front could solve the problem of finding the best values for a given objective. It is difficult to foresee whether such a solution would actually provide a better sampling of the non-dominated front.

8.4.4 Load-balancing objectives

The last scenario we evaluate in this chapter is the same as the previous one except that the long-term objective is load balancing. This scenario should be the easiest one for what concerns the long-term traffic objective (see Chapter 6), but we have no a priori knowledge of its relationship with short-term load balancing, because of the short-term dynamics of the traffic for prefixes. A long-term cost function and a short-term load balancing objective might appear at first conflicting since minimizing cost tends to move as much traffic towards the lower cost providers, hence making the load among providers uneven. Load-balancing as the two traffic objectives on the other hand might appear harmonious, so at least improving the long-term balance should improve the short-term one on average too. In the same way, one might think that improving the short-term balance is likely to improve the long-term balance. In practice, the validity of these intuitions depends to a large extent on the dynamics of the traffic over the short-term, which varies from one network to another.

Figure 8.11 shows the non-dominated front for the two load-balancing objectives. We do not have a nicely looking front, but there are few non-dominated points for each value of the number of BGP route changes. Most of the gain is once more achieved through the first 20 BGP route changes. The next twenty BGP route changes still provide a non-negligible gain in terms of both traffic objectives. However, after 40 BGP route changes the search appears to be stuck. It could be due to an actual difficulty of improving the traffic objectives, or to the size of the

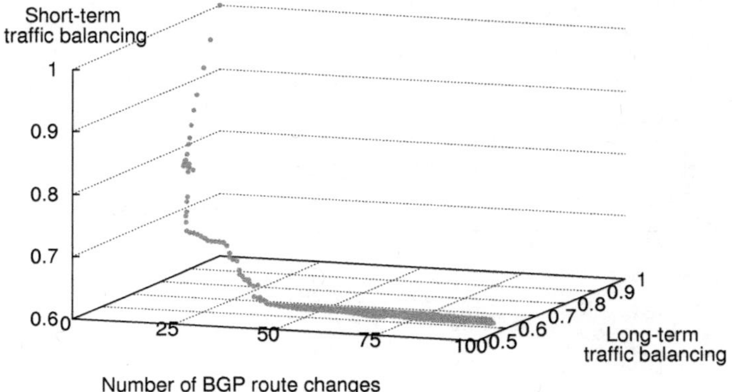

Figure 8.11: Solutions of the search for load balancing objectives. Reprinted from the International Journal of Computational Intelligence and Applications, Vol. 5, No. 2 (2005) 215-230.

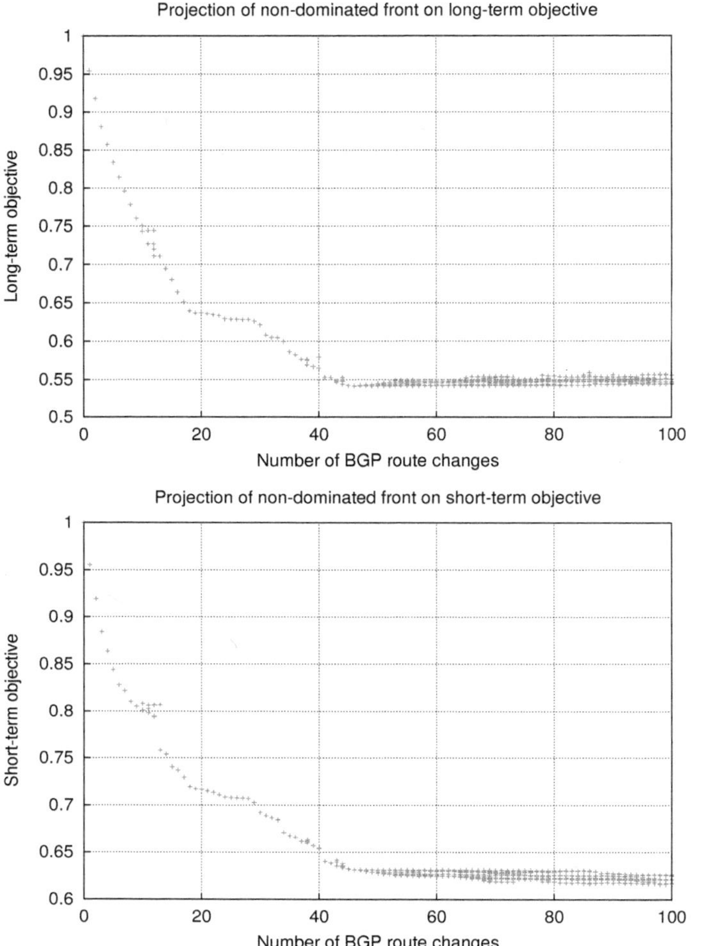

Figure 8.12: Projection of non-dominated front on each traffic objective for load balancing objectives.

search space that prevents our algorithm to find improvements as well.

Figure 8.12 provides the projections of the non-dominated front on the two traffic objectives. The absence of a well-spread non-dominated front indicates that the two load-balancing traffic objectives are indeed not conflicting. If they were conflicting we would be able to find a non-dominated front since we sample both objectives without biasing the search towards any of them. This simulation thus indicates that load balancing on both the short and longer timescales is possible even through a limited number of BGP route changes.

8.5 Conclusion

In this chapter we proposed to rely on evolutionary algorithms to evaluate multiple-objectives interdomain traffic engineering. The search consisted in finding the BGP routing changes in order to optimize two objective functions defined on the interdomain traffic.

To evaluate this problem, we relied on a real interdomain traffic trace and BGP routing tables from Oregon route-views. We studied two simultaneous traffic objectives, one short-term objective and one long-term objective, and evaluated different instances of the traffic objectives.

The search showed that the problem is difficult. The different objectives of this problem can be conflicting and relate to different timescales of Internet traffic. In addition, realistic objectives can be stochastic, depend on the dynamics of the Internet traffic and be interdependent. This chapter confirmed the results of the Chapter 6 showing that traffic balancing objectives are easily optimized. We showed in section 8.4.4 that in theory it would be possible to balance the traffic both on a daily and a hourly timescale at the same time, relying on a limited number of BGP route changes (about 40 in our simulations). For daily traffic objectives like a volume-based billing cost or percentile-based billing cost on the other hand, these objectives might be conflicting with short-term traffic balancing. Furthermore, realistic traffic objectives like a percentile-based billing cost are difficult to optimize due to their dependence on the short-term dynamics of the traffic.

This chapter confirmed the issue of interdomain traffic engineering, namely the BGP decision process and especially the tie-breaking rules that make the balance of the traffic over several BGP neighbors particularly uneven. This aspect is an issue for traffic engineering tools because it means that the optimization of some traffic objective is likely to spend valuable time trying to counteract "bad" choices made by the tie-breaking rules of the BGP decision process.

Part IV

The future of interdomain traffic engineering

In Chapter 9, we evaluate the current state of interdomain traffic engineering and intent to foresee its future. We discuss the current limitations of the interdomain Internet that impede on the deployment of interdomain traffic engineering. We point to further work on the Internet architecture that would ensure that the interdomain Internet be better designed for tomorrow's requirements. We also point for some further work in interdomain traffic optimization. Chapter 10 closes this book by summarizing the content.

Chapter 9

Evaluation and future work

Throughout this book, we have tried to look at interdomain traffic engineering from completely different perspectives. In the first part, we took the protocol-centered viewpoint of BGP. We presented in this first part the different aspects of the working of BGP that were important to perform interdomain traffic engineering. In the second part, we took the viewpoint of the interdomain traffic by analyzing its properties. Finally, in the third part we took the viewpoint of optimization by trying to evaluate the optimization of the interdomain traffic with BGP. The purpose of this chapter is to discuss the interrelationships between the different chapters of this book and their implications on the feasibility and the future of interdomain traffic engineering.

9.1 BGP

Part 1 discussed the difficulty of tweaking BGP to perform interdomain traffic engineering. It illustrated the complexity of the BGP decision process (see Figure 1.2) and how the interactions between its different rules make the tweaking of the BGP routes a non-trivial problem. Several issues concerning the BGP decision process must be discussed here. Chapter 1 discussed the difficulty of interdomain traffic engineering but only with respect to the BGP attributes and the BGP decision process. In this section we give this discussion a broader perspective.

The first problem with the BGP decision process, which is also the most important feature of BGP, is related to policy routing. The first rule of the BGP decision process emphasizes the importance of local preferences on the choice of the best BGP route. Each AS decides which next hop AS it will use (among those possible by its available BGP routes) for IP packets entering the local domain and having as destination a prefix outside the local AS. This choice in BGP implies that the end-to-end AS path between a source and a destination depends highly on

which ASes on the path modify the `local-pref` attribute of the BGP routes. The impact of `local-pref` on the choice of the best BGP route by each intermediate AS on the end-to-end path is significant [172] and highly affects the routing diversity inside ASes [201].

The second rule of the BGP decision process uses the shortest `AS path` length to choose the best route. While this might seem an appropriate distance metric for the routes, the number of AS hops have a limited meaning, especially in a dense network like the Internet. First, some networks do not have an AS number. To obtain an AS number, one must be connected to the Internet through two different providers. Networks who are not multi-connected cannot obtain an AS number and the AS path to attain them from other ASes will henceforth be shorter than AS paths of networks that do have an AS number. Some large transit providers have several AS numbers. ISPs can merge or split. Finally, the path across each AS depends on the diameter of the AS at the IP layer (in terms of router hops). Using the AS path length as a distance metric is thus questionable [106].

Next, the two rules concerning MED and IGP cost to the next-hop rely on local metrics. As explained in Chapter 1, these metrics represent an intradomain cost. For the IGP cost, it is the cost of the path from the local router towards to the exit point in the local AS. For the MED attribute, it is often the IGP cost of the path that crosses the neighboring AS to reach the destination. These costs concern the local AS or its next AS hop, so these costs have only a local meaning. Recent studies have shown that a large fraction of the routes in a tier-1 AS are selected based on the IGP cost rule [183].

The seventh and last rule of the BGP decision process (Figure 1.3) ensures that only one route will be selected for each destination prefix. In the context of large transit providers for which the previous rules are due to decide the best route [183], the last rule of the BGP decision process should not be heavily used. Generally speaking, we do not know which rules of the BGP decision process are most heavily used to decide about the best routes in the Internet. It depends on how much ASes actually tune their BGP routes to follow the best intradomain path according to the IGP routing or how they implement policy routing [117]. For stub ASes on the other hand, this last rule as it is currently implemented systematically biases the choice of the best route towards the one whose `BGP next-hop` has the smallest IP address [156, 50, 54]. This rule implies that if the set of BGP egress points is relatively small and the set of different routes for the same destination prefix that are still non-differentiated by the BGP decision process is often the same, then this last rule will not distribute evenly the best routes among the various equivalent egress points. A better choice for this last rule of the BGP decision process would have been to rely on a hash function that would choose the best route in a more "random" way in order to prevent a

systematic bias towards the same `BGP next-hop`. The choice of the last rule of the BGP decision process however requires care to prevent unwanted transitions between external peers [50].

From the previous remarks, we have realized that the attributes of the BGP routes and the rules of the BGP decision process were questionable from an interdomain traffic engineering viewpoint. What is often considered as the really important features of BGP are policy routing and the route filtering mechanism. The BGP decision process aims at ensuring that only one route for each prefix will be chosen in every domain. The purpose of policy routing is to permit to prefer some routes over others for each prefix [174, 115, 156, 45]. This feature was explicitly considered as important for all interdomain routing protocols [174, 115, 156, 45]. This mechanism must be differentiated from route filtering and route redistribution [174, 115]. [45] explicitly discussed the question of restricting route information distribution. In [45], the reasons given for limiting the redistribution of the routes were to hide information and limit resource consumption. However, no study evaluated these aspects to the best of our knowledge. Route filtering allows the local AS not to accept some BGP routes. This possibility also means that the discarded routes cannot be redistributed to other peer ASes hence a potential information loss about the AS-level topology. In addition, the process of the redistribution of the BGP routes currently permits that only one route be announced to each BGP peer router for each prefix. This further restricts the spread of the topological information contained in the BGP routes and limits the diversity of the routes known inside an AS [201]. The problems solved by policy routing and route filtering are different. Policy routing is aimed at influencing the local choice of what the local AS considers to be the best route to reach a given destination. For this problem, nothing else than locally changing the attributes of the BGP routes is required. Filtering on the other hand solves the problem of topological information hiding and BGP attributes manipulation (AS path prepending, MED, communities). These two problems need not be coupled, at least not in the working of the interdomain routing protocol as it is the case currently with BGP [78].

9.2 Interdomain traffic dynamics

Chapter 2 studied the timescale-dependent dynamics of the interdomain traffic of stub ASes. Self-similarity and scaling have been quite extensively studied in the context of networking [145, 71]. After more than a 15 years of research since the first paper on network traffic self-similarity [120], this topic is still misunderstood. There seems to be a gap between the understanding of the mathematical concepts and the physical emergence of scaling in network traffic [209]. [59]

was the first to propose a physical explanation for self-similarity in Web traffic through the distributional properties of the flows. [182] then showed that a physical explanation could be found in the rise of self-similarity in network traffic through an aggregation process with independent ON/OFF sources and heavy-tailed ON/OFF distribution times. [182] formally proved the possibility for the presence of self-similarity in the traffic without dependence among the traffic sources. This however did not prove that self-similarity in the traffic **is** due to heavy-tails in the ON/OFF times distribution of the sources. However, [209] showed that the ON/OFF model is able to generate self-similar processes of different types like fractional Brownian motion and alpha-stable processes [163, 111, 179], that seem to match the behavior of network traffic [181, 134].

[194] showed that over timescales between minutes and hours, the sample path properties of the process of the number of IP sources, prefixes and ASes that are active at any given instant also constitute a self-similar process on a one week trace of all the incoming traffic of a stub AS. The implication of [194] is that independent of the assumptions on the dependence between the traffic sources and their ON/OFF times durations, the simple fact that the time evolution of the number of sources (at different aggregation levels) is a self-similar process is sufficient for self-similarity to arise in the total traffic. [198] and Chapter 2 also showed that the process of the TCP flow arrivals had properties that could also explain the presence of self-similarity in the traffic. It is important to understand that [194] does not question the ON/OFF model. [210] confirmed that the ON/OFF model is likely to be right at the source level, i.e. for source-destination pairs at the IP level. What [194] showed is that independent of the validity of the ON/OFF model for source-destination pairs, the process of the number of network prefixes and autonomous systems that send traffic to the local network is sufficient to cause self-similarity in the traffic. This self-similarity could in turn be due to an ON/OFF model at the level of the network prefixes and autonomous systems. This is to be investigated in the near future, because it could imply that the Internet topology is partly involved in the emergence of self-similarity at the source level.

The "true" cause of self-similarity in the traffic might not exist. This might seem a disturbing answer but sometimes looking for physical explanations can be wrong at times [38]. Some properties of complex systems can be "emerging", in the sense that they are properties of the system itself as a whole, not of some identifiable parameters of the system. Scientists tend to think in terms of causes and effects, but this way of thinking can be simplistic. Whenever some protocol or control variable impacts the system in some way, then one can study the relationship between this protocol or variable and the dynamics of the system. Causes and effects have a meaning in that case, since there can be a functional relationship between the system and its variables. In the case of a protocol, one can study the impact of the state machine defining the behavior of the protocol and the

behavior of the system. This is because the state machine acts according to well-defined rules. In the Internet on the other hand, the traffic is generated by users (humans or machines) that do not follow precise rules or whose interactions are too complex to analyze. Therefore, the reason for the presence of self-similarity in the traffic might be the distributional properties of the flows as well as simply the behavior of the users, as we found in [194]. If such is the case, then it is possible that self-similarity in network traffic is an invariant that does not depend on the distributional properties of the flows but rather on the behavior of the users whose ensemble behavior generates a self-similar process. This discussion is not mere philosophical discussing about self-similarity, but has far-reaching engineering implications. If self-similarity in the traffic is actually an emerging property of the way users are active as a whole, it means that self-similarity is not an engineering problem because it is not rooted in the protocols or the properties of the various applications. This would mean that controlling the protocols to prevent self-similarity to arise will not work. Self-similarity would then have to be accepted as an inevitable fact for traffic engineering.

9.3 Topological distribution of interdomain traffic

Chapter 3 contributed to our understanding of how the traffic is distributed on the AS-level topology, as seen from two stub ASes. It showed that the ASes with which the stubs communicated were not far away in terms of the AS hop distance, between two and four hops away. We also saw that the traffic gets splitted in a tree-like manner, with limited traffic aggregation occurring in the AS-level topology, mainly at the peerings between the local stub and its provider, but not beyond. The results of the traffic analysis provided in Chapter 3 could be regarded as highly dependent on the particular traffic traces we used.

To assess the representativeness of the results of Chapter 3, we used the AS-level topology of [176] based on several BGP routing tables and computed the shortest AS path length between any two stub ASes of the topology. Figure 9.1 presents the distribution of the AS path length for the $11,976$ stubs among the $14,695$ ASes of the topology built with BGP tables dating from January 9, 2003. There were $30,815$ distinct inter-AS links in this topology. Among the $11,976$ stub ASes that the topology included, 4889 were single-homed and 7127 were multi-homed. The bars labeled "shortest path" on Figure 9.1 show the distribution of the AS path length of the shortest AS paths from every stub AS of the AS-level topology to every other stub AS. The "shortest path" in the topology does not take into account the routing policies applied by the intermediate ASes, so the shortest AS path from one stub to another is the shortest AS path that could be found given the available inter-AS edges available in the topology.

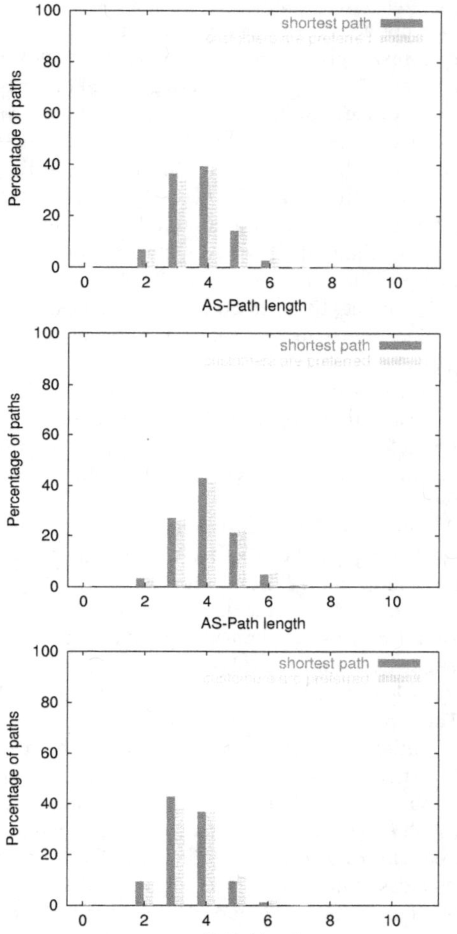

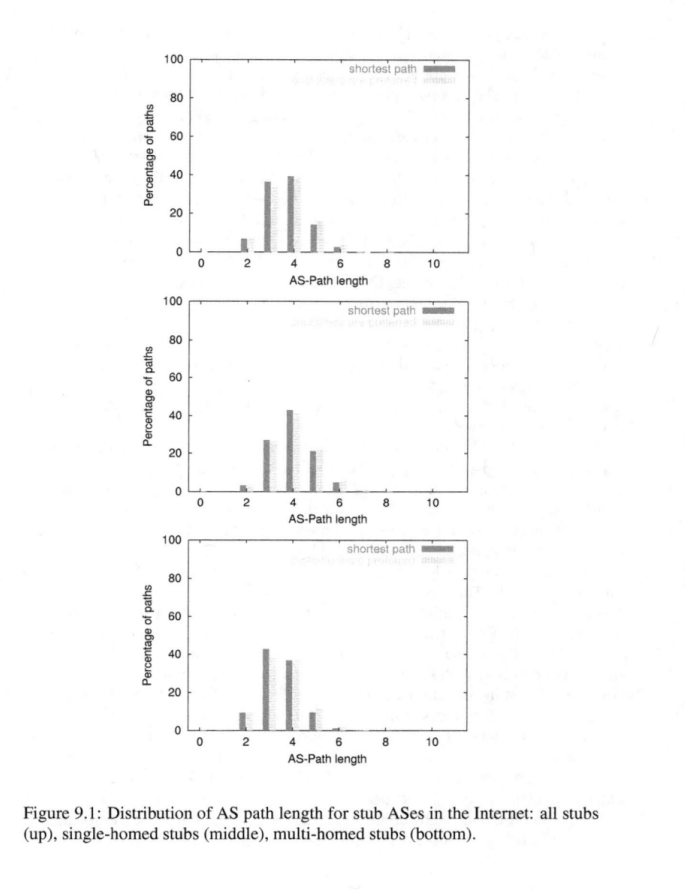

Figure 9.1: Distribution of AS path length for stub ASes in the Internet: all stubs (up), single-homed stubs (middle), multi-homed stubs (bottom).

The topology provided by [176] corresponds to the AS paths seen in the BGP routing tables gathered from several vantage points mainly in the core of the AS-level topology, these are not necessarily the real AS paths that would be used by each stub [136]. The AS paths given in the BGP tables are only valid for traffic sent by the ASes from which the BGP tables were gathered towards the destination prefixes of the routes of the BGP table. The reverse path however might not be valid. Because we cannot hope to obtain the BGP routing tables from all stubs of the Internet, we simulated the use of a common routing policy by all ASes [100]: a BGP route learned through a customer is preferred to a route learned through a peer, the latter being preferred to a route learned through a provider. [176] classified the peering relationships between the ASes in the AS-level topology as either "customer-provider" or "peer-to-peer", which is a simplification of realisty but still roughtly correct [137]. Based on this classification, we simulated the preference of customer routes to peer routes to provider routes. Each stub announced a single prefix to its neighbors. The advertisement of the BGP routes across the whole topology were simulated with the whole BGP decision process that chooses its best route according to the previously mentioned preferences. The bars labeled "customer preferred" on Figure 9.1 provide the distribution of the AS path length from each stub to each other stub of the topology from [176] when applying the "customer preferred" routing policy.

The top graph of Figure 9.1 shows the distribution of the AS path length for all stubs of the topology, while the middle graph shows it for the 4889 single-homed stubs and the bottom graph for the 7127 multi-homed stubs. For the middle and bottom graphs of Figure 9.1, we show the distribution of the AS path length for the AS paths used to attain every single-homed and multi-homed stub from every other stub AS (both single and multi-homed).

The typical AS path length is between 2 and 6, with most AS paths having length 3 or 4. The top graph of Figure 9.1 shows the limited effect of the "customer-preferred" routing policies on the distribution of the AS path lengths. The distribution of the "shortest paths" shows that the typical AS path length is 3 or 4. AS path lengths smaller than 2 or larger than 6 are rare. The effect of the routing policies is to slightly increase the AS path length compared to the shortest AS path, with an increase in the percentage of the AS paths of length 5 and 6 and decreases in the percentages of the AS paths of length 3 and 4.

The middle and bottom graphs of Figure 9.1 show the difference between single-homed (middle) and multi-homed stubs (bottom). Compared to the distribution for all stubs (top), single-homed stubs have a smaller percentage of AS paths of length 3 while a larger percentage of AS paths of length 5. The distribution for multi-homed stubs on the other hand is similar to the one of all stubs (top), with only slightly more AS paths of length three and less AS paths of length 4 and 5.

The previous simulation shows that the results of Chapter 3 are not just a bias due to the choice of our stub ASes. The structure of the Internet is mainly responsible for this topological distribution of the traffic. What happens is that stub ASes get their Internet connectivity through tier-1, tier-2 or tier-3 providers, and it is mainly how stubs are connected to the Internet that explains the distribution of the AS path lengths seen by the interdomain traffic. Most AS paths contain two parts, the first part goes from the source AS to the core of the Internet (tier-1 providers), and then from the core to the destination AS. Since most ASes in the Internet are stubs, the interconnection of the stubs to the core prevails over the interconnection structure of the core.

We computed for these simulations the fraction of the AS paths due to the different levels of the hierarchy. We removed from the AS paths the first and last AS, and computed how many times an AS from each level of the hierarchy appeared in this "transit" part of the AS path, when the routing policies preferring customer routes over peer routes over provider routes. Since there were $14,695$ ASes in the topology, we had $14,695 \times 14,694$ AS paths. We had on average 2.73 ASes in this transit part of the AS paths, among which about 42 % of tier-1's, 23 % of tier-2's, 24 % of tier-3's, and 10 % of tier-4's. Almost half of the transit part of the AS paths belong to tier-1's.

To illustrate on an example the impact of the hierarchical structure of the AS-level topology on the AS path length between ASes, Figure 9.2 presents a simplified view of the AS-level hierarchy. We show on Figure 9.2 four levels of the hierarchy. The core is made of tier-1's, to them are connected tier-2's, then the next level are tier-3's and finally stubs. This view is of course a simplification of reality, as a significant fraction of AS relationships are not observed [47, 101, 139].

Consider the stub AS $S1$ as the local AS. $S1$ is directly connected to a tier-1, $P1$. $P1$ is connected to three providers: another tier-1 $P2$, two tier-2 providers $P3$ and $P4$. $P1$ has also a customer stub AS, $S2$. Stub AS $S3$ is connected to the Internet through tier-2 provider $P3$. $S4$ is a dual-homed stub AS connected to the Internet through a tier-1 ($P2$) and a tier-2 provider, $P4$. Finally, stub AS $S5$ is connected to the Internet through a small tier-3 provider, $P5$. If we compute the shortest AS path length between stub $S1$ and the other stub ASes, we have the following:

- $S1 - S2$: AS path is S1-P1-S2, length = 2,
- $S1 - S3$: AS path is S1-P1-P3-S3, length = 3,
- $S1 - S4$: AS path is S1-P1-P2-S4, length = 3,
- $S1 - S5$: AS path is S1-P1-P4-P5-S5, length = 4.

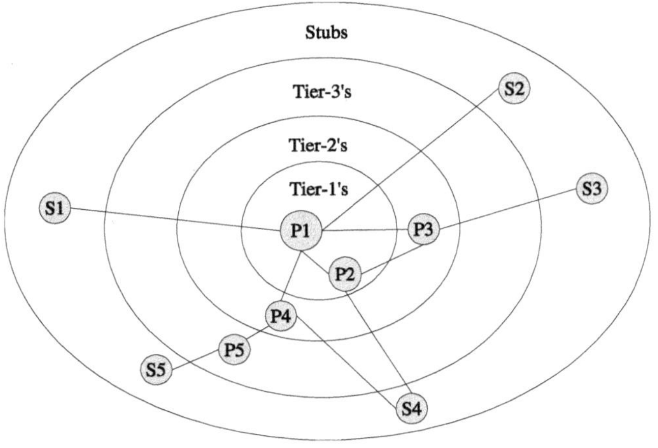

Figure 9.2: Impact of hierarchical structure of the AS-level topology on typical AS path length.

On the simple example of Figure 9.2, we can see why based on the AS-level topology of [176] most stub ASes are reachable within 2 to 4 AS hops depending on how they are connected to the Internet.

The previous simulation indicates that if the structure of the Internet does not change fundamentally and the ASes with whom a stub communicates with samples randomly the other stub ASes of the topology, then on average the topological distribution of the traffic will have the same look as the distribution of the AS path length shown on Figure 9.1. It appears that the structure of the Internet is slightly changing the last few years, with more ASes trying to bypass the core [95].

The topological properties of the traffic are not simply facts that are interesting to know. For outgoing traffic engineering, the control of the routes is performed locally, so the topological properties are not overly important even if we shall see in the next section that it has implications on the tweaking of BGP routes. For incoming traffic engineering on the other hand, the properties of the topology are extremely important. Influencing the incoming traffic requires that one be able to influence the source as well as the intermediate ASes on the path up to the local AS [152]. For incoming traffic engineering to be feasible, one must understand where the traffic is coming from over the interdomain topology and how policy routing and AS paths determine how the interdomain traffic crosses the AS-level topology. First, the limited AS hop distance between the local AS and interdomain traffic sources requires that the control information aimed at tweaking

the path of the incoming traffic travels across a few domains. This relatively short distance between the local AS and those to be influenced is good news. However, this short AS path distance is also a problem when one wants to rely on AS path prepending for inbound traffic engineering [152].In AS path prepending, a multi-homed AS will announce different routes to its different peers. AS path prepending consists in inflating the AS path announced to a particular peer with the objective of switching part of the incoming traffic from this link. Take the example of AS $S4$ on Figure 9.2. $S4$ has two providers, $P2$ and $P4$. Suppose that $S4$ performs AS path prepending by adding its AS number one more time in its route announced to $P2$ hoping to reduce the load of this link. The effect of this prepending is that $P2$ announced to $P1$ and $P3$ that it can reach $S4$ through a route with an AS path length of 2. The shortest AS paths for each stub is indicated on Table 9.1. Before prepending, $S1$ and $S2$ had similar AS path lengths through

Table 9.1: Effect of AS path prepending on shortest AS path and provider used to attain S4.

Source	Shortest AS path before prepending	Shortest AS path after prepending
S1	S1-P1-{P2|P4}-S4	S1-P1-P4-S4
S2	S2-P1-{P2|P4}-S4	S2-P1-P4-S4
S3	S3-P3-P2-S4	S3-P3-{P2-S4|P1-P4}-S4
S5	S5-P5-P4-S4	S5-P5-P4-S4

$P2$ and $P4$. The choice of the actual provider used to attain $S4$ hence depends on the choice of the next AS hop by $P1$. After prepending, $S1$, $S2$ and $S5$ send their traffic to $S4$ through $P4$ while for $S3$ it depends on the choice of the next AS hop by $P3$. So in this case, the effect of prepending is that at least three among the four sources attain $S4$ through $P4$.

If $P1$ prefers the routes learned from $P4$ to those by $P2$ because $P4$ is a customer while $P2$ is a peer, then $S4$ receives the traffic from $S1$, $S2$ and $S5$ from $P4$ while the traffic from $S3$ from $P2$. The net effect of prepending is then either null if $P3$ prefers $P2$ to $P1$ or to force all traffic to be received through $P4$ if $P3$ prefers $P1$ to $P2$.

If on the other hand $P1$ does not prefer $P4$ to $P2$ (if both are peers) but the BGP decision process decides that the routes announced by $P2$ are better, then before prepending the traffic from $S1$, $S2$ and $S3$ is received though $P2$ and the traffic from $S5$ through $P4$. After prepending, the traffic from $S1$, $S2$ and $S5$ is received through $P4$ and the traffic from $S3$ through $P2$.

The simplicity of the example contrasts with the complexity of the effect of AS path prepending due to the BGP decision process. Incoming traffic engineering depends on routing policies, on the particular values of the route attributes (for the tie-breaking rules) and the set of available alternative paths known by each AS along the path between the AS that performs prepending and the traffic sources.

The most important issue with BGP is the filtering mechanism that hides much of the information about the AS-level topology to BGP routers. This information is necessary to predict the effect of BGP tweaking on the incoming traffic distribution of stubs. It is an open issue whether relying on current models of the choice of the paths is sufficient for inbound traffic engineering [126, 136].

The limited aggregation of the traffic on the AS-level topology also raises the question of the granularity of the route control. Relying on AS path prepending or not announcing routes to some peers have a binary effect: either it does not work at all or it works by switching almost all traffic from one upstream provider to another [31, 152]. If one requires a finer control of the incoming traffic, a better way to go could be through communities [30]. As explained in section 1.5.3, the community attribute is an optional attribute of the BGP routes that allows an AS to influence the filtering process of the BGP routes of the BGP neighbors. This mechanism however requires mutual agreement between the ASes. This mechanism could in theory provide a finer control of the incoming traffic, by allowing ASes to indicate to their upstream providers the routing policy to be applied or the tweaking of the BGP attributes to be performed upon redistribution of these routes to their BGP neighbors. Nevertheless, communities are not a panacea. Communities merely defer the route control problem one hop away. It allows the BGP peers of the local AS to perform the BGP route manipulations on behalf of the local AS. If the BGP peer performs the BGP routes manipulation defined by the values present in the communities of the routes, this will have no effect on the routing policies applied upstream in the AS-level topology. So even if communities provide a finer control of the inbound traffic, it is not certain that they constitute a solution to this problem. For fine-grained interdomain traffic engineering, cooperation between ASes should be the way, for example as proposed in [151].

9.4 Topological dynamics of interdomain traffic

Chapter 4 presented a study of the dynamics of the traffic on the AS-level topology, in the case of two stub ASes. Generalizing the results of this chapter is not possible, because it could depend on the context of the studied traces like the users profile, the applications, the upstream peers of the stub,... Obtaining representative results in this area is unlikely to be possible, because traffic trace collection is

very time and resource consuming. The three main points of this chapter are the following:

1. the confirmation of [157] concerning the stability of the BGP routes for the traffic,

2. the observation of Chapter 3 concerning the limited traffic aggregation on the AS-level topology,

3. the observation that a significant percentage of the total traffic has stable AS paths while another significant part has unstable AS paths.

The first point requires little comments. The reason for observing stable AS paths probably does not go deeper than the relative stability of the routing policies that do not to change very fast. ASes do not dynamically change their best route towards a destination unless some failure happened in the network. Dynamic traffic engineering is not common, especially through BGP, even though per-prefix path choices by certain ASes are becoming more common [137].

Point 2 has already been discussed quite extensively in the previous sections. The limited aggregation of the traffic observed by stub AS on the AS-level topology is a consequence of the highly structure of the AS-level Internet. The dense core and the apparently less dense edge of the AS-level topology explain why stub ASes do not see much traffic aggregation except on the peering links. ASes in the core on the other hand see more aggregated traffic due to their centrality in the topology.

Point 3 requires some more explanations, especially to properly understand its implications on traffic engineering. "Stability" as mentioned in point 3 refers to the concept of "stability in presence" defined in section 4.6.3. In that section we counted the number of one hour intervals during which any AS path was present among the largest ones, for instance those AS paths that capture 90 percent of the hourly traffic. Given that the AS-level topology is assumed to be stable for the traffic, it is important to know what fraction of the traffic one can rely on to perform traffic engineering without too much uncertainty on the amount of traffic one actually is able to influence. To perform traffic engineering, one needs to know what amount of traffic will cross which outgoing or incoming link. If one has a high confidence on the distribution of the traffic that will be received on the incoming links for, say 80percent of the total traffic, but one has no idea of through which incoming links the remaining twenty percent of the traffic shall come, then information about from what part of the AS-level topology this traffic comes can be useful. Depending on the probability that some traffic might arrive through some AS path, one can compute the likelihood of the distribution of the remaining 20 percent of the traffic on the incoming links. If only long-term

traffic engineering is performed, traffic uncertainty might not be a problem. If on the other hand traffic engineering focuses on strict QoS guarantees or simply on optimizing a particular cost function that requires that a target traffic distribution among the ingress or egress links be ensured, the problem is different. In a best-effort Internet, these issues do not matter. Whenever tight traffic engineering objectives need to be met, a precise understanding of the behavior of the traffic is required.

9.5 Optimizing traffic objectives

Throughout Part III, we tried to assess the feasibility of optimizing traffic objectives by tweaking BGP, for stub ASes. That part uncovered the extent of the issues that lie within BGP if it is to be used to perform interdomain traffic engineering. Because the various chapters of Part III already discussed in details the implications on traffic engineering with BGP, we only add a few comments in this section. In this thesis, we only evaluated the optimization of objectives for outgoing traffic engineering in stub ASes. In the case of outgoing traffic, the local AS has full control over the manipulation of the BGP routes. For incoming traffic, the lack of control on the traffic makes us feel uncomfortable with the term "control". If traffic engineering means tweaking the path followed by the traffic, then inbound traffic engineering is possible. If however one uses the term traffic engineering as "traffic optimization" as we tried to do in Part III, then one should realize that even for outgoing traffic this task is far from being easy. Chapter 7 however provided encouraging results concerning the feasibility of on-line traffic engineering the outbound traffic, in the case of a traffic balancing objective. Note that in Part III, we have dealt with stub ASes only. Similar outbound traffic engineering was performed on a small transit AS (GEANT) in [200].

Throughout Part III we observed that optimizing a simple traffic objective while trying to keep the burden on BGP low was not simple. The tie-breaking rules of the BGP decision process were shown to have an important influence on the "inoptimality" of the default way BGP chooses its best routes and distributes the traffic among the available egress points. BGP is not to blame since the BGP decision process works on the set of routes towards a given prefix, not on the total traffic. Our feeling after having carried out the studies of Part III is that BGP should not be used to perform long-term traffic engineering. If BGP is to be used for traffic engineering, then either the purpose should be to compensate the uneven distribution of the traffic over multiple links due to the BGP decision process or to perform on-line traffic engineering.

BGP was designed as a reachability protocol that provided various features among which policy routing is one. Most metrics present in the BGP routes are

local administrative costs. `Local-pref` enforces strict preference among the routes. MED and IGP cost allow intradomain metrics to be taken into account. The `AS path` length has a limited meaning [106]. After spending much time trying to perform traffic engineering with BGP, we can safely claim that BGP is certainly not a proper tool to perform traffic engineering, even though it is possible to use it for that purpose.

It must be noted that the techniques used in part Part III are relatively weak from a pure optimization viewpoint, in that the optimality of the solutions found might, in theory, be relatively far from the true optima. In practice, we are confident that the solutions found by our algorithms are indeed good, because the main limitation in the optimization of traffic objectives comes from the distribution of the traffic over the interdomain sources or destinations, not the search technique. Our aim however was not at determining the optimal solutions but rather to assess how difficult it was to improve objective functions by tweaking BGP. Even though further work can be done in this area, we believe that the importance of optimality is marginal compared to the networking-related issues.

9.6 The future of interdomain traffic engineering

Before discussing the future of interdomain traffic engineering, let us first deal with its present. Except for a few commercial solutions whose internal working is opaque [15], interdomain traffic engineering is a very immature field. Contrary to intradomain traffic engineering that deals largely with technical objectives [87], interdomain traffic engineering deals mostly with economical and strategic objectives. Traffic billing and peering agreements are complex and difficult to evaluate [48]. If in the future well-defined technical objectives need to be achieved at the interdomain level, we expect that interdomain traffic engineering will become more popular. Having to rely on BGP though is an impediment from which interdomain traffic engineering might suffer.

If interdomain traffic engineering is a requirement for the future Internet, then either BGP must be replaced with an interdomain routing protocol that provides the necessary information, or tools must be developed to ease the deployment of interdomain traffic engineering. Admittedly, designing a routing protocol is not a simple task [105], and particularly for a large system like the interdomain Internet [103]. Failing to take into account the problems that the protocol is aimed to solve must be prevented. Our opinion is thus that if interdomain traffic engineering is not considered as a requirement for the future interdomain Internet, then the interdomain routing architecture should not be bothered by this objective [70, 69]. The absence of apparent progress in interdomain traffic engineering during the last years is not a good sign.

Few tools exist today for interdomain traffic engineering because few providers ask for it and the few ISPs who do such things do not share such knowledge. Is there a way out? Perhaps. The practice of multi-homing could be this way out. [10] noticed a trend towards more multi-homing during the last few years. Multi-homed ASes need to rely on interdomain traffic engineering. An AS multi-connected to the Internet needs to manage the load and performance of the traffic on its multiple link with its BGP peers. These ASes will ask for better interdomain traffic engineering tools since BGP on its own does not allow to automatically engineer the traffic. Multi-homed stub ASes represent less than half of all stubs in the Internet according to BGP tables from large transit providers [176]. If multi-homed ASes are able to put pressure on the demand for interdomain traffic engineering, then the latter may have a future.

Whatever the way the interdomain routing architecture evolves, we believe that traffic engineering in general has a future. One of the important drivers behind the evolution of the Internet is economics [130]. The viability of traffic engineering depends on its ability to answer the needs of ISPs. Nowadays, many ISPs still rely on over-provisioning as a strategy to prevent changes in the traffic demand to create congestion and increase the delay of the IP packets. This strategy works because the cost of the infrastructure is low enough. If infrastructure cost becomes more critical or changes in the traffic demand cannot be followed by network provisioning, then traffic engineering will become crucial to map the traffic demand to the network infrastructure.

9.7 Further work

As emphasized in the previous section, the two most important further works we envision for the close future are

1. the requirements for the future interdomain routing architecture,
2. developing interdomain traffic engineering tools.

Working on the future interdomain routing architecture is crucial because it will be both the main constraint for interdomain traffic engineering tools as well as its main basis. What information will be available through the interdomain routing protocol will guide the choice of the techniques upon which the interdomain traffic engineering tools will rely. On the one hand, the requirements for the interdomain routing architecture must be carefully studied to know what are the goals and what are the non-goals. Currently, policy routing is a goal of the interdomain routing architecture while accurate dissemination of the interdomain topology is not. Should the interdomain routing architecture be designed in such a way that

better topological information is disseminated and meaningful traffic-related information is included in the topological information? Unless better information is available to the routing plane, traffic engineering will remain a set of ad-hoc tweaks applied by few ISPs.

Working on routing alone is not enough. The four chapters of Part III were nothing more than the first step towards a more thorough evaluation of interdomain traffic engineering techniques. We relied on evolutionary algorithms due to their robustness and flexibility to easily be adapted to the problem at hand. These chapters on the other hand provide only weak optimality bounds on the gains that could be achieved through interdomain traffic engineering. The importance of evaluating the efficiency of various interdomain traffic engineering techniques is probably as important as working on the interdomain routing architecture, because it is by evaluating a system that one learns its strengths and weaknesses.

The amount of information and the ability to control the interdomain routing should ideally depend on the expected gains of more information and more control provided by interdomain routing. A trade-off between optimality, robustness and flexibility should be found. Bringing a system to optimality requires a perfect information about its state and perfect control on its behavior. Optimality cannot be ensured under uncertainty, nor can one expect robustness to be possible if global and centralized control of the system is performed. The situation can be illustrated through comparison of the intradomain and the interdomain routing architectures. At the intradomain level, the knowledge of the topology is good, the link state routing protocols distribute the whole intradomain topology to all routers. Globally optimizing the weights of the links is hence possible at the intradomain level [88, 87]. At the interdomain level on the other hand, a decentralized approach has been used, so that BGP routers only know the next hop AS to be used to reach a given destination and most metrics used for traffic control with BGP have a local meaning only. Globally optimizing the interdomain routing is thus not possible today, but the interdomain Internet is extremely robust. Even though testing the status of reachability is difficult [40], the current impression of the community is that reachability is working fine today. Today the interdomain routing architecture is designed to allow each AS to decide which routes it prefers to attain a destination. If advanced interdomain traffic engineering need to be performed, then ASes will need to allow other ASes to influence their choice of the best route.

Chapter 10

Conclusion

Based on many real traffic traces, we studied the implications of the characteristics of the interdomain traffic on traffic engineering. We evaluated the complexity and feasibility of interdomain traffic engineering with BGP, focusing on non-transit ASes.

Part 1 provided a detailed presentation of the working of the BGP protocol and the interactions between BGP and the control of the interdomain traffic. It discussed the complexity of interdomain traffic engineering with BGP and its causes: the BGP decision process and the uncertainty for controlling the traffic, especially in the inbound direction.

Part II studied the characteristics of the interdomain traffic from a stub AS viewpoint. It showed that interdomain traffic exhibits wide fluctuations and that its variability reduces only on timescales larger than about one hour. The topological distribution of the interdomain traffic for two stub ASes showed that the traffic sent by a stub splits on the AS-level topology within a few AS hops. Traffic aggregation on the AS-level graph occurs only with the direct peers of the stub. Finally, studying the dynamics of the topological distribution of the traffic showed that the AS-level graph seems stable for the traffic sent by a stub. On the one hand, the AS paths for an important fraction of the total traffic had a stable presence among the largest AS paths on a hourly basis. On the other hand, a significant fraction of the traffic had AS paths whose hourly presence among the largest AS paths is unstable in presence.

Part III evaluated the feasibility of interdomain traffic engineering by tweaking the BGP routing protocol. It relied on evolutionary algorithms to optimize objective functions defined on the interdomain traffic. Trying to optimize objective functions defined on the interdomain traffic for stub ASes was shown to be not trivial. We showed that the tie-breaking rules of the BGP decision process were largely responsible for the large distance between the default solution found by BGP and the optimum to be attained. Then, we showed that trying to min-

imize the number of BGP route changes to optimize the cost or traffic balance of the daily traffic over the available providers was difficult. Trying to leverage the aggregation of the interdomain traffic on the AS-level topology to minimize the number of BGP route changes was shown to provide a marginal gain mostly because of the limited aggregation of the traffic on the AS-level topology. Our evaluation of on-line interdomain traffic engineering provided the most positive results, indicating that BGP could be used to traffic balance the traffic while requiring a very limited number of BGP route changes every time interval. Finally, we studied the optimization of two traffic objectives, one defined on the short-term (ten minutes) and the other on the longer-term (one day). This study showed that traffic objectives working on these different timescales could be conflicting. Furthermore, we showed that percentile-based billing is a difficult objective to optimize due to its statistical nature and its dependence on the short-term traffic dynamics which is known to contain a lot of variability.

Part IV evaluated the current state of interdomain traffic engineering, discussed its issues and envisioned its future. It also provided a few guidelines for further work on the interdomain Internet. We discussed the limitations of BGP to perform interdomain traffic engineering. We then discussed the implications of the traffic characteristics on interdomain traffic engineering and showed that the topological distribution of the traffic was largely linked to the structure of the AS-level topology. Hence the topological distribution of the traffic is not due to change in the future unless the AS-level interconnection structure changes. Finally, we discussed two most important further work, namely the interdomain routing architecture and developing optimization methods to perform interdomain traffic engineering.

Bibliography

[1] Abilene NetFlow Statistics. Available at `http://www.itec.oar.net/abilene-netflow/`.

[2] GNU Zebra routing software. `http://www.zebra.org`.

[3] Internet Traffic Archive. `http://www.acm.org/sigcomm/ITA/`.

[4] P. Abry, R. Baraniuk, P. Flandrin, R. Riedi, and D. Veitch. The Multiscale Nature of Network Traffic: Discovery, Analysis, and Modelling. *IEEE Signal Processing Magazine*, May 2002.

[5] P. Abry, P. Flandrin, M. Taqqu, and D. Veitch. Wavelets for the Analysis, Estimation, and Synthesis of Scaling Data. *In Park and Willinger (editors) "Self-Similar Network Traffic and Performance Evaluation", Wiley-Interscience*, 2000.

[6] P. Abry, P. Flandrin, M. Taqqu, and D. Veitch. Self-similarity and Long-Range Dependence through the Wavelet Lens. *In P. Doukhan and G. Oppenheim and M. Taqqu (editors), "Theory and Applications of Long-Range Dependence", Birkhäuser, Boston*, 2002.

[7] P. Abry, P. Gonçalvès, and P. Flandrin. Wavelets, spectrum estimation and $1/f$ processes. In *Wavelets and Statistics, Lecture Notes in Statistics*. Springer Verlag, 1995.

[8] S. Agarwal, C. Chuah, S. Bhattacharyya, and C. Diot. The Impact of BGP Dynamics on Intra-Domain Traffic. Sprint ATL Research Report Nr. RR03-ATL-080677, August 2003.

[9] S. Agarwal, C. Chuah, S. Bhattacharyya, and C. Diot. The impact of BGP dynamics on intra-domain traffic. In *Proc. of ACM SIGMETRICS*, 2004.

[10] S. Agarwal, C. Chuah, and R. Katz. OPCA: Robust Interdomain Policy Routing and Traffic Control. In *Proc. of IEEE OPENARCH*, April 2003.

[11] S. Agarwal and T. Griffin. BGP proxy community community. Internet draft, draft-agarwal-bgp-proxy-community-00.txt, work in progress, January 2004.

[12] A. Akella, B. Maggs, S. Seshan, A. Shaikh, and R. Sraman. A measurement-based analysis of multihoming. In *Proceedings of ACM SIGCOMM*, August 2003.

[13] A. Akella, S. Seshan, and A. Shaikh. Multihoming performance benefits: An experimental evaluation of practical enterprise strategies. In *Proc. of USENIX Annual Technical Conference 2004, Boston, MA*, 2004.

[14] B. Alberts, A. Johnson, J. Lewis, M. Raff, K. Roberts, and P. Walter. *Molecular Biology of the Cell, Fourth Edition.* Garland Science, 2002.

[15] D. Allen. NPN: Multihoming and route optimization: Finding the best way home. *Network Magazine*, February 2002.

[16] C. Association for Internet Data Analysis (CAIDA). CoralReef Software Suite. `http://www.caida.org/tools/measurement/coralreef`.

[17] R. Atkinson and S. Floyd. IAB Concerns and Recommendations Regarding Internet Research and Evolution. Internet draft, draft-iab-research-funding-02.txt, work in progress, October 2003.

[18] P. Aukia, M. Kodialam, P. Koppol, T. Lakshman, H. Sarin, and B. Suter. RATES: A server for MPLS Traffic Engineering. *IEEE Network Magazine, pages 34–41*, March/April 2000.

[19] D. Awduche, A. Chiu, A. Elwalid, I. Widjaja, and X. Xiao. Overview and Principles of Internet Traffic Engineering. Internet Engineering Task Force, RFC3272, May 2002.

[20] D. Awduche, A. Elmalid, I. Widjaja, and X. Xiao. A framework for Internet traffic engineering. Internet draft, draft-ietf-tewg-framework-05.txt, work in progress, May 2001.

[21] D. Awduche, J. Malcom, B. Agogbua, M. O'Dell, and J. McManus. Requirements for Traffic Engineering Over MPLS. Internet RFC 2702, September 1999.

[22] J. E. Baker. Reducing bias and inefficiency in the selection algorithm. In *Proceedings of the Second International Conference on Genetic Algorithms. Lawrence Erlbaum Associates (Hillsdale)*, 1987.

[23] J.-M. Bardet, G. Lang, G. Oppenheim, A. Philippe, and M. Taqqu. Generators of Long-Range Dependent Processes. *In P. Doukhan and G. Oppenheim and M. Taqqu (editors), "Theory and Applications of Long-Range Dependence", Birkhäuser, Boston*, 2002.

[24] S. Bartholomew. The art of peering. *BT Technology Journal*, 18(3), July 2000.

[25] J. Bartlett. Optimizing multi-homed connections. *Business Communications Review*, 32(1):22–27, January 2002.

[26] T. Bates, E. Chen, and R. Chandra. BGP Route Reflection - An Alternative to Full Mesh IBGP. Internet Engineering Task Force, RFC4456, April 2006.

[27] S. Bellovin, R. Bush, T. Griffin, and J. Rexford. Slowing routing table growth by filtering based on address allocation policies. Preprint, `http://www.research.att.com/~jrex`, June 2001.

[28] J. Bendat and A. Piersol. *Random data analysis and measurement procedures (third edition)*. Wiley Series in Probability and Statistics, Wiley InterScience, 2000.

[29] J. Beran. *Statistics for Long-Memory Processes*. Monographs on Statistics and Applied Probability, Chapman & Hall, 1994.

[30] O. Bonaventure, S. D. Cnodder, J. Haas, B. Quoitin, and R. White. Controlling the redistribution of BGP routes. Internet draft, draft-ietf-ptomaine-bgp-redistribution-01.txt, work in progress, August 2002.

[31] O. Bonaventure, P. Trimintzios, G. Pavlou, B. Quoitin, A. Azcorra, M. Bagnulo, P. Flegkas, A. Garcia-Martinez, P. Georgatsos, L. Georgiadis, C. Jacquenet, L. Swinnen, S. Tandel, and S. Uhlig. Internet Traffic Engineering. Book chapter of Cost Action 263 final report, LNCS 2856, Springer-Verlag, pp. 118-179, September 2003.

[32] O. Bonaventure, S. Uhlig, and B. Quoitin. The case for more versatile BGP route-reflectors. Internet draft, draft-bonaventure-bgp-route-reflectors-00, work in progress, July 2004.

[33] S. Borthick. Will Route Control Change the Internet ? *Business Communications Review*, September 2002.

[34] G. Box, G. Jenkins, and G. Reinsel. *Time Series Analysis: Forecasting and Control*. Prentice-Hall, 1994.

[35] J. Branke. *Evolutionary Optimization in Dynamic Environments.* Kluwer Academic Publishers, 2002.

[36] A. Broido, E. Nemeth, and K. Claffy. Internet expansion, refinement and churn. *European Transactions on Telecommunications*, January 2002.

[37] A. Bryzon. *Dynamic Optimization.* Prentice-Hall, 1999.

[38] M. Buchanan. *Ubiquity: the science of history...or why the world is simpler than we think.* Phoenix, London, 2001.

[39] L. Burgstahler and M. Neubauer. New Modifications of the Exponential Moving Average Algorithm for Bandwidth Estimation. In *Proc. of the 15th ITC Specialist Seminar*, July 2002. Wüerzburg, Germany.

[40] R. Bush, J. Hiebert, O. Maennel, M. Roughan, and S. Uhlig. Testing the reachability of (new) address space. In *Proc. of the 2007 SIGCOMM workshop on Internet network management (INM'07), Kyoto, Japan*, 2007.

[41] R. Caceres, N. Duffield, A. Feldmann, J. Friedmann, A. Greenberg, R. Greer, T. Johnson, C. Kalmanek, B. Krishnamurthy, D. Lavelle, P. Mishra, K. Ramakrishnan, J. Rexford, F. True, and J. van der Merwe. Measurement and analysis of IP network usage and behavior. *IEEE Communications Magazine*, May 2000.

[42] M. Caesar, D. Caldwell, N. Feamster, J. Rexford, A. Shaikh, and J. van der Merwe. Design and implementation of a routing control platform. In *Proceedings of the 2nd Symposium on Networked Systems Design and Implementation (NSDI)*, Berkeley, CA, USA, 2005.

[43] CAIDA. cflowd: Traffic Flow Analysis Tool. `http://www.caida.org/tools/measurement/cflowd/`, 1998.

[44] A. Carlisle. Applying the Particle Swarm Optimizer to Non-stationary Environments. Ph.D. thesis, Auburn University, 2002.

[45] I. Castineyra, N. Chiappa, and M. Steenstrup. The Nimrod Routing Architecture. Internet Engineering Task Force, RFC1992, August 1996.

[46] S. Cerav-Erbas, O. Delcourt, B. Fortz, and B. Quoitin. The interaction of IGP weight optimization with BGP. In *Proceedings of the International Conference on Internet Surveillance and Protection*, 2006.

[47] H. Chang, R. Govindan, S. Jamin, S. Shenker, and W. Willinger. Towards Capturing Representative AS-Level Internet Topologies. *Computer Networks*, 44(6), April 2004.

[48] H. Chang, S. Jamin, and W. Willinger. To peer or not to peer: Modeling the evolution of the Internet's AS-level topology. In *Proc. of IEEE INFOCOM*, April 2006.

[49] C. Chatfield. *The Analysis of Time Series: An Introduction, Fifth Edition.* Chapman & Hall, 1996.

[50] E. Chen and S. Sangli. Avoid BGP Best Path Transition from One External to Another. Internet draft, draft-chen-bgp-avoid-transition-00.txt, work in progress, June 2003.

[51] Cisco. BGP Case Studies Section 1. Available at `http://www.cisco.com/warp/public/459/13.html`.

[52] Cisco. Sample Configurations for Load Sharing with BGP in Single and Multihomed Environments. Available at `http://www.cisco.com/warp/public/459/40.html`.

[53] Cisco. NetFlow services and applications. White paper, `http://www.cisco.com/warp/public/732/netflow`, 1999.

[54] Cisco. BGP Best Path Selection Algorithm. http://www.cisco.com/warp/public/459/25.shtml, 2002.

[55] K. Claffy, H. Braun, and G. Polyzos. Traffic characteristics of the T1 NSFNET backbone. In *Proc. of IEEE INFOCOM*, 1993.

[56] K. Claffy, G. Miller, and K. Thompson. The nature of the beast : recent traffic measurements from an Internet backbone. In *INET98*, 1998.

[57] C. A. C. Coello. An updated survey of GA-based multiobjective optimization techniques. *ACM Computing Surveys*, 32(2):109–143, 2000.

[58] N. Z. Computer Science Dept., University of Waikato. The DAG project. Available from `http://dag.cs.waikato.ac.nz/`.

[59] M. Crovella and A. Bestavros. Self-similarity in world wide web traffic evidence and possible causes. In *Proc. of* ACM SIGMETRICS'96, pages 160–169, May 1996.

[60] T. Dang and S. Molnar. On the Effects of Non-stationarity in Long-Range Dependence Tests. *Periodica Polytechnica Ser. El.*, 43(4):227–250, 1999.

[61] M. Daniel and A. Willsky. Modelling and Estimation of Fractional Brownian Motion using Multiresolution Stochastic Processes. *In Lévy Véhel, Lutton and Tricot (editors) "Fractals in Engineering: From Theory to Applications"*, 1997.

[62] I. Daubechies. *Ten Lectures on Wavelets*. Number 61 in CMBS-NSF Series in Applied Mathematics, SIAM, Philadelphia, 1992.

[63] K. Deb. Multi-objective genetic algorithms: Problem difficulties and construction of test problems. *Evolutionary Computation*, 7(3):205–230, 1999.

[64] K. Deb. *Multi-objective Optimization using Evolutionary Algorithms*. Wiley Interscience series in systems and optimization, 2001.

[65] K. Deb, S. Agrawal, A. Pratab, and T. Meyarivan. A Fast Elitist Non-Dominated Sorting Genetic Algorithm for Multi-Objective Optimization: NSGA-II. In *Proc. of the Parallel Problem Solving from Nature VI Conference*, pages 849–858, Paris, France, 2000. Springer-Verlag (LNCS 1917).

[66] S. V. den Bosch, F. Poppe, and G. Petit. Single-Path Traffic Engineering with Explicit Routes in a Differentiated-Services Network. In *IEEE International Conference on Communications*, Helsinki, Finland, June 2001.

[67] L. Deri. nProbe: an Open Source NetFlow probe for Gigabit Networks. In *Proc. of Terena TNC 2003*, May 2003.

[68] B. Donnet and O. Bonaventure. On BGP communities. *ACM Computer Communication Review*, 38(2), April 2008.

[69] A. Doria and E. Davies. Analysis of IDR requirements and History Group B contribution. Internet draft, draft-irtf-routing-history-00.txt, work in progress, February 2002.

[70] A. Doria and E. Davies. Future Domain Routing Requirements Group B contribution. Internet draft, draft-irtf-routing-reqs-groupb-00.txt, work in progress, February 2002.

[71] P. Doukhan, G. Oppenheim, and M. T. (editors). *Theory and Applications of Long-Range Dependence*. Birkhäuser, Boston, 2002.

[72] N. Duffield and C. Lund. Predicting resource usage and estimation accuracy in an IP flow measurement collection infrastructure. In *Proceedings of the 3rd ACM SIGCOMM conference on Internet measurement*, 2003.

[73] T. Ernst, N. Montavont, R. Wakikawa, E. Paik, C. Ng, K. Kuladinithi, and T. Noel. Goals and benefits of multihoming. Internet draft, draft-ernst-generic-goals-and-benefits-02, work in progress, October 2005.

[74] E. Falkenauer. A hybrid grouping genetic algorithm for bin packing. *Journal of Heuristics*, 2:5–30, 1996.

[75] E. Falkenauer. Applying genetic algorithms to real-world problems. In L. D. Davis, K. De Jong, M. D. Vose, and L. D. Whitley, editors, *Evolutionary Algorithms*, pages 65–88. Springer, New York, 1999.

[76] W. Fang and L. Peterson. Inter-AS traffic patterns and their implications. In *IEEE Global Internet Symposium*, December 1999.

[77] N. Feamster, J. Borkenhagen, and J. Rexford. Guidelines for interdomain traffic engineering. *ACM SIGCOMM Comput. Commun. Rev.*, 33(5):19–30, 2003.

[78] N. Feamster, R. Johari, and H. Balakrishnan. Implications of autonomy for the expressiveness of policy routing. In *Proc. of ACM SIGCOMM*, 2005.

[79] N. Feamster, Z. Mao, and J. Rexford. Borderguard: detecting cold potatoes from peers. In *Proceedings of the 4th ACM SIGCOMM conference on Internet measurement*, 2004.

[80] N. Feamster and J. Rexford. Network-Wide BGP Route Prediction for Traffic Engineering. In *Proc. of SPIE ITCOM 2002, Boston, MA*, August 2002.

[81] A. Feldmann. Characteristics of TCP connection arrivals. *In Park and Willinger (editors) "Self-Similar Network Traffic and Performance Evaluation", Wiley-InterScience*, 2000.

[82] A. Feldmann, A. Gilbert, P. Huang, and W. Willinger. Dynamics of IP Traffic: a Study of the Role of Variability and the Impact of Control. In *Proc. of ACM SIGCOMM*, pages 301–313, 1999.

[83] A. Feldmann, A. Gilbert, and W. Willinger. Data Networks as Cascades: Investigating the Multifractal Nature of Internet WAN Traffic. In *Proc. of ACM SIGCOMM*, 1998.

[84] A. Feldmann, A. Gilbert, W. Willinger, and T. Kurtz. The changing nature of network traffic: scaling phenomena. *ACM Computer Communication Review*, 28(2):5–29, 1998.

[85] A. Feldmann, A. Greenberg, C. Lund, N. Reingold, J. Rexford, and F. True. Deriving traffic demands for operational IP networks: methodology and experience. In *Proc. of ACM SIGCOMM*, September 2000.

[86] C. Fonseca and P. Fleming. An overview of evolutionary algorithms in multiobjective optimization. *Evolutionary Computation*, 3(1):1–16, 1995.

[87] B. Fortz, J. Rexford, and M. Thorup. Traffic engineering with traditional IP routing protocols. *IEEE Communications Magazine*, October 2002.

[88] B. Fortz and M. Thorup. Internet traffic engineering by optimizing OSPF weights. In *Proc. of IEEE INFOCOM*, March 2000.

[89] C. Fraleigh, C. Diot, B. Lyles, S. Moon, P. Owerzarski, D. Papagiannaki, and F. Tobagi. Design and deployment of a passive monitoring infrastructure. In *Passive and active measurements workshop*, Amsterdam, April 2001.

[90] L. Gao. On inferring autonomous system relationships in the Internet. In *Proc. of IEEE Global Internet Symposium*, November 2000.

[91] L. Gao and F. Wang. The Extent of AS Path Inflation by Routing Policies. In *Proceedings of IEEE Global Internet Symposium*, 2002.

[92] R. Gao, C. Dovrolis, and E. Zegura. Interdomain ingress traffic engineering through optimized AS-path prepending. In *Proc. of IFIP/TC6 Networking Conference*, May 2005.

[93] M. Garey and D. Johnson. *Computers and Intractability - A Guide to the Theory of NP-completeness*. W.H. Freeman & Co., San Francisco, USA, 1979.

[94] A. Gilbert, W. Willinger, and A. Feldmann. Scaling Analysis of Conservative Cascades, with Applications to Network Traffic. *IEEE Trans. on Information Theory*, 45(3):971–992, 1999.

[95] P. Gill, M. Arlitt, Z. Li, and A. Mahanti. The flattening internet topology: Natural evolution, unsightly barnacles or contrived collapse? In *Proc. of the 2008 Passive and Active Measurement conference (PAM)*, pages 1–10, April 2008.

[96] D. Goldberg. *Genetic Algorithms in Search, Optimization and Machine Learning*. Addison-Wesley, 1989.

[97] J. Grefenstette. Genesis: A system for using genetic search procedures. In *Proceedings of the 1984 Conference on Intelligent Systems and Machines*, pages 161–165, 1984.

[98] S. D. Gribble. UC Berkeley Home IP HTTP Traces. Available at `http://www.acm.org/sigcomm/ITA/`, July 1997.

[99] T. Griffin and G. Wilfong. An analysis of BGP convergence properties. In *Proc. of ACM SIGCOMM*, September 1999.

[100] B. Halabi. *Internet Routing Architectures (2nd edition)*. Cisco Press, 2000.

[101] Y. He, G. Siganos, M. Faloutsos, and S. Krishnamurthy. A systematic framework for unearthing the missing links: Measurements and impact. In *Proc. of the 4th USENIX Symposium on Networked Systems Design and Implementation (NSDI)*, Cambridge, MA, USA, 2007.

[102] S. Hergarten. *Self-Organized Criticality in Earth Systems*. Springer-Verlag, 2002.

[103] R. Hinden. Internet Routing Protocol Standardization Criteria. Internet Engineering Task Force, RFC1264, October 1991.

[104] J. Holland. *Adaptation in Natural and Artificial Systems*. University of Michigan Press, Ann Arbor, 1975.

[105] G. Holzmann. *Design and validation of computer protocols*. Prentice-Hall, 1991.

[106] B. Huffaker, M. Fomenkov, D. Plummer, D. Moore, and K. Claffy. Distance Metrics in the Internet. In *Proc. of IEEE International Telecommunications Symposium (ITS)*, September 2002.

[107] G. Huston. Analyzing the Internet's BGP routing table. *Internet Protocol Journal*, 4(1), 2001.

[108] V. Jacobson, C. Leres, and S. McCanne. tcpdump. Available at ftp://ftp.ee.lbl.gov, 1989.

[109] S. Jaffard. Multifractal formalism for functions Part I : Results valid for all functions. *SIAM J. Math. Anal.*, 28(4):944–970, July 1997.

[110] Juniper. Junos software release 5.6 : New features list. `http://www.juniper.net/products/ip_infrastructure/junos/105012.html`.

[111] A. Karasaridis and D. Hatzinakos. Network Heavy Traffic Modeling using alpha-Stable Self-Similar Processes. *IEEE Transactions on Communications*, 49(7):1203–1214, July 2001.

[112] K. Keys, D. Moore, R. Roga, E. Lagache, M. Tesch, and K. Claffy. The architecture of Coralreef: an Internet traffic monitoring software suite. In *Proc. of Passive and Active Measurement workshop*, April 2001.

[113] L. Kleinrock and W. Naylor. On measured behavior of the ARPA network. In *AFIS Proceedings, 1974 National Computer Conference*, volume 43, pages 767–780. John Wiley & Sons, 1974.

[114] B. Krishnamurthy, C. Wells, and Y. Zhang. On the use and performance of content distribution networks. In *Proc. of the first ACM SIGCOMM Internet Measurement Workshop*, November 2001.

[115] C. Kunzinger. Protocol for the Exchange of Inter-Domain Routing Information among Intermediate Systems to Support Forwarding of ISO 8473 PDUs. Internet Draft ISO/IEC 10747, draft-kunzinger-idrp-ISO10747-00.txt, April 1994.

[116] C. Labovitz, G. Malan, and F. Jahanian. Internet Routing Instability. In *Proc. of ACM SIGCOMM*, September 1997.

[117] A. Lange. Issues in Revising BGP-4. Internet draft, draft-ietf-idr-bgp-issues-01.txt, work in progress, June 2003.

[118] S. Leinen. Evaluation of candidate protocols for IP flow information export (IPFIX). Internet draft, draft-leinen-ipfix-eval-contrib-02, work in progress, January 2004.

[119] W. Leland, M. Taqqu, and W. Willinger. On the Self-Similar Nature of Ethernet Traffic. *IEEE/ACM transactions on networking*, 2(1):1–15, 1994.

[120] W. Leland and D. Wilson. High time-resolution measurement and analysis of LAN traffic: Implications for LAN interconnection. In *Proc. of* IEEE INFOCOM, 1991.

[121] R. Mahajan, D. Wetherall, and T. Anderson. Understanding BGP Misconfigurations. In *Proc. ACM SIGCOMM*, September 2002.

[122] B. B. Mandelbrot. Intermittent turbulence in self-similar cascades: divergence of high moments and dimension of the carrier. *Journal of Fluid Mechanics*, 62:331–358, 1974.

[123] B. B. Mandelbrot. *Multifractals and 1/f noise*. Springer-Verlag, 1997.

[124] B. B. Mandelbrot. *Fractals and Scaling in Finance*. Springer-Verlag, 1999.

[125] B. B. Mandelbrot. *Gaussian Self-Affinity and Fractals*. Springer-Verlag, 2001.

[126] Z. Mao, L. Qiu, J. Wang, and Y. Zhang. On as-level path inference. In *Proceedings of the ACM SIGMETRICS*, pages 339–349, 2005.

[127] Z. M. Mao, J. Rexford, J. Wang, and R. Katz. Towards an accurate AS-level traceroute tool. In *Proc. of ACM SIGCOMM, Karlsruhe, Germany*, 2003.

[128] S. McCreary and K. Claffy. Trends in wide area IP traffic patterns : a view from Ames Internet Exchange. Proceedings of the 13th ITC Specialist Seminar on Internet Traffic Measurement and Modelling, Monterey, CA, 2000.

[129] T. McGregor, H.-W. Braun, and J. Brown. The NLANR Network Analysis Infrastructure. *IEEE Communications Magazine*, May 2000.

[130] L. McKnight and J. Bailey. *Internet Economics*. MIT Press, 1997.

[131] P. McManus. A passive system for server selection within mirrored resource environments using AS path length heuristics. Available from `http://www.gweep.net/~mcmanus/proximate.pdf`, April 1999.

[132] D. Meyer. University of Oregon Route Views Project. Available at `http://antc.uoregon.edu/route-views/`.

[133] Z. Michalewicz. *Genetic Algorithms + Data Structures = Evolution Programs*. Springer-Verlag, 1996.

[134] T. Mikosch, S. Resnick, H. Rootzén, and A. Stegeman. Is Network Traffic Approximated by Stable Lévy Motion or Fractional Brownian Motion? *The Annals of Applied Probability*, pages 23–68, 2002.

[135] T. Mikosch and C. Starica. Long-Range Dependence Effects and ARCH Modeling. *In P. Doukhan and G. Oppenheim and M. Taqqu (editors), "Theory and Applications of Long-Range Dependence", Birkhäuser, Boston*, 2002.

[136] W. Mühlbauer, A. Feldmann, O. Maennel, M. Roughan, and S. Uhlig. Building an AS-topology model that captures route diversity. In *Proc. of ACM SIGCOMM*, 2006.

[137] W. Mühlbauer, S. Uhlig, B. Fu, M. Meulle, and O. Maennel. In search for an appropriate granularity to model routing policies. In *Proc. of ACM SIGCOMM*, 2007.

[138] U. of Waikato. The DAG project. `http://dag.cs.waikato.ac.nz/`.

[139] R. Oliveira, D. Pei, W. Willinger, B. Zhang, and L. Zhang. In Search of the elusive Ground Truth: The Internet's AS-level Connectivity Structure. In *ACM SIGMETRICS*, Annapolis, USA, June 2008.

[140] J. Org, M. Ian, and G. Brownlee. The auckland data set: an access link observed. In *Proceedings of the* 14th *ITC Specialist Seminar*, April 2001.

[141] C. Papadimitriou and K. Steiglitz. *Combinatorial Optimization: algorithms and complexity*. Prentice-Hall, 1982.

[142] V. Pareto. *Cours d'Economie Politique, volume I et II*. F. Rouge, Lausanne, 1896.

[143] K. Park, G. Kim, and M. Crovella. On the relationship between file sizes, transport protocols, and self-similar network traffic. In *Proc. of IEEE International Conference on Network Protocols (ICNP)*, 1996.

[144] K. Park, G. Kim, and M. Crovella. On the Effect and Control of Self-Similar Network Traffic. In *Proc. SPIE International Conference on Performance and Control of Network Systems*, pages 296–310, 1997.

[145] K. Park and W. Willinger. *Self-Similar Network Traffic and Performance Evaluation*. Wiley-Interscience, 2000.

[146] V. Paxson. Empirically-Derived Analytic Models of Wide-area TCP Connections. IEEE/ACM *Transactions on Networking*, 2(4):316–336, 1994.

[147] V. Paxson and S. Floyd. Wide-Area Traffic: The Failure of Poisson Modeling. IEEE/ACM *Transactions on Networking*, 3(3):226–244, 1995.

[148] M. Priestley. *Spectral Analysis and Time Series*. Academic Press, 1981.

[149] R. Purshouse and P. Fleming. Conflict, Harmony, and Independence: Relationships in Multi-criterion Optimisation. In *Proc. of the Second International Conference on Multi-Criterion Optimization (EMO2003), Portugal*, pages 16–30, April 2003.

[150] Y. Qiao, J. Skicewicz, and P. Dinda. Multiscale predictability of network traffic. Technical Report NWU-CS-02-13, Nothwestern University, October 2002.

[151] B. Quoitin and O. Bonaventure. A cooperative approach to interdomain traffic engineering. In *Proc. of Next Generation Internet Networks Conference (NGI)*, April 2005.

[152] B. Quoitin, C. Pelsser, O. Bonaventure, and S. Uhlig. A performance evaluation of BGP-based traffic engineering. *Int. J. Netw. Manag.*, 15(3):177–191, 2005.

[153] B. Quoitin, S. Tandel, S. Uhlig, and O. Bonaventure. Interdomain Traffic Engineering with redistribution communities. *Computer Communications*, 27, March 2004.

[154] B. Quoitin, S. Uhlig, and O. Bonaventure. Using redistribution communities for Interdomain Traffic Engineering. In *Proc. of the third COST 263 workshop on Quality of future Internet Services, Zurich, Switzerland*, October 2002.

[155] B. Quoitin, S. Uhlig, C. Pelsser, L. Swinnen, and O. Bonaventure. Interdomain traffic engineering with BGP. *IEEE Communications Magazine, Internet Technologies Series*, May 2003.

[156] Y. Rekhter, T. Li, and S. Hares. A Border Gateway Protocol 4 (BGP-4). Internet draft, draft-ietf-idr-bgp4-22.txt, work in progress, October 2003.

[157] J. Rexford, J. Wang, Z. Xiao, and Y. Zhang. BGP Routing Stability of Popular Destinations. In *Proc. of the second ACM SIGCOMM Internet Measurement Workshop*, pages 197–202, November 2002.

[158] R. H. Riedi. An improved multifractal formalism and self-similar measures. *J. Math. Anal. Appl.*, 189:462–490, 1995.

[159] S. Romig, M. Fullmer, and R. Luman. The OSU Flow-tools package and Cisco NetFlow logs. In *Proc. of* USENIX LISA, December 1995.

[160] M. Roughan and D. Veitch. Measuring Long-Range Dependence under Changing Traffic Conditions. In *Proc. of IEEE INFOCOM*, 1999.

[161] M. Roughan and D. Veitch. Some remarks on unexpected scaling exponents. *ACM Computer Communication Review*, 37(5), October 2007.

[162] S. Sahni and T. Gonzalez. P-complete approximation problems. *J. ACM*, 23:555–565, 1976.

[163] G. Samorodnitsky and M. Taqqu. *Stable Non-Gaussian Random Processes: Stochastic Models with Infinite Variance*. Chapman & Hall, 1994.

[164] A. Sang and S. Li. A Predictability Analysis of Network Traffic. In *Proc. of IEEE INFOCOM*, 2000.

[165] S. Sangli, D. Tappan, and Y. Rekhter. BGP Extended Communities Attribute. Internet draft, draft-ietf-idr-bgp-ext-communities-06.txt, work in progress, August 2003.

[166] S. Saroiu, P. Gummadi, and S. Gribble. A Measurement Study of Peer-to-Peer File Sharing Systems. In *Proc. of Multimedia Computing and Networking Conference*, January 2002.

[167] S. Savage, A. Collins, E. Hoffman, J. Snell, and T. Anderson. The end-to-end effects of internet path selection. In *Proceedings of ACM SIGCOMM*, 1999.

[168] S. Sen and J. Wang. Analyzing peer-to-peer traffic across large networks. In *Proc. of the second ACM SIGCOMM Internet Measurement Workshop*, November 2002.

[169] G. Siganos and M. Faloutsos. BGP Routing: A Study at Large Time Scale. In *Proc. of IEEE Global Internet Symposium*, November 2002.

[170] P. Smith. Weekly routing table report. Weekly reports from APNIC's router in Japan sent to `bgp-stats@lists.apnic.net`.

[171] D. Sornette. *Critical Phenomena in Natural Sciences*. Springer-Verlag, 2000.

[172] N. Spring, R. Mahajan, and T. Anderson. Quantifying the causes of path inflation. In *Proceedings of ACM SIGCOMM*, August 2003.

[173] N. Srinivas and K. Deb. Multi-Objective function optimization using non-dominated sorting genetic algorithms. *Evolutionary Computation*, 2:221–248, 1995.

[174] M. Steenstrup. Inter-Domain Policy Routing Protocol Specification: Version 1. Internet Engineering Task Force, RFC1479, July 1993.

[175] J. Stewart. *BGP4 : interdomain routing in the Internet*. Addison Wesley, 1999.

[176] L. Subramanian, S. Agarwal, J. Rexford, and R. Katz. Characterizing the Internet hierarchy from multiple vantage points. In *INFOCOM 2002*, June 2002.

[177] P. D. Surry and N. J. Radcliffe. Formal algorithms + formal representations = search strategies. In *Parallel Problem Solving from Nature*, pages 366–375, 1996.

[178] H. Tangmunarunkit, R. Govindan, S. Shenker, and D. Estrin. The Impact of Routing Policy on Internet Paths. In *Proc. of IEEE INFOCOM 2001*, pages 736–742, 2001.

[179] M. Taqqu. The modeling of Ethernet data and of signals that are heavy-tailed with infinite variance. *Scandinavian Journal of Statistics*, 29(2):273–295, 2002.

[180] M. Taqqu and V. Teverovsky. On Estimating the Intensity of Long-Range Dependence in Finite and Infinite Variance Time Series. *"A Practical Guide to Heavy Tails: Statistical Techniques and Applications"*, 1998.

[181] M. Taqqu, V. Teverovsky, and W. Willinger. Is network traffic self-similar or multifractal? *Fractals*, 5:63–73, 1997.

[182] M. Taqqu, W. Willinger, and R. Sherman. Proof of a Fundamental Result in Self-Similar Traffic Modeling. *ACM Computer Communication Review*, 27, 1997.

[183] R. Teixeira, A. Shaikh, T. Griffin, and J. Rexford. Dynamics of Hot-Potato Routing in IP Networks. In *Proc. of ACM SIGMETRICS*, 2004.

[184] A. Tewfik and A. Kim. Correlation structure of the discrete wavelet coefficients of fractional brownian motion. IEEE *Trans. Inform. Theory*, 38:904–909, 1992.

[185] K. Thompson, G. Miller, and R. Wilder. Wide-Area Internet traffic patterns and characteristics. *IEEE Network magazine*, 11(6), November/December 1997.

[186] P. Traina, D. McPherson, and J. Scudder. Autonomous System Confederations for BGP. Internet Engineering Task Force, RFC3065, February 2001.

[187] S. Uhlig. 3D-LD : a Graphical Wavelet-based Method for Analyzing Scaling Processes. Infonet technical report Infonet-2002-04, available at `http://www.infonet.fundp.ac.be/doc/tr/`, April 2002.

[188] S. Uhlig. Interdomain Traffic Engineering: an Evolutionary Approach. Université catholique de Louvain, Research Report RR 2002-10 INFO, December 2002.

[189] S. Uhlig. Conservative cascades: an Invariant of Internet traffic. In *Proc. of the 2003 IEEE International Symposium on Signal Processing and Information Technology, Darmstadt, Germany*, December 2003.

[190] S. Uhlig. High-order Scaling and Non-stationarity in Flow Arrivals. *ACM Computer Communication Review*, 34(2):9–24, Apr 2004.

[191] S. Uhlig. *Implications of the traffic characteristics on interdomain traffic engineering*. PhD thesis, Computer Science Dept., University of Louvain-la-neuve, Belgium, 2004.

[192] S. Uhlig. A Multiple-objectives Evolutionary Perspective to Interdomain Traffic Engineering. *International Journal of Computational Intelligence and Applications (IJCIA)*, 5(2):215–230, June 2005.

[193] S. Uhlig and O. Bonaventure. On the cost of using MPLS for interdomain traffic. In *Proc. first COST263 workshop on Quality of future internet services (QoFIS)*, September 2000.

[194] S. Uhlig and O. Bonaventure. Understanding the long-term self-similarity of internet traffic. In *Proceedings of the Second International Workshop on Quality of Future Internet Services (QoFIS)*, September 2001.

[195] S. Uhlig and O. Bonaventure. Implications of Interdomain Traffic Characteristics on Traffic Engineering. *European Transactions on Telecommunications*, January 2002.

[196] S. Uhlig and O. Bonaventure. Designing BGP-based outbound traffic engineering techniques for stub ASes. 34(5):89–106, October 2004. ACM Computer Communication Review.

[197] S. Uhlig, O. Bonaventure, and B. Quoitin. Interdomain Traffic Engineering with minimal BGP Configurations. In *Proc. of the* 18th *International Teletraffic Congress, Berlin*, September 2003.

[198] S. Uhlig, O. Bonaventure, and C. Rapier. 3D-LD : a Graphical Wavelet-based Method for Analyzing Scaling Processes. In *Proc. of the* 15th *ITC Specialist Seminar, Würzburg, Germany*, July 2002.

[199] S. Uhlig, V. Magnin, O. Bonaventure, C. Rapier, and L. Deri. Implications of the Topological Properties of Internet Traffic on Traffic Engineering. In *Proc. of the* 19th *ACM Symposium on Applied Computing, Special Track on Computer Networks, Nicosia, Cyprus*, March 2004.

[200] S. Uhlig and B. Quoitin. Tweak-it: Bgp-based interdomain traffic engineering for transit ass. In *Proc. of Next Generation Internet Networks Conference, Rome, Italy*, April 2005.

[201] S. Uhlig and S. Tandel. Quantifying the Impact of Route-Reflection on BGP Routes Diversity inside a Tier-1 Network. In *Proc. of IFIP Networking*, Coimbra, Portugal, May 2006.

[202] D. Veitch, P. Abry, P. Flandrin, and P. Chainais. Infinitely divisible cascade analysis of network traffic data. In *Proc. of ICASSP*, 2000.

[203] M. Vetterli and J. Kovacevic. *Wavelets and subband coding*. Prentice-Hall, 1995.

[204] D. Walton, D. Cook, A. Retana, and J. Scudder. Advertisement of multiple paths in BGP. Internet draft, draft-walton-bgp-add-paths-01.txt, work in progress, November 2002.

[205] L. Wang, X. Zhao, D. Pei, R. Bush, D. Massey, A. Mankin, S. Wu, and L. Zhang. Protecting BGP Routes to Top Level DNS Servers. In *Proc. of* 23rd *International Conference on Distributed Computing Systems (ICDCS)*, May 2003.

[206] Y. Wang, Z. Wang, and L. Zhang. Internet traffic engineering without full mesh overlaying. In *Proc. of IEEE INFOCOM*, April 2001.

[207] Z. Wang. Internet traffic engineering. *Special section of IEEE Network magazine*, March-April 2000.

[208] M. Wickerhauser. *Adapted wavelet analysis: from Theory to Software*. A K Peters, 1994.

[209] W. Willinger, V. Paxson, R. Riedi, and M. Taqqu. Long-Range Dependence and Data Network Traffic. *In P. Doukhan and G. Oppenheim and M. Taqqu (editors), "Theory and Applications of Long-Range Dependence", Birkhäuser, Boston*, 2002.

[210] W. Willinger, M. Taqqu, R. Sherman, and D. Wilson. Self-similarity through high-variability: statistical analysis of Ethernet LAN traffic at the source level. *IEEE/ACM Transactions on Networking*, 5(1):71–86, 1997.

[211] C. Wills, M. Mikhailov, and H. Shang. Inferring relative popularity of internet applications by actively querying DNS caches. In *Proceedings of the 3rd ACM SIGCOMM conference on Internet measurement*, 2003.

[212] D. Wolpert and W. Macready. No free lunch theorems for optimization. *IEEE Transactions on Evolutionary Computation*, 1(1):67–82, April 1997.

[213] J. Xu, J. Fan, M. Ammar, and S. Moon. On the Design and Performance of Prefix-Preserving IP Traffic Trace Anonymization. In *Proc. of the first ACM SIGCOMM Internet Measurement Workshop*, November 2001.

www.ingramcontent.com/pod-product-compliance
Lightning Source LLC
LaVergne TN
LVHW022306060326
832902LV00020B/3299

www.ingramcontent.com/pod-product-compliance
Lightning Source LLC
LaVergne TN
LVHW022306060326
832902LV00020B/3298

* 9 7 8 9 3 8 7 2 8 4 5 0 0 *